MIDDLE EAST OIL

MIDDLE EAST OIL

*A Study in Political
and Economic Controversy*

GEORGE W. STOCKING

Vanderbilt University Press
1970

COPYRIGHT © 1970
VANDERBILT UNIVERSITY PRESS
Standard Book Number 0–8265–1156–2
Library of Congress Catalogue
Card Number 73–115095
Printed in the United States of America by
Kingsport Press, Inc.
Kingsport, Tennessee

To my Sons and Daughters
G. W. S., Jr.
M. R. S.
S. S. W.
C. S. S.

Preface

OIL production in the Middle East, which contains the world's most prolific oil reserves and supplies Western Europe with most of its crude oil, has been under the control of a half dozen of the world's largest corporations. These companies are fully integrated, skillfully managed, and possess the latest techniques for finding, recovering, transporting, and refining petroleum; they are equipped with far-flung facilities for marketing petroleum products. Their leaders have manifested a propensity for risk-taking consistent with the finest traditions of business enterprise. They have evidenced a paternalistic concern for the welfare of their employees and have not been unaware of their social responsibilities. But their major concern is making money. They have found in the Middle East a gratifying source for the realization of their aims.

In achieving this goal they have mobilized thousands of untutored workers and taught them skills. They have transformed nomadic tribesmen living under the simplest conditions in rude tents into urban dwellers with fixed and comfortable abodes possessing many of the amenities to which Western workers have long been accustomed. They have brought business opportunities to local merchants, suppliers, and contractors and have engendered thereby a prosperity never before realized. They have supplied the host countries with a large and increasing proportion of their total revenues—revenues that have grown faster than the region's oil output—and have thereby afforded the means for extensive programs of social betterment and industrial planning. On its face it looks as though a mutually satisfactory and profitable relationship should have been established between the host countries and the producing companies. Looks are deceiving.

While a visiting professor at the American University of Beirut in 1960, I attended the Second Arab Petroleum Congress. My familiarity with the oil industry covers a span of more than half a century intermittently as an employee, student, author, and government regulator. But I was unprepared for the hostility, bitterness, and suspicions manifested by congress spokesmen for the host countries or the feeling of insecurity, uncertainty, and resentment manifested by representatives of the oil industry. The tone of the congress was set in the address of welcome by its secretary general, Sayed Mohammad Salman, when he said, "I simply do not want to let this opportunity pass without making a point of the extent of bitterness felt by the Arabic public towards the companies' position, which has borne until now the mark of obscurity as well as that of challenge." It was accentuated in a paper by the legal adviser on petroleum affairs to the Kingdom of Saudi Arabia, who developed the theme that a sovereign nation cannot alienate by contract its responsibility to its people, and that when it believes that these responsibilities have not been adequately fulfilled it has a right, indeed an obligation, to modify unilaterally the petroleum concessions. It was carried to a crescendo by Sheik Abdullah Tariki, then Saudi Arabia's director and soon to be designated minister of petroleum and mineral resources, in accusing the oil companies through their pricing policy of having defrauded the oil countries of billions of dollars during the preceding decade.

I had accepted the invitation to teach at A.U.B. partly because it afforded an opportunity to do research on the Middle East oil industry. The congress afforded the inspiration for the scope of my research. Believing that the present cannot be understood nor the future forecast without an understanding of the past, I began the study with a review of the political and economic environment in which the original concessions were granted and in which the seeds of controversy were sown. There follow in succession an analysis of the terms of the concessions and how they eventually led to conflict; a study of the specific conflicts and the role that national and collective action have played in trying to resolve them; an analysis of the influences that oil pricing, cost of production, and profits have exerted on the controversies; and some observations on the security and future of the concessions.

To encompass the breadth and depth of study that the outline demanded would have taxed the capacities of any individual working alone and lacking a knowledge of Arabic. Fortunately, I had opportunities commensurate with the task and compensating in a measure for my inadequate linguistic equipment. I have made three trips to the Middle East over a span of seven years. Vanderbilt University, with funds supplied by the Rockefeller Foundation, made two of these possible; the first in 1960–61, the second in 1963–64. The Ford Foundation through a personal grant made the third possible in 1966–67. During my first visit, Professor Zuhayr Mikdashi of A.U.B. and I conducted a seminar on the Middle East oil industry in which representatives of the oil industry participated. Later, Mikdashi and I visited Saudi Arabia, Kuwait, Iraq, Iran, and northern Africa, conferring with company and government officials. On my last sojourn I revisited Kuwait and Saudi Arabia, the latter as a guest of the minister of petroleum. En route to and from the Middle East I conferred with company officials in the United States, England, Germany, France, Italy, and Japan.

In the early winter of 1966–67 I participated in the Kuwait oil seminar under the auspices of the Institute for Social and Economic Planning, organized by Salaiman Mutawa, co-director of the institute, and Professor Mikdashi, who at that time was serving as adviser to the Kuwaiti minister of finance. In the late winter of 1967 I participated in the seminar at A.U.B. on petroleum economics organized by Professor A. J. Meyer of Harvard University and Professors Sayegh and Mikdashi of A.U.B.

These several contacts were helpful, and I am grateful for them. But to the Middle East Research and Publishing Center, publishers of the *Middle East Economic Survey* my debt is greatest. The *MEES*, with a professionally competent staff, fluent in English and Arabic, with access to a wide range of sources and contacts, including governmental publications, legal documents, concession agreements, and industry and governmental personnel, presents weekly a realistic, comprehensive, and skillfully edited review of the news and views on oil industry developments in the Middle East and Northern Africa. On two of my visits, extending through more than a year, the center supplied me with office space and made available to me a mine of information, including access to English translations of important articles. More important, its files supplied me with a moving account of political and economic developments in the several oil-producing countries and English translations of the news and views as published in Arabic journals.

I am also grateful to the many individuals, including representatives of the oil industry and staff members of the *MEES*, who read critically several chapters of the study. I am particularly grateful to Fuad Itayem, publisher and editor, and to Ian Seymour, news editor, of the *Survey;* to Professor Mikdashi, Professor James McKie of Vanderbilt University, and to Professor M. A. Adelman of Massachusetts Institute of Technology for their critical comment on certain chapters. That their comments have improved the study goes without saying. They are in no way responsible for my interpretation of the facts nor for my opinions expressed in the study.

I completed the study as a visiting scholar at Stanford University, where I had access to the Hoover Institution's library. I wish to express my appreciation to Arline Paul of the reference department of this library and to her staff. In locating documents they manifested a graciousness and generosity so frequently characteristic of those who work with books. I wish also to express my gratitude to Elizabeth Post who assisted me in the early phases of the study. But I am indebted most to my wife, Dorothé Reichhard Stocking, who deciphered my illegible longhand and typed the manuscript, edited it, read the page proofs and helped in the preparation of the index.

George W. Stocking

Portola Valley
California
February 1970

Contents

Contents

PART IV
The Economics of Middle East Oil Pricing

PART V
The Prospects for Middle East Oil

PART I

THE HISTORICAL AND INSTITUTIONAL BACKGROUND OF THE CONCESSIONS

The Iranian Oil
Concessions

IRAN, known as Persia until 1935, is the oldest oil-producing country in the Middle East. Production developed under a concession originally obtained by William Knox D'Arcy, a British subject whose interest in Persian oil arose through the efforts of Kitabji Khan, Persia's Commissioner General at the Paris Exposition of 1900. Shah Muzaffar ed-Din granted the concession, and on May 28, 1901, Alfred M. Marriot, D'Arcy's attorney, signed it on his behalf.[1]

To understand the significance of what was given and what was got under the D'Arcy concession, it will be helpful to review briefly recent Persian political history. Persia for a long time had been ruled by rapacious and cruel kings. The Qajar dynasty, ushered in by the predatory reign of Muhammad in the last quarter of the eighteenth century, ruled Persia for approximately one hundred and fifty years. Like their predecessors, they ruled with iron, frequently bloody, but uneasy hands. Theirs was an absolute despotism in which the people and the land were regarded as the property of the shah to do with as he chose. A position embodying such power was naturally coveted. There being no customary line of descent, frequently several contenders aspired to the throne. The ruling shah just as

1. The concession appears as an appendix to Annex 1419c in the League of Nations *Official Journal,* XIII (1932), 2305–2307.

frequently took strong measures to make his reign secure and to designate his successor.

Social and Economic Groups

That despotism as a form of government prevailed in Persia throughout the nineteenth century, as it did in most Asiatic countries, is an institutional fact that has a ready explanation. No middle class developed in Persia capable of curbing the power of the shahs. The chief social and economic groups were the peasants, the landowners, the nomadic tribes, the religious hierarchy, and the merchants. The peasants, who constituted the bulk of the population, made their living by farming on land generally belonging to an absentee landlord. They lived in small mud villages close to a subsistence level, the unfortunate victims of ignorance, disease, and poverty. Although legally free, for all practical purposes they were bound to the soil they cultivated. The landlords, who frequently measured their wealth by the number of villages they owned, generally lived in the larger cities, leaving the management of the villages to their stewards.

Although the most substantial and influential class in Persia, the landlords never developed a class consciousness strong enough to curb the absolutism of the throne. In truth, the shahs saw to it that they did not. As ultimate owners of the land and the people, the shahs did not hesitate to confiscate a landlord's property when the needs of the treasury or the security of their power demanded it.

The nomadic tribes, representing about one fourth of the total population, lived on the mountain ranges along Persia's eastern and western borders. Their nomadic habits gave them a strong sense of personal freedom and independence and engendered an attitude of haughty indifference toward the throne.

The religious hierarchy was composed of the mujtahids and the mullahs, distinguishable primarily on a basis of scholarship and personality and the stratum of society within which they worked. The mujtahids, learned in religious law and philosophical and literary culture, were held in esteem by the upper stratum of Persian society. The mullahs, less learned and cultured, occupied a similar position among the masses. Neither group ordinarily exercised any substantial restraint on the power of the shah.

The bazaar or merchant class represented the nearest approach to a middle class comparable to the Western bourgeoisie, but there were important differences. The bazaar class contained none of the liberal professions that played such an important role in transforming the essentially ecclesiastical civilization of Western Europe of the Middle Ages into the secular, scientific, democratic civilization of modern times. It also lacked the cohesion essential to an articulate and persistently politically influential group. It embraced all categories of those who sold goods—the itinerant peddler with his pack on his back, the shopkeeper catering to a local trade, and the importer-exporter whose commerical connections were worldwide.

Nor did Persia boast a nobility of sufficient coherence and with sufficiently well-defined rights and duties to be of influence in the affairs of the throne. There were, of course, large groups of individuals holding royal titles clustered around the throne and receiving their sustenance and social standing therefrom. Royal harems bred royal princes and their numbers grew as a dynasty aged. According to Haas, in the regime of Nasir ed-Din Shah (1848–1896) the descendants of his predecessor, Fath Ali Shah, could be counted by the thousands.[2] Some of them held official positions of responsibility, but most, as loafers, constituted a heavy drain on the shah's purse but exercised no restraint on his behavior.

Foreign Intervention

While the Persian shahs of the nineteenth century customarily were neither accountable to nor held in restraint by any coherent social group, they were not equally free from foreign influence. Foreign powers vied with each other in trying to obtain a preferred economic and geographic position in Persia and to utilize Persian military forces to strengthen their imperial dominions. Chief among the powers that looked on Persia as an instrument in the realization of their own imperial ambitions were Russia and Great Britain. Twice in the first quarter of the nineteenth century Russia went to war with Persia, and on each occasion Russian military successes were followed by a recognition of Russian claims to territory. Meanwhile Great Britain likewise had resorted to armed strength and peaceful diplomacy as the occasion suited in an effort to protect her imperial interests in India and her control over the Persian Gulf.

During the second half of the nineteenth century Russia and Britain vied for the confidence of the Persian ruler. At one time or another both

2. William S. Haas, *Iran* (New York: Columbia University Press, 1946), p. 99.

sent military missions to modernize the Persian army and resorted to various other devices to win the shah's friendship and confidence. Throughout this period their machinations in Persia were subordinate to their broader imperialist ambitions, and with the rising might of Germany near the close of the century they participated in a regrouping of power alignments. The failure of the Anglo-German naval agreement was followed by the Franco-British agreement respecting spheres of influence in Africa. Shortly thereafter Russia and Great Britain settled their rivalry in Persia by signing a treaty, the Anglo-Russian Convention of 1907, which, while piously recognizing Persia's integrity and independence, divided the country into spheres of Russian and British influence. Under this treaty each country bound itself to recognize any concession previously obtained by its nationals and to refrain from seeking public or private interests in the zone allotted to the other. The Russian zone included all of northern and central Persia; Britain's, the southeastern desert, with a neutral zone (embracing the new oil field) lying in between.

Persia's Fiscal Structure

Great Britain's concern with Persia was twofold: the protection of her interests in India—greatly facilitated by the Anglo-Russian Convention which gave her command of the Persian Gulf littoral—and the exploitation of such economic and commercial potentialities as Persia might possess. The instrument through which exploitation rights were acquired was the concession. It was an instrument well suited to the times. Persia lacked both capital and technological knowledge, and the ruling shah was frequently short of funds with which to support his court and his government. The shah's revenue came ultimately from Persian economic resources, but it came through an ingenious and corrupt administration that placed the burden on those who could least afford it. Customarily governmental posts from the highest to the lowest were bought and sold. Direct appointees of the shah made an appropriate monetary contribution to the shah's purse. They, in turn, sold subordinate positions. This process continued until a hierarchy of officials was built up whose chief duties were providing revenues and appropriate military quotas and administering justice. Each official got from those below funds for those above with enough to spare for his own personal expenditures on such a scale as the system might allow. Lord Curzon characterized the whole arrangement as "an arithmetical progression of plunder from the sovereign to the subject, each unit on the descending scale remunerating himself from the unit next in rank below him, the helpless peasant being the ultimate victim."[3] Under such a system it is not surprising that income frequently was inadequate to meet the shah's governmental and personal expenses. During the latter part of the nineteenth century Nasir ed-Din Shah looked with increasing frequency to concessions as a means of supplementing the royal treasury, and usually he granted them with an irresponsible lack of prudence and foresight.

The de Reuter Concession

The de Reuter concession of 1872 illustrates the point. Covering all of Persia, it granted to a naturalized British subject, Baron Julius de Reuter, a seventy-year monopoly for the construction and operation of all Persian railroads and streetcar lines; the sole right to exploit all of Persia's mineral resources except gold, silver, and precious stones; a monopoly of government forests including all uncultivated lands; an option on all future enterprises connected with the construction of roads, telegraphs, mills, factories, workshops, and public works of every kind; and the right to collect all Persian customs duties for twenty-five years and payment to the shah of 16 percent of the net revenue for the remaining twenty years. In return de Reuter agreed to pay the Persian government 20 percent of the profits derived from the railways and 15 percent of those derived from all other sources. In the language of Lord Curzon, this concession represented "the most complete and extraordinary surrender of the entire industrial resources of a kingdom into foreign hands that has probably ever been dreamt of, much less accomplished in history."[4]

As the shah set out on a trip to England in 1873, his first foreign travel, news of this "amazing document . . . fell like a bombshell" on all Europe. According to Curzon, the British government, aware of the political complications likely to arise from the exercise of such an all-inclusive monopoly by one of its subjects, received the scheme coldly. Russian reaction was more positive and hostile, so menacing in fact that the shah on his return to Persia canceled the concession and confiscated de Reuter's £40,000 de-

3. George N. Curzon, *Persia and the Persian Question* (London: Longmans, Green, 1892), 2 vols.; I, 443n7.
4. *Ibid.*, p. 480.

posit on the grounds that work had not begun within the fifteen months stipulated in the contract.

However great may have been the British government's misgivings about the initial de Reuter concession, Britain did not accept its cancellation with equanimity. On the contrary, Britain refused to recognize the cancellation and brought sufficient pressure on the shah to cause him to grant de Reuter another less comprehensive concession in 1889. The new concession, which was to run for sixty years, granted de Reuter the privilege of establishing a bank and of exploiting certain minerals, including petroleum, from which the Persian government was to receive 16 percent of the net profits. Baron de Reuter organized the Imperial Bank of Persia and the Persian Bank Mining Rights Corporation. The former, having a capital of £1 million and the right to issue banknotes, soon became operative. But the mining company, despite a two-year search for oil, met with no success. After the Persian government declared the mineral concession void, the de Reuter interests liquidated the mining corporation in 1901.

The D'Arcy Concession is Granted

On the assassination of his father in 1896 Shah Muzaffar ed-Din began a reign which was to witness the dissipation of the state's revenue, an increase in burdensome foreign indebtedness, and the alienation of the country's most important source of national wealth, oil. Accustomed to extravagant living at home and expensive travel abroad, surrounded by a corrupt and pretentious court, faced with an empty treasury, lacking in appreciation of the economic potentialities of Iran's oil resources, confronted by concession-seekers willing to pay for the privilege of risking their capital in a search for petroleum, and responsible to no one for any mistakes that he might make, the shah signed the D'Arcy concession on May 28, 1901. Although the initial move in finding a concessionaire seems to have been taken by Persia's commissioner general, Kitabji Khan, at the Paris Exposition, once the British were interested, they pushed the matter aggressively. After Kitabji Khan had contacted Sir Henry Drummond Wolff, former British minister to Teheran, he was invited to London where he discussed a concession with William Knox D'Arcy. A geological investigation of Persia's oil potentialities by H. T. Burls eventuated in Alfred M. Marriot's visit to Persia accompanied by Kitabji Khan in search of a concession. Meanwhile Wolff requested Sir Arthur Hardinge, who had succeeded Wolff as minister to Teheran, to intervene on D'Arcy's behalf. Hardinge was advised that the offer of shares in the proposed enterprise to influential ministers might help in obtaining the concession, and apparently he followed this advice. He gave this account of developments:

> The first important duty which confronted me a few months after my arrival at Tehran was that of securing for a British Company an important concession of Persian oil-fields. . . . Sir Henry Wolff accordingly wrote a letter of introduction, brought by a Mr. Marriott, whom he recommended in it to my good offices, and who would, he went on to say, explain to me fully the character and objects of his mission.
>
> Its main end, which appeared to me well deserving of support, was to win the goodwill of the Persian Government by assigning shares in the proposed development of the rich oil-fields believed to exist in Western Persia to some of its most influential Ministers, including the Grand Vizier [prime minister] himself.[5]

His efforts were successful. After ingeniously contriving to conceal his actions from the Russian Legation until it was accomplished,[6] the grand vizier awarded a concession to D'Arcy.

Terms of the Concession

The concession gave D'Arcy a sixty-year exclusive privilege to "search for, obtain, exploit, develop, render suitable for trade, carry away and sell

5. Sir Arthur H. Hardinge, *A Diplomatist in the East* (London: Jonathan Cape, 1928), p. 278.

6. Hardinge gives the following account of the grand vizier's machinations: "The Grand Vizier declared himself prepared to fall in with the project, but he suggested that a letter—to be written by me, in the Persian language, embodying its main features—should be immediately drawn up for submission to the Russian Legation. He was aware that M. Argyropulo could not read Persian, more especially in the written or 'shikaste' character, which is illegible, owing to its peculiar abbreviations, even to scholars familiar with the printed language. He also knew from his own spies that the Russian Oriental Secretary, M. Stritter, who alone could read it, was about to leave Zergendeh, the summer residence of the Russian Legation, for a short sporting excursion in the neighboring hills. He therefore sent the letter to Zergendeh, where it lay several days untranslated, awaiting M. Stritter's return, and as no objection to the proposal contained in it was made by the Russian Minister, who could not read it, and never suspected the importance of its contents, all the Persian members of the Government supported the Grand Vizier's decision to sign the concession to Mr. D'Arcy. M. Argyropulo was far from pleased when he learnt what actually happened; but the Grand Vizier could not be blamed for the accidental and temporary absence of his Legation's Persian translator, and the Russian Minister accordingly adopted the sensible course of accepting the accomplished fact." *Ibid.*, p. 279.

natural gas, petroleum, asphalt and ozocerite throughout the whole extent of the Persian Empire"[7] with the exception of the five northern provinces that were Russia's particular concern. The concession embraced 500,000 square miles, an area somewhat larger than half the United States east of the Mississippi River. It granted D'Arcy the exclusive right to lay pipelines to the Persian Gulf and made available all land within the concession, except holy places, that was necessary to D'Arcy's operations. D'Arcy was to get without charge the use of any state-owned uncultivated lands and was to purchase cultivated lands at fair prices. He could buy private lands by agreement with their owners, but they could not charge prices out of line with those prevailing in the locality. The concession exempted all lands and all materials imported for the maintenance and development of the concession from all taxes and imposts throughout the life of the concession. The government bound itself to take all necessary measures for protecting the property and personnel of the exploiting company; but having done this, the government denied the company "the power, under any pretext whatever, to claim damages from the Persian Government." Other provisions also aimed to protect the government's interests. Under the terms of the concession the government could appoint an imperial commissioner to advise with D'Arcy and to be paid by him, who should establish by agreement with him such supervision as might be deemed "expedient to safeguard the interests of the Imperial Government." D'Arcy agreed to employ Persians exclusively except for managerial and technical staff, including oil-drillers. When the concession expired, D'Arcy was to turn over to the government without charge all materials, buildings, and equipment used by him. The concession gave D'Arcy the right to organize one or more companies to work the concession with all the privileges and responsibilities that D'Arcy enjoyed.

In return for these rights the exploiting company or companies were to pay the Persian government within a month of their formation £20,000 in cash, £20,000 in stock, and 16 percent of the annual net profits, plus a fixed sum of 2,000 tomans ($1,800) per year. All disputes arising over the meaning of the concession regarding the rights and responsibilities of either party were to be submitted to two arbitrators, one appointed by each of the parties, and an umpire selected by the arbitrators. The decision of the arbitrators or, in case they disagreed, of the umpire was to be final.

7. League of Nations, *Official Journal*, XIII (1932), Annex 1419c, Appendix, 2305–2307.

In June 1901, one month after the signing of the concession, the shah issued a royal decree to be added to the concession agreement.

Pursuant to the concession granted to Mr. William Knox D'Arcy, as a result of the particularly friendly relation which unites powerful Great Britain and Persia, it is accorded and guaranteed to the Engineer William D'Arcy, and to all his heirs and assigns and friends, full powers and unlimited liberty for a period of sixty years to probe, pierce and drill at their will the depths of Persian soil; in consequence of which all the subsoil products wrought by him without exception will remain the property of D'Arcy. We declare that all the officials of this blessed kingdom and our heirs and successors will do their best to help and assist the honorable D'Arcy, who enjoys the favor of our splendid court.[8]

When the concession was granted, no one could foresee the pecuniary costs and physical difficulties of developing Persia's oil lands, the vast riches that were to flow from them, or the political turmoil to which their exploitation would eventually lead. The circumstances under which the concession was negotiated and its terms indicate that an irresponsible and despotic ruler, unsophisticated in business affairs, was more interested in meeting his and the royal family's needs than in protecting the patrimony of his people.

First Steps in Exploiting the Concession

At the outset the effort to exploit Persia's oil resources was strictly a commercial enterprise, but a commercial venture with important political implications. After having conducted explorations, hired drillers, imported drilling equipment and technical and domestic supplies essential to drilling operations in a bleak and inhospitable environment, G. B. Reynolds, D'Arcy's field commander, began drilling near an oil seepage at Chiah Surkh, an area on the Persian side of the Persian-Iraqian frontier. After May 1903, the First Exploitation Company, organized with a capital of £600,000 in shares of £1 each, took control of the drilling program. The first venture met with failure. Although oil trickled from the first well and flowed more generously from the second, remote as they were from the seaboard, they did not promise production in commercial quantities. Undeterred, Reynolds, a tough-fibered man of experience and tenacity, in

8. Ketab-Khanah-i Saltanati, Majmouah-i-Farmin (Royal Library, Teheran), quoted in Nasrollah Saifpour Fatemi, *Oil Diplomacy* (New York: Whittier Books, 1954), p. 357.

1905 shifted operations to the south, where transportation to the Persian Gulf would be simpler. At Mamatain in the province of Khuzistan, a barren area with a torrid climate, devoid of roads, in the midst of hostile and thieving tribesmen, with primitive equipment, Reynolds and his crew began drilling late in 1906.

Under the concession contract the central government at Teheran assumed responsibility for protecting the company operations, but weak and corrupt as it was, it lacked control of the independent Bakhtiari tribesmen. To protect operations from their raids D'Arcy formed the Bakhtiari Oil Company, Ltd., with 400,000 shares of £1 each, and presented 3 percent of the share capital to the Bakhtiari khans. In addition, the company agreed to pay the tribesmen £2,000 annually for safeguarding the drilling equipment and supplies and £1,000 annually for safeguarding the pipeline when built.

Oil for the British Navy

Meanwhile D'Arcy, whose funds were practically exhausted, tried to obtain additional capital in London. Failing in this, he turned to foreign sources. At this juncture the British Admiralty took a step that was to convert oil production in Persia from a private commercial enterprise into a quasi-public venture rich with promise to Great Britain and with political implications for the whole Middle East. As long ago as 1882 Captain John Arbuthnot Fisher, with prophetic vision, had contended that "the general adoption of oil instead of coal as the fuel for ships would immediately increase the fighting capacity of every fleet by at least 50 percent."[9] By the time D'Arcy had run short of funds in his Persian oil operations, Fisher had become first sea lord of the Admiralty. His ideas of 1882 were his obsession of 1904, and he was in a position to make them a reality.

One of Fisher's first acts was to appoint an oil committee with E. G. Pretyman, a member of Parliament and civil lord of the Admiralty, as chairman. Years later Pretyman, speaking before Parliament, had this to say about the committee:

It was in the time of the late Board of Admiralty that it first became evident that oil fuel would be necessary for the British Navy. . . . That being so, it became obviously a question of the greatest import as to whence our supplies

9. Quoted in Anton Mohr, *The Oil War* (New York: Harcourt, Brace, 1926), p. 3.

were to be drawn. . . . With that object in view a Committee was appointed by the First Lord of the Admiralty, Lord Selborne, to take charge of that particular question. . . . [The Committee decided that what was wanted was] an independent source of supply which is, as far as possible, uncontrolled by any agency which can exact undue prices. . . . When we surveyed the whole of the oil-fields, it appeared . . . that practically the whole area was covered by certain large concessions. . . . It was not only a question of the magnitude of those concerns, it was also a question of whether they were under British control or foreign or cosmopolitan control.[10]

Pretyman explained that the Admiralty had learned that D'Arcy was negotiating with foreign capitalists for the sale of his concession. The Admiralty approached Lord Strathcona of the Burmah Oil Company, which then had an agreement with the Admiralty regarding emergency fuel oil for the navy, with a suggestion that he form a syndicate with D'Arcy and provide the funds necessary to develop the Persian concession. In the parliamentary debates, Pretyman gave this account of subsequent developments:

It is only due to them [Strathcona and Burmah Oil] to point out that this enterprise originated at the request of the Admiralty, and not from a purely commercial purpose. Lord Strathcona, whom I saw personally—I think it was characteristic of him—only asked me one question. He asked, is it in the interest of the British Navy that this enterprise should go forward, and that I should take part in it? I said, "It is," and Lord Strathcona, without any further questions, agreed to do what he had been asked.

Thereafter Strathcona, representing Burmah Oil, began negotiations with D'Arcy, who had been asked by the Admiralty to break off his negotiations with foreign capitalists. Strathcona's negotiations bore fruit in May 1905, when the Concessions Syndicate Ltd. was formed to take over the assets of the First Exploitation Company, with D'Arcy as a director and with adequate capital to continue operations in Persia.

Oil Strike and APOC

It was at this juncture that Reynolds shifted his operations to the south. After two unsuccessful attempts at Mamatain, Reynolds moved his drilling rigs to the site of an ancient fire temple known locally as Masjid-i-Sulaiman

10. *Parliamentary Debates, House of Commons* (hereafter P.D.C.), Vol. 63, cols. 1189–1191, June 17, 1914.

or "Mosque of Solomon." He spudded in the first well in January 1908 and the second in March. With funds nearing exhaustion and orders drafted to abandon the whole enterprise, on May 26, 1908, the drill punctured the cap rock that sealed one of the world's largest oilfields.

With oil in abundance, D'Arcy's financial difficulties were at an end. Appropriate to the larger scale of operations made possible by the discovery, the Concessions Syndicate in April 1908 formed the Anglo-Persian Oil Company; the Burmah Oil Company bought most of its £1 million of common stock, and its preferred stock, open to the public, was quickly oversubscribed. At the Admiralty's suggestion, Lord Strathcona, then in his eighty-ninth year, became its chairman, with D'Arcy a member of its board of directors.

British Government Protects APOC

The British government, indirectly responsible for the birth of the Anglo-Persian Oil Company, took steps to provide an environment favorable to its growth. In truth, it gave extended prenatal care. Since the protection that the Bakhtiari khans had been paid to provide proved inadequate,[11] the Indian government in 1907 dispatched Lt. Arnold T. Wilson with twenty men of the 18th Bengal Lancers to the drilling area. Stationed in Ahwaz, he could guard the British consulate and at the same time protect the drillers from predatory tribesmen.

Lt. Wilson was later—the Indian guard went back to India in 1909—to play an important part in the activities of both his country and the Anglo-Persian Oil Company in the Middle East. He became acting consul

11. Sir Arnold Wilson's description of the lack of connection between the payments made to the Bakhtiari khans and the payments to the guards who were to protect the concession is revealing. In a letter dated August 5, 1908, when he was a lieutenant of the Indian Army stationed in Persia, Wilson wrote: "The Company pay the tribal chiefs a large sum for these 'guards'; it is of course a sort of blackmail; the tribal Khans pay them nothing. The guards levy toll upon the local labourers and are changed from time to time so as to give everyone a chance." *SW. Persia: A Political Officer's Diary,* 1907–1914 (London: Oxford University Press, 1941), p. 53. The discovery of oil not unnaturally increased the British government's willingness to help. Wilson wrote on September 8, 1908, "As oil has been found and as prospects are now good the Home Government may wish to ensure that the D'Arcy Exploration Company may continue their work without hindrance. They are ready to send more men if need be. But for troubles in Persia I should be back in India." *Ibid.,* p. 55.

at Mohammareh, where the company was "my particular charge";[12] and in
the years 1918–1920 he served as Acting Civil Commissioner for Mesopotamia, when he may have determined the boundary line between Turkey
and Mesopotamia after the Armistice. E. Thurtle, M. P., speaking in the
House of Commons in 1926, stated that at the time of the Armistice British
military forces were south of Mosul. Thurtle surmised that the military
commander, unfamiliar with the territory, would be likely to take the
advice of the highest civil authority.

The Acting Civil Commissioner was at the military headquarters, and the
boundary line was moved on beyond Mosul and then for a hundred miles
beyond that. This was done on the authority of the Acting Civil Commissioner
in consultation with the chief military officer. . . . The sequel . . . is that Sir
Arnold Wilson is now the general manager of the Anglo-Persian Oil Company
for the Persian and Mesopotamian area.[13]

The Arab tribes, occupying an area in southern Khuzistan bordering on
Iraq and the Persian Gulf, presented a more complex problem. A pipeline
to the gulf must cross their province and a projected refinery on the gulf
coast be located in it. If the Anglo-Persian Oil Company were to carry out
these operations with a reasonable degree of security, it was necessary to
obtain the co-operation of Sheik Khazal, the legal owner of the area, who
as a semi-independent emir had with the Bakhtiari openly opposed the
Teheran government during the Revolution of 1906.

British government agents took steps to insure the necessary co-operation. Within a month after the organization of the Anglo-Persian Oil
Company, the British resident at Bushire, Sir Percy Cox, journeyed to
Mohammareh to open negotiations with Sheik Khazal on behalf of the
company. Sir Percy succeeded. After prolonged discussion Sheik Khazal on
July 16, 1909, signed an agreement granting a pipeline right of way over
his territory and permitting the Anglo-Persian Oil Company to buy one

12. *Ibid.,* p. 133. An early service to the company consisted of successful intervention in 1910 when the Persian government claimed the company's barges were
dutiable on the Upper Karun River, "under a convention or declaration older by
some years than the Company's concession." (p. 125.) Wilson could not recall what
he did or how he did it, but he preserved an expression of the Viceroy of India's
recognition of "the great credit due to Lt. Wilson for the excellent manner in which
he championed the Oil Company's cause." *Idem.*
13. P.D.C., Vol. 191, col. 2223, February 18, 1926. On March 10, 1921, it was
stated in the House of Commons that Sir Arnold Wilson "is now on leave preparatory to retirement and will not take up duties with or receive salary from the
Anglo-Persian Oil Company until he has retired." *Ibid.,* Vol. 139, col. 652.

square mile on Abadan Island for a refinery, a depot, warehouses, and other paraphernalia essential to refinery operations. At the same time Sheik Khazal agreed to protect the company property from marauding tribesmen. For these services the company agreed to pay Sheik Khazal an annual rental of £650 for ten years, the whole ten years' rental to be paid in advance. After ten years the annual rental was to be increased to £1,500. In addition, Anglo-Persian lent the sheik £10,000 with the British government officiating as the nominal lender. Of greater significance, the British not only recognized Sheik Khazal as the lawful ruler of Mohammareh but guaranteed his and his successors' rights to Mohammareh. That such an agreement was necessary is a reflection of the weakness of the central government, but the terms negotiated by a representative of the British government plainly disregarded the letter and spirit of the D'Arcy concession. The company deducted the payments to the local tribal chiefs from its royalty payments to the Persian government.[14]

Pipeline, Refinery, and F7

Shortly after the signing of the agreement the company began the construction of a pipeline designed to carry approximately three million barrels of oil a year. With primitive facilities, untrained labor, and a difficult terrain, the construction of the pipeline took more than two years. To build a refinery on an uninhabited mudflat between the Shatt-al-Arab and Bahmeshire Rivers required even longer. Although a single distillation unit was completed by August 1912, the refinery was not completed until early in 1913. Meanwhile drilling had proceeded apace. By 1914 some thirty wells had been drilled just short of pay rock.[15] One of these, F7, completed in November 1911 and producing more than 17,000 barrels a day in March 1914, was to become one of the most celebrated single wells in the world, with a total output of more than 52 million barrels (7 million tons) before it was sealed in April 1926.

British Government Buys Control of APOC

By this time the dreams of Admiral Fisher were fast coming true. When Winston Churchill became first lord of the Admiralty in 1911, the Admi-

14. "Anglo-Iranian Answers Iran with Facts," *Oil Forum*, April 1952, Special Insert, p. iv.
15. Stephen Hemsley Longrigg, *Oil in the Middle East* (London: Oxford University Press, 1954), 19n1.

ralty had or was building 56 destroyers and 74 submarines powered by oil.[16] Nevertheless, the British Navy was not yet so dependent on oil that it had a serious supply problem. But under Churchill's leadership it was on the eve of a three-year expansion program that "comprised the greatest additions in power and cost ever made to the Royal Navy." With the exception of the battleships built in 1913 (later converted to oil), these did not contain a single coal-burning ship. As Churchill put it,

Submarines, destroyers, light cruisers, fast battleships—all were based irrevocably on oil. The fateful plunge was taken when it was decided to create the Fast Division. Then, for the first time, the supreme ships of the Navy, on which our life depended, were fed by oil and could only be fed by oil. The decision to drive the smaller craft by oil followed naturally upon this. The camel once swallowed, the gnats went down easily enough.[17]

With no appreciable oil production anywhere in the British Empire, prudence seemed to dictate that the Admiralty seek a supply that it could control. With Anglo-Persian a going enterprise in whose survival the British government had already interested itself, with production from fast-flowing wells, a completed pipeline, and a refinery in full operation, the Admiralty turned to it. According to Sir Winston, "An unbroken series of consequences conducted us to the Anglo-Persian Oil Convention." The first step was to create a Royal Commission on Oil Supply. The commission sent Admiral Sir E. J. W. Slade, destined to become a vice-chairman of the Anglo-Persian Oil Company,[18] and a committee of experts, including John Cadman, professor of mining at Birmingham University and later chairman of the company, to the Persian Gulf to examine at first hand Anglo-Persian's operations. What the committee saw impressed it greatly. On the basis of its enthusiastic report the Admiralty made first a twenty-year supply contract with Anglo-Persian and then an agreement to buy a controlling interest in it for £2,200,000. Less than two months before the outbreak of World War I, on June 17, 1914, Churchill laid the agreement before Parliament and moved its approval.

In the debate that followed, Churchill outlined the advantage to the Admiralty and Great Britain that this purchase would bring. It would free the navy from exclusive reliance on a world market for oil dominated by "two gigantic corporations," Standard Oil and Royal Dutch-Shell. It would help to insure supplies in peace and war at reasonable prices. Britain would

16. Winston S. Churchill, *The World Crisis* (New York: Charles Scribner's Sons, 1923), p. 134.
17. *Ibid.*, p. 136.
18. *P.D.C.*, Vol. 63, col. 1137, 1140, June 17, 1914.

obtain control for nearly fifty years of the oil potentialities of a petroliferous area of about half a million square miles in extent—that is to say, roughly speaking, nearly as big as France and Germany put together.

. . . .
Over the whole of these enormous regions we obtain the power to regulate developments according to naval and national interests, and to conserve and safeguard the supply of existing wells pending further development.[19]

Nor would the acquisition of control in the company "produce any untoward effect on our foreign relations," or aggravate the political difficulties that confronted Persia. On the contrary, the additional "investment of capital, the development of roads, railways, and industries in which the tribesmen and the Persian Government are both interested, and from which both profit ought to tend to make the Persian Government strong and the tribesmen tame."

Commenting on the fact that a supply contract was a counterpart of the agreement, Churchill said,

We are not only, of course, proposing to be the largest shareholder and the predominant partner in this concern, but we are also its principal and most regular customer. What we do not gain at one end of the process we recover at the other.

The proposal was sound, he argued, not only as a means of assuring the navy an adequate and continuing oil supply, but as a business proposition. The Admiralty would gain control over a "proved proposition, a going concern." Having negotiated a supply contract that would "confer upon the Anglo-Persian Company an immense advantage which, added to their concession, would enormously strengthen the company and increase the value of their property,"[20] why should the government not share in the advantage that it had created? If the House did not approve the agreement, Churchill argued, Anglo-Persian would doubtless merge with one of the world's dominant firms and the navy would have to pay more for its oil.

Not all House members agreed with Churchill. George Lloyd pointed out that the oil wells, with perhaps two exceptions, lay outside the zone of British influence, where defense would be difficult. He argued that the oil properties would be exposed to attack by the fierce and turbulent Bakhtiari and Arab tribes and threatened by the Russian Cossacks in the north, and that their purchase would burden the Indian government with "new and immense responsibilities."

19. *Ibid.*, col. 1140, 1145.
20. *Ibid.*, col. 1146, 1140.

Ramsay MacDonald felt that the agreement's political significance was more important than its economic meaning. He feared that action by British troops against hostile tribes in the neutral zone would invite a Russian countermove. He feared that reliance on local tribes for the protection of the properties would tend to weaken the central government. He feared that ownership by the British government of a controlling interest in the Anglo-Persian Oil Company might ultimately lead to a demand for the repartition of Persia. "All commercial concessions, especially with Government money in them," he argued, "have a very unhappy knack, in the course of time, of becoming territorial acquisitions."[21]

A. W. H. Ponsonby was even more articulate in his belief that the venture threatened the independence of Persia.

We know that ever since the Anglo-Persian Convention of 1907 we have been weakening the Government of Persia.
There can . . . be no manner of doubt that we are working towards a partition of Persia.[22]

Spokesmen for the Shell group, among them Samuel Samuel, a director of Shell Transport and Trading Company, protested the government's acquiring a majority interest in a competitor. They rejected Churchill's assertion that high oil prices were caused by monopolies and labeled his attack on the Shell group as Jew-baiting.[23]

Despite this frank and forceful opposition, Churchill's views prevailed. After a debate lasting only slightly more than seven hours, the House approved the agreement by a vote of 254 to 18.[24]

Benefits to British Government

About the entire naval oil program Churchill had this to say in 1923:

And so it all went through. Fortune rewarded the continuous and steadfast facing of these difficulties by the Board of Admiralty and brought us a prize from fairyland far beyond our brightest hopes.[25]

Manifesting a bent for mathematical calculation and optimistic forecasting, he estimated the government's gains from the appreciation in value of

21. *Ibid.*, col. 1164.
22. *Ibid.*, col. 1175–1176.
23. *Ibid.*, col. 1221, 1229.
24. *Ibid.*, col. 1250.
25. Churchill, 137n42.

its original investment of £2,200,000 (subsequently raised to £5 million), dividends, lowered prices charged by other suppliers, and savings on the purchase of navy oil under the supply contract through the life of the contract to be £40 million, and from this he concluded,

The aggregate profits, realised and potential, of this investment may be estimated at a sum not merely sufficient to pay for all the programme of ships, great and small of that year and for the whole prewar oil fuel installation; but are such that we may not unreasonably expect that one day we shall be entitled also to claim that the mighty fleets laid down in 1912, 1913 and 1914, the greatest ever built by any power in an equal period, were added to the British Navy without costing a single penny to the taxpayer.[26]

Arthur C. Millspaugh, an American who served as administrator general of the finances of Persia from 1922 to 1927 and from 1943 to 1945, said of the British government's interest in Anglo-Persian,

While the government undertook not to interfere in the commercial management of the company, the latter became to all intents and purposes an arm of the British Admiralty and of British strategic policy.[27]

Benefits to APOC

But the gain was not one-sided. The Anglo-Persian Oil Company obtained additional capital with which to carry out an extensive program of expansion and obtained an outlet for most of its oil without the necessity of competing for it in world markets. What is perhaps of greater importance, the welfare of Great Britain and the Anglo-Persian Oil company became identified, an identity that was to stand the company in good stead in the turbulent years ahead.

With new funds and assured markets Anglo-Persian rapidly increased its production. In 1913 it had produced only 1,857,000 barrels of crude oil; in the closing year of World War I it produced 8,623,000 barrels. During the war it rendered important services to the British government by supplying fuel oil to ships in the Mediterranean and the Far East and oil, gasoline, and kerosene in other war areas, particularly Mesopotamia.

In the postwar decade Persia attained the position of a major oil

26. *Ibid.*, p. 140.
27. Arthur C. Millspaugh, *Americans in Persia* (Washington: Brookings Institution, 1946), p. 162.

producer. In 1920 it produced 12,230,000 barrels and in 1930, 45,833,000 barrels, becoming the fourth largest among oil-producing countries of the world, exceeded only by the United States, Russia, and Venezuela. But with mushrooming production came an intensification of Anglo-Persian's political difficulties.

Anglo-Persian Agreement of 1919

A brief review of relations between Persia and Great Britain in the war and early postwar period will aid in understanding the difficulties that the Anglo-Persian Oil Company encountered during the 1920s. Although Persia tried to maintain her neutrality during the war, her fear of and hostility toward Russia and Great Britain made it difficult to do so. Encouraged by Persia's attitude towards Great Britain and Russia and with an eye on the main chance, Germany attained considerable influence in Persia and particularly in Teheran before and during the early war years. Russia, to protect its sphere of influence, speedily occupied northern and northwestern Persia, and to protect Anglo-Persian's Abadan refinery, British forces occupied the Turkish port of Basrah. As a result of the Bolshevik Revolution, Russian forces withdrew from Persia, and by the end of the war Persia was under British protection and receiving £225,000 monthly to meet regular governmental expenses and help maintain the Cossack Division.

Aware of Russia's long-standing efforts to subjugate Persia to its territorial ambitions and of the necessity for a stable and friendly government to protect its own territorial and economic interests, Great Britain after the war sought to establish on a lasting basis control over the Persian government and Persia's armed forces. Lord Curzon, the chief architect of this postwar plan, stated its objectives:

Now that we are about to assume the mandate for Mesopotamia, which will make us coterminous with the western frontiers of Persia, we cannot permit the existence, between the frontiers of our Indian Empire . . . and those of our new Protectorate, of a hotbed of misrule, enemy intrigue, financial chaos, and political disorder. Further . . . we possess in the south-western corner of Persia great assets in the shape of the oilfields, which are worked for the British Navy and which give us a commanding interest in that part of the world.[28]

28. Memorandum by Earl Curzon on the Anglo-Persian Agreement of August 9, 1919, *Documents on British Foreign Policy, 1919–1939,* ed. E. L. Woodward and Rohan Butler (London: Her Majesty's Stationery Office, 1952), First Series, IV, 1119.

The chosen instrument for stabilizing a government responsive to its need was the so-called Anglo-Persian Agreement of 1919.[29] Negotiated in secrecy and eased along by substantial "subsidies" to important Persian officials, yet avowing "the close ties of friendship which have existed between the two Governments in the past" and the conviction that "these ties should be cemented," and reiterating "in the most categorical manner" the British government's intention "to respect absolutely the independence and integrity of Persia," the treaty provided that Great Britain should furnish Persian governmental departments with advisers "endowed with adequate powers," and that it should supply the Persian army with British officers and such munitions and equipment as would be necessary to preserve order within the country and protect its frontiers. All of this was to be done at Persia's expense but to be financed by a British loan secured by customs revenue or "other sources of income at the disposal of the Persian Goverment."

The agreement was stillborn. Its publication brought forth violent criticism from all sides, both domestic and foreign. Under the Persian constitution the agreement required ratification by the Majlis (national assembly), but before elections to it were completed, British forces withdrew in the face of Soviet military action on Persian soil, shaking Persian confidence in the ability of the British to defend the country and causing the fall of the prime minister who had negotiated the agreement. His successor refused to carry out the agreement until it was ratified, and it remained "in a state of suspended animation until its final repudiation"[30] after the Reza Khan coup of February 21, 1921. Unfortunately the hostility and suspicion that the agreement engendered did not die with it. They remained to intensify Persia's historic distrust of the British and to aggravate disputes that subsequently arose over the relative rights and responsibilities of the Anglo-Persian Oil Company and the Persian government, plaguing the British in their effort to resolve them.

Disputes over Concession Terms

Article 9 of the D'Arcy concession authorized the concessionaire "to found one or several companies for the working of the concession," pro-

---

29. The agreement is reproduced in H. W. V. Temperley, "The Liberation of Persia," pt. v. of Chap. 1 in *A History of the Peace Conference of Paris,* ed. H. W. V. Temperley (London: British Institute of International Affairs, 1924), VI, 212–213.
30. Sir Arnold T. Wilson, *Persia* (London: Ernest Benn, Ltd. 1932), p. 143.

vided the concessionaire gave the government official notice of the formation of each company, a copy of its charter ("statutes"), and information "as to the places at which such company is to operate." The article also gave such company or companies "all the rights and privileges granted to the concessionaire" but required them to "assume all his engagements and responsibilities."[31] Article 10, it will be recalled, provided that the concessionaire should pay the Persian government annually "a sum equal to 16 percent of the annual net profits of any company or companies that may be formed in accordance" with Article 9. Under Article 14 the Persian government bound itself to take "all and any necessary measures" to insure the safety of the exploitation company in its operation. Having done so, the Persian government absolved itself from any claim for damages.

The provisions came into dispute during World War I. As previously indicated, early in the war both Britain and Russia violated Persia's neutrality. This heightened the sympathy of the Persian people toward the Central Powers, and unruly Bakhtiari tribesmen ignored the obligation they had assumed to protect Anglo-Persian's operations and, incited by the Turks, severed its pipeline in February 1915.[32] The oil company, relying on Article 14 of the concession, refused to pay royalties due the Persian government and presented a claim for damages. The government, contending that the damages were occasioned by factors beyond its control, denied responsibility and requested arbitration as provided in Article 17 of the concession. The company was willing to submit to arbitration the amount of damages[33] but not the question of responsibility.

31. League of Nations, *Official Journal,* XIII (1932), Annex 1419c, Appendix 2306.

32. The Anglo-Iranian Oil Company's description of the incident was as follows: "Iranian tribesmen, incited by enemy propaganda and gold, cut the Company's pipeline, thereby bringing all its operations to a standstill for a number of months and so causing serious loss to it." "Anglo-Iranian Answers Iran with Facts," *Oil Forum,* April 1952, Special Insert, p. iv. The Persian government's view of the incident is reflected in a statement made by M. Davar, the Persian government's representative, before the Council of the League of Nations in 1933: "Our neutrality was violated by the *company* . . . whereupon other belligerents came and organized armed bands, and it was these which cut the pipelines. And we were expected to pay for it! We could not do so, since the losses sustained were due to the war and to acts at variance with Persian neutrality." League of Nations, *Official Journal,* XIV (1933), 208. Emphasis supplied.

33. The parties' estimates of the amount of damage varied by a wide margin, the company claiming more than £600,000 and the government declaring that "the real amount of the loss sustained" was about £20,000. Memorandum from the Imperial Government of Persia, January 18, 1933, Annex 1422b, *ibid.,* p. 290. According to

Meanwhile, a dispute had arisen with regard to the calculation of the 16 percent net profits due annually to the Persian government. The government contended that the profit-sharing provision applied to all companies formed by Anglo-Persian regardless of the branch of the industry or the country in which it operated. The company contended that it applied only to companies operating in Persia under the control of the Anglo-Persian Oil Company.

Armitage-Smith Agreement

After the Anglo-Persian Agreement of 1919 was signed, but before the Reza Khan administration repudiated it, the British Treasury lent Sydney A. Armitage-Smith to the Persian government as financial adviser. On May 6, 1920, he and six assistants took up their duties in Teheran. On August 29, 1920, the undersecretary of the Persian ministry of finance addressed a letter to Armitage-Smith notifying him of his appointment by the Persian government "to finally adjust all questions in dispute between the Anglo-Persian Oil Company and the Imperial Government of Persia."[34]

On December 22, 1920, Armitage-Smith on behalf of the Persian government reached an agreement with the Anglo-Persian Oil Company which was subsequently signed by the parties.[35] The agreement accepted the Persian government's contention that the profit-sharing provision in the concession applied to all companies formed by Anglo-Persian. When Articles 9 and 10 are read in conjunction with Article 1, the Persian government's contention seems not unreasonable. As previously indicated, Article 1 gives an "exclusive privilege" to the concessionaire "to search for, obtain, exploit, develop, *render suitable for trade, carry away and sell* natural gas petroleum, asphalt and ozocerite. . . ." While accepting in principle the Persian government's contention on the scope of profit-sharing, the Armitage-Smith agreement provided for deductions from the net profits of

---

Davar, the £20,000 figure was set by a British chartered accountant who had been commissioned by the Persian government to audit the company's books and who the United Kingdom's representative before the Council had said was "above suspicion." *Ibid.,* p. 208. See also p. 210.

34. Appendix No. 3 to Annex 3 "Statement of Relevant Facts up to May 1, 1951" to Memorial Submitted by the Government of the United Kingdom (hereafter United Kingdom Memorial), International Court of Justice *Pleadings, Anglo-Iranian Oil Co., Case (United Kingdom* v. *Iran,* 1952) (hereafter I.C.J. Pleadings), p. 228.

35. Appendix No. 4, *Ibid.,* pp. 229–235.

"subsidiary companies refining, distributing or dealing with Persian oil outside Persia,"[36] based on the volume of petroleum products handled, before the calculation of the 16 percent due the Persian government. This in effect provided the subsidiary companies with capital at the Persian government's expense. The agreement also exempted tanker companies from the profit-sharing principle and held that time-charter rates, not actual transportation costs, should apply in calculating the freight costs of companies using Persian oil—a second modification of the concession at the Persian government's expense. It also provided that any dispute regarding royalties should be referred not to arbitrators sitting at Teheran, as provided in Article 17 of the concession, but to a chartered accountant nominated by the president of the Institute of Chartered Accountants in England, whose decision should be final.

In return for these concessions the oil company dropped its claim for losses suffered by the cutting of its pipeline and paid the Persian government £1 million in settlement of all past claims.[37] This settled the issue only temporarily.

Reza Khan's Revolution

Shortly after the signing of the Armitage-Smith agreement a bloodless revolution took place in Persia. On the night of February 20, 1921, Colonel Reza Khan marched into Teheran at the head of a Cossack detachment and with Sayyid Zia ed-Din Tabatabai, a young political reformer, took control of the central government on the following day. Thereafter Sayyid Zia became prime minister and Reza Khan his minister of war. Clashing with vested interests in the speed with which he endeavored to institute reforms and unable to retain the support of Reza Khan, Sayyid Zia fled the country barely three months after assuming the prime ministership. Reza Khan as commander-in-chief of the army and minister of war remained as undisputed master of the revolution. In 1923 he became prime minister. In 1925 the Majlis voted to depose Ahmad Shah, the last of the Qajar dynasty, who since the revolution had functioned primarily as a figurehead, and named as shah Reza Khan, who had suppressed tribal rebellions and saved the central government. Reza Shah

36. Article 3 of Armitage-Smith agreement, Appendix No. 4 to Annex 3 to United Kingdom Memorial, *I.C.J. Pleadings*, p. 230.
37. Annex 3 to United Kingdom Memorial, *ibid.*, p. 190.

Pahlevi remained virtual dictator of Persia until forced to abdicate by British and Russian pressure in World War II.

Shortly after Reza Khan assumed control of the Persian government, Lumley & Lumley, a London law firm that had been retained by the Persian government to examine the documents related to the Armitage-Smith agreement, expressed the opinion that the agreement constituted not an interpretation of the concession agreement but a modification of it and hence required ratification by the Persian government to make it binding.[38] Lumley & Lumley also expressed the opinion that "Mr. Armitage-Smith may have exceeded the powers conferred upon him" in negotiating a settlement on such broad terms. After analyzing the agreement they concluded with these remarks:

> As we understand the matter, it is a question entirely for the Imperial Government whether they adopt the alleged Agreement or not, but we venture to think that the various matters we have indicated above point to the conclusion that the adoption of the Agreement would mean the Imperial Government giving up important rights and interests under the Concession.

They warned, however, that "if . . . it is intended to repudiate the Agreement . . . steps should be taken to do so as promptly and with as little delay as possible."

Failures in Negotiation

Although the agreement was never ratified, neither was it properly repudiated. In accordance with the complicated procedure it set up, the company made annual payments to the Persian government totaling £8,579,195 in the ten years after Armitage-Smith made it.[39] During the early years of the decade disputes arose relative to their adequacy. Mild at the outset, with the passing of time they became more and more acrid. As

38. Letter of July 27, 1921, to the Imperial Commissioner attached to the Persian Legation in London, League of Nations, *Official Journal*, XIV (1933), Annex 1422b, Appendix III, 296–297.

39. Annex 3 to United Kingdom Memorial, *I.C.J. Pleadings*, p. 188. The Armitage-Smith agreement was signed December 22, 1920; but the figure given for total royalties paid to the Persian government in the following decade is computed from amounts given for the periods April 1 to March 31 for the years 1921–1928 and January 1 to December 31 for the years 1929–1930 and so understates them by the amount paid for the period January 1 to April 1, 1921.

early as 1924 the government challenged the adequacy of the payments, and it repeated the challenge for each of the next three years.

On May 9, 1928, the Persian oil commissioner addressed a letter to the Anglo-Persian Oil Company in London calling attention to a private letter that on December 22, 1926, he had sent to Sir John Cadman, chairman of the company, asking for an early settlement of the disputes pertaining to the royalties paid in 1924 and thereafter.[40] "The company making no movement towards a settlement," in February 1927 the commissioner had proceeded to London, where he "took the initial steps" toward arbitration. Thereafter matters had dragged. In his letter of May 9, 1928, the commissioner charged that Armitage-Smith had exceeded his instructions, that the deductions his agreement authorized to be made before calculating the 16 percent of profits due the government vitiated Article 10 of the concession agreement, and that the Persian government had never officially ratified the agreement but had "unofficially acquiesced in this arrangement as a *modus vivendi* for the time being."[41] He closed his letter with the request that the company's legal advisers get in touch with his government in order that a "final and complete settlement" might be effected.

On August 12, 1928, Prince Teymourtache, minister of the Persian court, wrote to Sir John Cadman lamenting the unsatisfactory progress made in negotiations.[42] He alluded to "the obsolescence of the D'Arcy Concession, which was obtained at a time when the Government of the Kadjars did not realise what was being taken from it and what it was giving," and he contended that the "concession needs to be remade, recast. . . ." Sir John Cadman in the same year suggested to the minister of court that if the Anglo-Persian Oil Company were to obtain adequate capital to develop the concession properly, it would be necessary that its life be extended.

Negotiations looking to a new concession agreement continued intermittently throughout the next three years. They were discontinued in 1931 because, in the language of the British government, "the demands of the Persian Government were greatly in excess of anything which the Company could accept."[43]

---

40. League of Nations, *Official Journal*, XIV (1933), Annex 1422b, Appendix IV, 298.

41. *Ibid.*, p. 299.

42. *Ibid.*, Annex 1422b, p. 291.

43. In April 1931 the Persian government proposed that any new concession contain a guarantee by the company of not less than £2,700,000 annual payments to

The Persian government was not content to let the matter rest, however, and in November 1931 the parties opened negotiations anew directed specifically toward the question that most troubled the Persian government, the calculation of the annual payments due it under Article 10 of the concession. The negotiators made progress, reaching a preliminary agreement on principles and concluding a provisional agreement which the Council of Ministers in Teheran approved in February 1932. The parties then referred the provisional agreement to lawyers and accountants with instructions to draft a final agreement. The latter forwarded a final agreement approved by the negotiators to Teheran for ratification by the Persian government. It arrived on May 29, 1932.[44] The Persian cabinet undertook examination of the agreement preparatory to submitting it for parliamentary sanction.

It was never submitted. Five days after its arrival, the company completed its accounting for the year 1931 and cabled the results to Teheran. The figures made disconcerting reading. Whereas the company had paid the Persian government £1,288,312 in 1930, it calculated the royalty payable for 1931 at £306,872.[45] The decline in the Anglo-Persian Oil Company's revenues upon which payments to the Persian government largely depended reflected the world economic depression and more particularly the depressed condition of the oil industry, suffering not only from a decline in demand but from a glut in world markets. With a decline of about 3 percent in Persian production, royalty payments to the Persian government based on the company's profits fell, as indicated above, by about three-fourths.

The Persian government was greatly disturbed and on June 29, 1932, refused to accept the royalty as calculated. Long suspicious of the oil company and its accounting methods, the government sought not a rational explanation but more revenue. The cabinet accordingly subjected the new royalty agreement to careful and skeptical scrutiny. Finding it "complicated and obscure" and likely to "lead to differences of interpretation,"[46] it

the government whether the company made any profits or not. Statement by Sir John Simon, United Kingdom representative, before the Council of the League of Nations, January 26, 1933, League of Nations, *Official Journal*, XIV (1933), 202.

44. Memorandum by His Majesty's Government in the United Kingdom, submitted to the Council on December 19, 1932 (hereafter United Kingdom Memorandum), League of Nations, *Official Journal*, XIII (1932), Annex 1419c, 2300.

45. United Kingdom Memorandum, League of Nations, *Official Journal*, XIII (1932), Annex 1419c, 2300.

46. Memorandum from the Imperial Government of Persia, January 18, 1933, League of Nations, *Official Journal*, XIV (1933), Annex 1422b, 291.

requested the company to send its experts to Teheran for further discussion. The company replied that unfortunately this was "impossible" and suggested that talks be resumed in London. On July 7, 1932, the Persian government indicated that it was preparing a new set of proposals "on an entirely different basis."[47] There the matter seemed to rest for the next few months. On November 16 the minister of finance informed the company's representative in Teheran that the proposals were almost complete.

Eleven days later, on November 27, 1932, the minister of finance delivered formal notice to the resident director of the Anglo-Persian Oil Company that the Persian government, having "lost hope" of protecting Persia's interests under the D'Arcy concession, was canceling the concession. On November 29, 1932, the company's resident director notified the minister of finance that the company did not acknowledge the right of the Persian government to cancel the concession and that the company hoped the Persian government would reconsider. Three days later the minister of finance notified the resident director that the government's decision to cancel was final.[48]

British Government Acts

The British government, whose interests in the Anglo-Persian Oil Company were those of both a sovereign and a shareholder, took charge of negotiating withdrawal of Persia's cancellation of the concession.[49] On December 2, 1932, the British minister in Teheran notified the Persian government in writing that the cancellation of the concession was "an inadmissible breach of its terms." The British government, he stated, took "a most serious view of the conduct of the Persian Government" and had intructed him to "demand the immediate withdrawal of the notification" of cancellation. Expressing the hope that an amicable settlement might still be reached through direct negotiations, the British minister indicated that his government would "not hesitate, if the necessity arises, to take all legitimate measures to protect their just and indisputable interests" and would "not tolerate any damage to the Company's interests. . . ."

47. United Kingdom Memorandum, League of Nations, *Official Journal,* XIII (1932), Annex 1419c, 2300.

48. The correspondence referred to in this section appears as Appendices 5–7, Annex 3 to United Kingdom Memorial, *I.C.J. Pleadings,* pp. 235–237.

49. The correspondence referred to in this section appears as Appendices 8–10, Annex 3 to United Kingdom Memorial, *I.C.J. Pleadings,* pp. 237–239.

In reply to these warnings the Persian minister for foreign affairs stood firm on cancellation but indicated a willingness to negotiate a new concession that would safeguard in an equitable manner the rights and interests of the Persian government. He denied responsibility for any damage that the company might suffer in the meanwhile.

On December 8, 1932, the British minister at Teheran advised the Persian government in writing that the cancellation was confiscatory and a clear breach of international law and that unless the Persian government withdrew its notification of cancellation within one week, his government would have no alternative but to refer the matter to the Permanent Court of International Justice at The Hague. With respect to Persia's denial of responsibility for any damage accruing to the company, the minister stated,

I have the honor to inform Your Excellency categorically that His Majesty's Government will hold the Persian Government directly responsible for any damage to the Company's interests, any interference with their premises or business activities in Persia, or any failure to afford the Company adequate protection, and, in the event of any such damage occurring, His Majesty's Government will regard themselves as entitled to take all such measures as the situation may demand for that Company's protection.[50]

Before the Council of the League of Nations

To this note the Persian minister for foreign affairs replied at length on December 12, 1932, setting forth in detail the reasons for the Persian government's dissatisfaction and denying the competence of the Permanent Court of International Justice to examine the differences between his government and the Anglo-Persian Oil Company. He again disavowed his government's responsibility for any losses the company might suffer, and he expressed its hope that the company would send an authorized representative to Teheran to open negotiations looking toward a new agreement in

50. When asked in Parliament whether this note meant armed interference, Anthony Eden, undersecretary of state for foreign affairs, replied: "I should have thought that the position is quite clear. We hold the Persian Government responsible for protecting the rights of this British Company." P.D.C., Vol. 272, col. 1793, December 8, 1932. Ten days later, when asked whether any warships had been sent to Persian waters for the purpose of protecting the interests of the Anglo-Persian Oil Company, the First Lord of the Admiralty replied, "No war vessels have been despatched to Persian waters *for this purpose.*" P.D.C., Vol. 273, col. 740, December 19, 1932. Emphasis supplied.

harmony with Persian interests. He deplored the "threats and intimidation" that the British government was resorting to and closed with these words:

The Persian Government consider this attitude of His Majesty's Government as incompatible with the spirit of uprightness and the desire for peace which should prevail amongst friendly Powers and Members of the League of Nations, and consider themselves within their rights in bringing to the notice of the Council of the League of Nations the threats and pressure which have been directed against them.[51]

Although the Persian government threatened to bring the matter before the Council of the League of Nations, it was the British government that did so. On December 19, 1932, Sir John Simon on behalf of the British government requested the secretary-general of the League of Nations to put the cancellation dispute on the council's agenda.

Thereafter each party submitted to the council a memorandum setting forth its position in the dispute; the representatives of the parties argued the case orally before the council. The council appointed Edward Benes of Czechoslovakia as *rapporteur*.

Persia's Grievances under the D'Arcy Concession

The proceedings before the Council of the League of Nations clearly revealed the sharpness of the cleavage between the company and the Persian government. They illuminated not only the immediate issue that precipitated the dispute but also the basic grievances that the concession had engendered.

The Persian government pointed out that the original concession had been granted by a government that had no constitutional basis and was under the influence of two foreign powers.[52] It alleged that although the concession covered an area larger than Germany and France combined, the

51. Appendix VI to Memorandum from the Imperial Government of Persia submitted to the Secretary-General of the League of Nations for transmission to the Council, January 18, 1933, League of Nations, *Official Journal*, XIV (1933), 303.

52. The allegations are made in a "Memorandum from the Imperial Government of Persia" submitted to the Secretary-General of the League of Nations for transmission to the Council, January 18, 1933, *ibid.*, XIV (1933), 289–295, and in a note sent December 12, 1932, by the Persian minister for foreign affairs to the British minister at Teheran, attached to the above-mentioned memorandum as Appendix VI, *ibid.*, pp. 300–303.

company had confined its exploitation to an area slightly in excess of one square mile. It contended that the company had restricted its operations in Persia while extending them abroad, that it had defaulted in royalty payments since 1919 and had refused to pay taxes due under Persia's 1930 income tax laws, that it had sold petroleum products in Persia at unwarrantedly high prices, and that it had failed to comply with its obligations to employ Persian subjects in its operations, to pay annually to the government 16 percent of its net profits, and to allow the Persian government to supervise expenditures that were to be deducted from gross income to determine net income. It argued that Armitage-Smith had exceeded his authority in negotiating the 1920 agreement and that it was invalid, not having been ratified by the Persian Parliament. It contended that if the government had allowed the company to work the oil deposits freely without any payments whatever, subjecting it only to customs charges on materials it imported, the receipts would have been approximately 100 percent greater than they had actually been—£19,998,509 between 1905 and 1932 instead of about £11 million. The government alleged that the company had been unwilling to negotiate a new agreement consistent with the interests of Persia. For these reasons, the Persian government contended, cancellation was the only course open to it.

The British Government's Position

The British government, having intervened as sovereign on behalf of the company of which it was the controlling owner as citizen, not only denied virtually all these allegations but declared that even if true they could not justify cancellation.[53] Conceding only that questions as to the meaning of "net profits" were not surprising in view of the fact that profits were first earned only "after some fifteen years' work on the concession" and in view of "the steady expansion in all directions of the company's business relations," the government noted that the concession contained no provision for termination before 1961 except for failure to form a company to work the concession, which company had been formed. The Persian government's cancellation of the concession was a breach of contract and

53. The British government's position is set forth in part in its memorandum submitted to the Council of the League of Nations on December 19, 1932, *ibid.*, XIII (1932), 2298–2305.

contrary to international law, a step taken without invoking the arbitration provision of Article 17 of the concession.

In oral argument before the Council of the League of Nations on January 26, 1933, Sir John Simon, the British government's representative, declared that there was no dispute as to whether the concession bound the Persian government at the time it was made or whether the government had recognized the Anglo-Persian Oil Company "for many years as the holder of the concession."[54] The Persian government's "real reason for canceling the concession is that it hopes by this means to dictate to the company a new concession while the company is in the adverse and unfair situation of having its concession cancelled." Simon stated that oil was being produced from an area "over 150 times" the area mentioned by the Persian government in its complaint (slightly in excess of one square mile). He said that the Anglo-Persian Oil Company had "examined in detail over 150,000 square miles of territory" and "geologically traversed another 250,000 square miles," carrying out "380 geological and geophysical surveys" and spending more than £3,500,000 on exploration. Rather than "extend oil works at random," the company was following a plan of development which took account of available markets for the oil—a manner of development in which "the interests of the company are identical with the interests of the Persian Government." Far from defaulting in royalty payments before 1919, the company had paid £325,000 in the years 1914 to 1919. If the Armitage-Smith agreement was invalid, "the Persian Government has received £1,000,000 from the company which it ought to pay back." Under the Armitage-Smith agreement the Persian government had had the company's records examined each year by a firm of chartered accountants; because of certain questions the accountants had raised, the calculations being complicated, royalty payments had been made without prejudice to later adjustments. As for the employment of Persian subjects in the company's operations, the great expansion necessitated by the war had led to the importation of skilled workers from India, but "since then the number of non-Persian employees has been consistently diminished, and 90 percent of the company's non-European employees in Persia are now Persian subjects." The company had spent more than £100,000 on education, more than £550,000 since 1924 on medical services, and not less than £22,000,000 in total expenditures, from which "the Persian Government has benefited directly and indirectly." The Per-

54. *Ibid.,* XIV (1933), 198–204.

sian government's assertion that it could have received more revenue from ordinary taxes than from royalty payments was "based on a rather obvious fallacy,"[55] since there was no tax on the exportation of oil, and any company whose importation of machinery would be subject to £19,000,000 in taxation would most likely not undertake the operation.

The company's unwillingness to discuss a new concession arose after the Persian government in April 1931 proposed that any new concession should contain a guarantee by the company of a minimum annual payment to the government of £2,700,000 "whether any profits were made or not."[56] Because such a payment in the year 1931 would have exhausted the company's profits and because the uncertainties of the future made such a proposal unacceptable, the company proposed that negotiations be confined to settling difficulties under the existing concession while deferring consideration of a new concession until "the chaotic condition of the oil industry" had returned to order. It was this suggestion that the Persian government had interpreted as a categorical refusal to contemplate further any revision of the concession.

Settlement and a New Concession

While proceedings were under way before the Council of the League of Nations, the Anglo-Persian Oil Company and the Persian government agreed to reopen negotiations directed toward working out a new concession agreement. Taking cognizance of this, the two governments agreed to suspend proceedings before the council until its May 1933 session. Negotiations extended over approximately three months, the negotiators meeting first in Geneva and later in Teheran. Eventually, through the direct participation of the Shah of Persia and Sir John Cadman, chairman of Anglo-Persian's board, the government and the company reached an accord. They signed a new concession agreement on April 29, 1933, and on May 26 Dr. Benes announced to the council that the dispute between the two governments was virtually settled.[57] On June 5 the Persian government informed the secretary-general of the League of Nations that the Persian Parliament had ratified the concession.

55. *Ibid.*, p. 209 (Simon's rejoinder to the oral argument made by M. Davar, representative of Persia).
56. *Ibid.*, p. 202.
57. League of Nations, *Official Journal*, XIV (1933), 827.

The new concession differed in important respects from the old.[58] It will be recalled that governmental revenues under the old concession depended primarily on the profits of the company. This not only made a major source of revenue dependent upon the vagaries of international commerce in oil but raised complicated questions of accounting that were a continuous source of controversy. The new concession provided for a more certain and precise governmental income. It obliged the Anglo-Persian Oil Company to pay to the government four shillings for each ton of oil sold in Persia or exported and, in lieu of all taxes (except custom duties and other taxes on imports for employees' exclusive needs), to pay during the first fifteen years ninepence a ton for the first six million tons of oil produced in any year and sixpence a ton for production in excess of six million tons. After fifteen years the payments were to go up to one shilling a ton for the first six million tons and ninepence a ton for production in excess of six million tons. In addition, Anglo-Persian agreed to pay the Persian government a sum equal to 20 percent of the distribution of earnings in excess of £671,250, whether the distribution was made to stockholders or as an addition to the company's reserves shown on its books on December 31, 1932. The company guaranteed that the payments to the government from these three sources, royalties, payments in lieu of taxes, and a percentage of distributed earnings, should not be less than £975,000 in any one year during the first fifteen years and not less than £1,050,000 during the next fifteen years.

The new concession was to run until 1993, and the parties agreed to negotiate before 1963 the annual payments to be made in lieu of taxes for the last thirty years of its term. On the expiration of the concession or its surrender by the company, the company agreed to pay the government 20 percent of the amount by which its reserves had increased since December 31, 1932, and 20 percent of the increase in its balance carried forward on the expiration or surrender date over its balance carried forward on December 31, 1932.

The company agreed to pay within thirty days from April 29, 1933, £1 million in settlement of all the government's claims except those with reference to taxation and to settle the payments due for the years 1931 and 1932 in accordance with the new rather than the old agreement. To settle the government's claim that the company had refused to pay income taxes

58. The English translation of the concession of April 29, 1933 (the official version was in French) appears as part of Appendix No. 16 to Annex No. 3 to United Kingdom Memorial, *I.C.J. Pleadings*, pp. 258–270.

under a law enacted in 1930, the company agreed to pay for the period March 21, 1930, to December 31, 1932, a sum calculated on the basis of the formula laid down for calculating payments in lieu of taxes.

The new concession tightened and made more comprehensive the provisions relating to the disposition of the company's property on the expiration of the concession. Whereas the old concession had provided that "all materials, buildings and apparatuses *then* used by the company for the exploitation of its industry shall become the property of the said Government,"[59] the new concession provided that "all the property of the Company in Persia shall become the property of the Government" at the end of the concession, and it prohibited the alienation of any company property during the last ten years of the concession or during the two years' period of notice required if the company surrendered it sooner.

The concession contained several new provisions designed to protect Persia's interests in an efficient and economical operation of its petroleum resources and to provide the Persian government with all scientific information bearing on the petroleum deposits which the company should obtain in exploiting the concession. It gave the government the right to appoint a "Delegate of the Imperial Government" at an annual salary of £2,000 to be paid by the company, who should have access to all information that stockholders were entitled to, who might attend all meetings of the board of directors, of its committees, and of stockholders where questions arising out of the relations between the government and the company were to be considered, and who might request special meetings of the board of directors at any time to consider any proposal the government submitted to it. The concession committed the parties to plan for "yearly and progressive reduction of the non-Persian employees with a view to replacing them in the shortest possible time and progressively by Persian nationals," and it made the company responsible for sanitary and public health services, "according to the requirements of the most modern hygiene practised in Persia," within the company's territory. The concession contained provisions designed to insure lower prices of petroleum products for Persian consumption than had prevailed under the old concession, and it provided for a reduction after 1938 in the area of the concession to 100,000 square miles, instead of the 500,000 square miles granted in the original concession. And finally, the company obliged itself "to have regard at all times and in all places to the rights, privileges and interests of the Government"

59. League of Nations, *Official Journal,* XIII (1932), Annex 1419c, Appendix, 2307. Emphasis supplied.

and to "abstain from any action or omission which might be prejudicial to them."

The government for its part exempted the company from all taxation on its operations by either central or local authorities for the first thirty years of the concession and agreed not to annul or alter the concession "either by general or special legislation in the future, or by administrative measures or any other acts whatever of the executive authorities."

Under this new concession no significant dispute arose until the middle of the century.

Institutional Factors in the Cancellation

It would be a task of supererogation at this late date to try to assess the validity and weigh the relative merits of all the arguments and counterarguments presented in the course of the D'Arcy concession disputes. That the disputes culminated in cancellation of the concession was owing not so much to their insoluble character as to the institutional changes that were taking place in Persia. Basically they represented a conflict between nineteenth-century business morality and British "dollar diplomacy" and a twentieth-century social revolution that was taking place in Persia and is not yet completed.

The revolution in Persia represented a clash between political and economic feudalism and twentieth-century nationalism. For a hundred years before World War I the weakness and corruption of the Qajars and the economic, social, and political backwardness of Persia had made her an easy prey to the imperial ambitions of Russia and Great Britain. During the first two decades of the twentieth century these two countries, fearful of Germany's encroachments in Persia and jealous of each other, had carved out spheres of influence for themselves and agreed on a neutral area barred to outsiders. During this period the rulers of Persia held uneasy sway. The Revolution of 1906, which was essentially an urban movement, culminated in the adoption of a constitution in December 1906. It was short lived. Within the year Shah Muhammad Ali had launched a countercoup and by jailing, murdering, or exiling his opponents brought a temporary end to constitutional government. A reign of terror was followed by civil war, Russian intervention, and dethronement of the shah on July 16, 1909. It is a reflection of Russia's and Great Britain's concern with and influence on

Persia's internal affairs that the shah was exiled and pensioned under terms to which Russia and Great Britain formally agreed.

The shah's thirteen-year-old son's regency was a troubled one. Within two years his exiled father, crossing Russia in disguise with arms and ammunition labeled "Mineral Waters," returned to Persia and set off a civil war in his efforts to recapture the throne—apparently with the sympathy of Russia and the opposition of Great Britain. The effort failed. In June 1912 the ex-shah was again ejected. During the period of near political chaos, an unstable central government gave the tribes another opportunity to assert their independence and authority. World War I, with Russian and British armies occupying most of Persia, brought further disunity.

Reza Khan, who had been swept into the political vacuum occasioned by a weak and inept regency and a war-engendered turbulence, was a fitting figure to bring strength to the central government and unity to the country not hitherto achieved. More than six feet tall, with a breadth of shoulders appropriate to his height, handsome and haughty, with a reticent manner that belied his ruthless determination and a military bearing that commanded respect and obedience, quick of perception and firm of decision, Reza Shah as dictator of Persia ruled with a ruthless and iron hand. With an opportunistic regard for such administrative talents as were available, he commandeered the intelligent, experienced, and energetic leaders of the old regime and utilized them in achieving the goals he had set for himself and his country. When they had served his purposes, he disposed of them without mercy. He made Firuz Mirza first minister of justice and later minister of finance; Abdul Hoseyn Timurtash, minister of court; and Ali Akbar Davar successively minister of agriculture, commerce, and public works, minister of justice, and minister of finance. The first two were later accused of bribery, arrested, and murdered, and the third committed suicide.

Reza Shah, determined to bring unity, stability, and prosperity to his country and if possible to recapture something of Persia's historical grandeur, set himself three major goals: to free Persia from foreign influence, to consolidate the power of the central government, and to bring economic self-sufficiency to his country. The first two called for a strong military establishment and the latter was dependent to a considerable extent upon the second. His program, which proceeded piecemeal and frequently uneconomically along several fronts rather than in the orderly manner that a centralized control theoretically makes possible, involved universal military conscription; reforms in taxation, financing, and credit; the establishment of state-owned or state-controlled monopolies in tea, tobacco, sugar, and twenty-four other commodities; the construction of glass, textile, and cement plants; improvements in communication by telegraph, roads, and railways; the regulation of imports; improvements in education and the status of women; and numerous other social and economic changes designed to free Persia from dependency upon the imperial powers in whose pattern Reza Shah was trying to remake his country.

It is only when viewed as an episode in a political drama, mercantilistic in aims, haphazard in method, and ruthless in character, that the real significance of the cancellation of the D'Arcy concession can be understood. The country celebrated the cancellation with a two-day national holiday. The new concession brought a truce, but not lasting peace.

Iraq and the Iraq Petroleum
Company's Concession

BEFORE Qasim by a decree in 1961 reclaimed more than 99 percent of its concession, the Iraq Petroleum Company, Ltd., (IPC) directly or through its subsidiaries, the Mosul Petroleum Company, Ltd., and the Basrah Petroleum Company, Ltd., had exclusive oil rights covering an area in Iraq of approximately 174,000 square miles, virtually the whole of the country.[1] How it got them is a complex story of diplomatic and commercial maneuvering in which Iraq played a largely passive role. The details have never been fully recorded, but enough is known to present a thumbnail sketch of the major factors shaping the contemporary corporate structure and ownership of Iraq's oil.

1. The Khanaquin Oil Company, a subsidiary of British Petroleum Company, owns a concession covering small areas on the eastern Iraq border. It was originally a part of Anglo-Persian's concession in Persia. In 1913 Persia transferred these areas to the Ottoman Empire with the understanding that the Ottoman Empire would honor existing concession arrangements. Iraq confirmed these in 1925. *Special Report by His Majesty's Government to the Council of the League of Nations on Progress of Iraq, 1920–1931*, British Colonial Office, 1931, p. 222. Hereafter referred to as *Special Report on Progress of Iraq*. See also, Stephen Longrigg and Frank Stoakes, *Iraq* (London: Ernest Benn, 1958), pp. 148–149.

Gulbenkian's Early Interest in Mesopotamia's Oil

The everlasting sacred fires of northern Mesopotamia, fueled by escaping gas from crevices in the earth's surface, still warm nomadic shepherds who camp at night in the Kirkuk area with their flocks nearby. Enterprising primitive entrepreneurs, long before modern invention had made oil a necessity to Western civilization, collected oil from seepages and sold it by the donkeyload to feed the lamps of Turkey.

But it remained for Calouste Sarkis Gulbenkian, member of a wealthy Armenian family and known in imaginative oil literature as "Mr. Five Per Cent,"[2] to recognize the commercial possibilities of Mesopotamian oil. Gulbenkian's comprehensive report on oil in Turkish Asia prompted the Turkish sultan, Abdul Hamid II, in 1904 to transfer Mesopotamian oil lands from state ownership to his own privy purse.

The Germans and D'Arcy Become Interested

Meanwhile a German technical commission reported to its sponsors that the oil resources of Mesopotamia warranted more intensive investigation. Three years later the Deutsche Bank, through the Anatolia Railway Company, obtained a concession from the Sultan's *civil liste* authorizing the railway company to explore for oil over an area in the vilayets of Mosul and Baghdad extending for twenty miles on either side of the proposed railway. Later, alleging that the railway company had failed to fulfill its exploration obligations under the concession, the sultan informed the Germans that the concession had expired. Because the sultan did not remunerate the company for the £20,000 it had spent on geological surveys, the Germans never acknowledged the invalidation of the concession. Nevertheless, in 1906 William D'Arcy, who had obtained for the D'Arcy Exploitation Company the concession covering most of Persia, petitioned the sultan for the "lapsed" German rights. He did so with the "full support of His Majesty's Ambassador at Constantinople."[3] The D'Arcy negotiations

2. Ralph Hewins, *Mr. Five Per Cent, The Story of Calouste Gulbenkian* (New York: Rinehart & Company, 1958).

3. Correspondence between His Majesty's Government and the United States Ambassador respecting Economic Rights on Mandated Territories, *Parliamentary*

dragged on for several years. Before their conclusion D'Arcy transferred (1908) both his Persian concession and such claim as he had to Mesopotamian oil to the British-owned Anglo-Persian Oil Company.

Gulbenkian Reconciles Conflicting Interests

The Young Turk Revolution swept Sultan Abdul Hamid out of power and placed all oil rights again in the hands of the government. This necessitated that all oil negotiations begin anew. In 1910 the British, alert to their new opportunities, under the leadership of Sir Ernest Cassel and Lord Revelstoke, organized the National Bank of Turkey in support of British enterprise within the country. Gulbenkian, long and continuously interested in Mesopotamian oil and "a known personality in the offices and ante-rooms of Stambul,"[4] where corruption, intrigue, and bargaining were established procedures, became a member of the Bank's board of directors and of its executive committee. Meanwhile Gulbenkian had interested the Royal Dutch-Shell group, with which he had been associated in various business ventures, in Mesopotamian oil. Believing that union of rival interests promised more than competition, he acted as a catalyst in precipitating the organization on January 31, 1911, of African and Eastern Concessions, Ltd. Sir H. Babington Smith became its first chairman. African and Eastern Concessions allotted 25 percent of its £50,000 capital stock to the Deutsche Bank. In return the Deutsche Bank surrendered to the National Bank such rights as it had to Mesopotamian oil. The National Bank, or interests connected with it, retained 75 percent of African and Eastern Concessions' stock. In the fall of 1912 African and Eastern Concessions revised its articles, increased its capital to £80,000, and changed its name to the Turkish Petroleum Company (TPC). The D'Arcy interests were yet to be brought in. The Deutsche Bank wanted to exclude them; the British, to bring them in. The Turkish grand vizier urged an accord.

Papers, Cmd. 675 (1921), Earl Curzon to Ambassador Davis, February 28, 1921, reproduced in *Foreign Relations of the United States,* Department of State, 1921, II, 80–84. Hereafter referred to as *For. Rel.*

4. Stephen H. Longrigg, "The Early History of the Iraq Petroleum Company," a 4-page printed pamphlet, undated, p. 1.

The British Government Steps In

Private commercial interests were the prime movers in the foregoing developments. From here on the British government played an increasingly important role. To make secure an adequate oil supply to fuel its navy, it had taken steps under Churchill's leadership to acquire a controlling interest in the Anglo-Persian Oil Company with its vast Persian concession. This effort, as recounted in Chapter I, culminated in May 1914, in the government's buying 51 percent of Anglo-Persian's stock.

While the British government was negotiating for control of Anglo-Persian, it was also making its influence felt in the annals of TPC. On March 19, 1914, the British Foreign Office sponsored an agreement with TPC fusing German, British, and Dutch interests. This agreement is remarkable not only because it united the Deutsche Bank, the Anglo-Saxon Petroleum Company, Ltd. (forerunner of the Royal Dutch-Shell combine) and the Anglo-Persian Oil Company in a common undertaking to develop the Mosul oil deposits, but it pledged the three groups to refrain from engaging in the production or manufacture of crude oil anywhere in the Ottoman Empire in Europe or Asia (excepting only that part under the administration of the Egyptian government or of the sheik of Kuwait) other than through TPC.

Representatives of the British and German governments, the National Bank of Turkey, the Anglo-Saxon Petroleum Company, the Deutsche Bank, and the D'Arcy group all signed the agreement. What they sought was not only a unification of diverse interests in Mesopotamian oil but a united front in the development of oil in the Middle East.[5]

The agreement assigned 50 percent of TPC's stock to the D'Arcy Group, 25 percent to the Deutsche Bank, and 25 percent to the Anglo-Saxon Petroleum Company. The D'Arcy group was to nominate four and the Deutsche Bank and Anglo-Saxon each two of the company's eight directors. Under the agreement the D'Arcy group and Anglo-Saxon obliged themselves to provide Gulbenkian with a 5 percent beneficiary, nonvoting interest in TPC.

Turkey's Grand Vizier Agrees to Grant TPC Concession

Within less than a month after the British government had unified conflicting rivalries for Mesopotamia's oil and only four days after it had

5. The agreement is reproduced in *For. Rel.,* 1927, II, 821–823.

obtained a controlling interest in Anglo-Persian Oil Company, the German and British ambassadors presented to the government of Turkey the draft of an agreement covering oil rights in Mosul and Baghdad. Six weeks later the grand vizier agreed in writing to grant TPC a lease covering the areas it wanted. In doing so, he stated,

> The Minister of Finance, which has taken over from the civil list matters concerning petroleum deposits already discovered or to be discovered in the vilayets of Mosul and Baghdad, agrees to lease them to the Turkish Petroleum Co., and reserves the right later on to fix its own share as well as the general terms of the agreement.[6]

World War I interrupted negotiations and the agreement never became final. Nevertheless, the war strengthened British control over both TPC and Mesopotamian oil. In December 1918, the British government expropriated the Deutsche Bank's 25 percent interest in TPC and placed it in the hands of a British trustee. This, with the government's ownership in the Anglo-Persian Oil Company, gave the British effective control of 75 percent of TPC's stock. Meanwhile Henri Deterding, signer on behalf of the Royal Dutch-Shell combine of the original TPC agreement, had become a naturalized citizen of Great Britain (1915). For his war and postwar services, reflected in Lord Fisher's earlier characterization of Deterding as "Napoleonic in his audacity and Cromwellian in his thoroughness,"[7] the British king conferred knighthood on Sir Henri Deterding. These several developments placed control of TPC firmly in British hands.

World War I Tightens British Control

Without control over Mesopotamian oil TPC would have been an empty corporate shell. Partly through chance and partly through design, events decreed that it should not become such. By the fall of 1918 British troops had occupied Baghdad and Mosul and, allegedly as a military necessity, had begun the construction of pipelines, railways, and refineries, and the operation of certain oil wells. Two years earlier Great Britain,

6. *History of IPC and Mr. Gulbenkian's Part in Its Foundation,* April 1944. Quoted in Federal Trade Commission, *The International Petroleum Cartel,* Select Committee on Small Business, Senate, 82nd Congress, 2nd Session (Washington, D.C.: Government Printing Office, 1952), p. 50.

7. E. M. Earle, "The Turkish Petroleum Company—A Study in Oleaginous Diplomacy," Political Science Quarterly, XXXIX, No. 2 (1924), 265–279.

France, and Russia through the so-called Sykes-Picot agreement had reached an accord on respective spheres of influence in their projected reconstruction of the Turkish Empire and Sir Edward Grey had obtained from M. Cambon a pledge that the French government would respect the validity of the economic rights of British nationals in all Ottoman territories acquired by France or in which French influence was predominant.

Specifically, M. Cambon asserted that "the French Government is ready to sanction the various British concessions dating with certainty from before the war, in the regions which may be attributed to it or which may depend on its actions."[8] On March 20, 1919, in a secret meeting in Lloyd George's London apartment, M. Pichon, French minister of foreign affairs, to make more secure the Sykes-Picot treaty, agreed to its revision, acquiescing to Lloyd George's demand that Mosul be transferred from the French to the British zone.[9] Reinforcing the British claim to Mosul oil, the Treaty of Sèvres, August 10, 1920, confirmed acquired rights by Allied nationals in territories severed from Turkey.[10]

San Remo Agreement

While thus fortifying the claims of TPC to Mosul oil and eliciting French co-operation in making them more secure, Great Britain in the San Remo Agreement of April 24, 1920, guaranteed the French government a

8. Letters from M. Cambem to Sir Edward Grey, May 9 and May 15, 1916. Reproduced in J. de V. Loder, *The Truth About Mesopotamia, Palestine & Syria* (London: George Allen & Unwin, 1923), pp. 161–164. Quotation from p. 164.

9. Ray Stannard Baker, *Woodrow Wilson and World Settlement* (New York: Doubleday, Page & Co., 1922), 70–72. See also Zeine N. Zeine, *The Struggle for Arab Independence* (Beirut: Khayat's, 1960), pp. 76–77.

10. Article 311 of the Treaty of Sèvres reads as follows: "In territories detached from Turkey to be placed under the authority or tutelage of one of the Principal Allied Powers, Allied nationals and companies controlled by Allied groups or nationals holding concessions granted before October 29, 1914, by the Turkish Government or by any Turkish local authority shall continue in complete enjoyment of their duly acquired rights, and the Power concerned shall maintain the guarantees granted or shall assign equivalent ones." Quoted in Earle, 272n4. The Ottoman delegation signed the treaty, but the Ottoman government never ratified it. The government later ratified the (Lausanne) Treaty of Peace (July 24, 1923) under Article 16 of which Turkey renounced all rights to territories situated outside the frontiers delimited in the Lausanne Treaty. J. C. Hurewitz, *Diplomacy in the Near and Middle East, A Documentary Record: 1914–56* (Princeton, N.J.: D. Van Nostrand and Company, 1956), 120.

role in its development. Should the British government itself develop the Mosul fields, Great Britain pledged itself to grant "the French Government or its nominee 25 percent of the net output of crude oil at current market rates." Should a private company develop the fields, the British government agreed to grant the French government a 25 percent share in such company. Both parties agreed that "the said petroleum company" should "be under permanent British control."[11]

In return the French government agreed to the construction of two separate pipelines, and railways necessary for their construction and maintenance, for the transport of oil from Mesopotamia and Persia through French spheres of influence to ports in the Mediterranean.[12]

Husain Negotiates with British for Arab Independence

These agreements were made before it was known what the final arrangements for the government of the fragments of the Turkish Empire would be, and hence necessarily without the knowledge or consent of such governments. In truth, while the British and French were conducting their secret negotiations on the political controls to be established over the dismembered parts of the Turkish Empire, the British were making commitments to the Arabs not wholly compatible with their commitments to the French. In June 1915 the British high commissioner in Egypt had begun negotiations on behalf of the British government with Sharif Husain, emir of Mecca, on terms with which to secure Arab co-operation with the Allies in their war with Turkey.[13] As a price of his co-operation Husain asked that Great Britain acknowledge the future independence of a vast Arabic region extending from the Persian frontier to the Red Sea and the Mediterranean. Aware of its own and French claims to portions of this area, the British high commissioner wrote Husain accepting certain designated portions of Husain's territorial claims and stating,

With the above modification and without prejudice to our existing treaties with Arab chiefs, we accept these limits.

11. Paragraph 6 of San Remo Agreement, April 24, 1920. Emphasis supplied. *For. Rel.,* II (1920), 655–658.

12. *Ibid.,* Arts. 10, 11, 12. France also agreed that oil so exported should be exempt from export and transit duties.

13. Known as the McMahon-Sharif Husain Correspondence, the exchange extended from July 14, 1915, to March 10, 1916. Reproduced in Hurewitz, pp. 13–17. See also Zeine, pp. 6–10.

As for those regions lying within those frontiers wherein Great Britain is free to act without detriment to the interest of her ally, France, I am empowered in the name of the Government of Great Britain to give the following assurance and to make the following reply to your letter:—

"Subject to the above modifications, Great Britain is prepared to recognize and support the independence of Arabs in all the regions within the limits demanded by the Sharif of Mecca."

In making these commitments Great Britain reserved the right to advise the Arabs and assist them in establishing "what may appear to be the most suitable forms of government in these various territories." McMahon further stipulated that the Arabs would seek only the advice and guidance of Great Britain in forming a sound administration. He also stipulated that the established position and interests of Great Britain in the vilayets of Baghdad and Basrah necessitated special administrative measures "in order to secure these territories from foreign aggression, to promote the welfare of local populations, and to safeguard our mutual economic interests."[14]

Mandates and Arab Independence

This is not the place to recount in detail the postwar political maneuvering that prevented the Arabs from realizing in full their nationalistic aspirations.[15] Suffice it to say that the Allied Supreme Council at the conference in San Remo in April 1920 allotted to Great Britain the mandate over Iraq and Palestine and to France the mandate over Syria and Lebanon. It should also be noted that the mandate system represented a political device under which the Allied powers would exercise control over the former German colonies and the detached territories of the Turkish Empire. The mandate system introduced two new concepts into the relationships of the mandatory power and the mandated state. The mandatory was obliged to act as a guardian and tutor of the mandated state—to assist, educate, and guide it until it was ready for independent statehood. The mandatory power assumed its responsibility under the authority and control of an international body, the League of Nations.

When the political infant under the mandatory's beneficent guidance reached political maturity, it was to be granted independent statehood. Meanwhile the mandatory promised to maintain an open-door policy with

14. Hurewitz, p. 15.
15. For a lucid and remarkably objective account of these developments, see Zeine, *passim.*

regard to commercial, intellectual, and religious matters and obliged itself to see that concessions for the development of national resources were granted without distinction as to nationality and without monopolization.[16]

Arab leaders found it difficult to reconcile the mandatory system with the French and British declaration immediately following the armistice, in which they promised political autonomy to Syria and Mesopotamia. They had jointly declared that

> The object aimed at by France and Great Britain in prosecuting in the East the war let loose by German ambition is the complete and definite emancipation of the people so long oppressed by the Turks, and the establishment of national governments and administrations deriving their authority from the initiative and free choice of the indigenous governments and administrations in Syria and Mesopotamia, now liberated by the Allies, and in territories the liberation of which they are engaged in securing, and in recognizing these as soon as they are actually established. Far from wishing to impose on the populations of these regions any particular institutions, they are only concerned to ensure by their support and by adequate assistance the regular working of governments and administrations freely chosen by the populations themselves. . . . Such is the policy which the two Allied Governments uphold in the liberated territories.[17]

France Established Mandate over Syria

Following the armistice, events in the Middle East moved rapidly and in a way to thwart the national aspirations of the Arabs and to challenge the integrity of this declaration. It soon became evident that France would insist on a mandate over Syria and Lebanon and that Great Britain had no intention of surrendering its political interests in Palestine and Iraq. Faisal, Sharif Husain's son, who had spoken for the Arab nationalists at the peace conference, after a fruitless effort to prevent the partition of Arab territory in any form, left Paris after the French government had drawn up a proposed agreement incorporating the principle of the mandate system, an agreement which according to Zeine left France "in possession" of Syria.[18]

16. Article 11 of the Mandate for Syria and Lebanon, for example, provides that "Concessions for the development of . . . natural resources shall be granted without distinction of nationality between the nationals of all State Members of the League of Nations, but on condition that they do not infringe upon the authority of the local government. Concessions in the nature of a general monopoly shall not be granted." Norman Bentwich, *The Mandates System* (London: Longmans, Green, 1930), p. 169.

17. *Ibid.*, pp. 52–53.

18. Zeine, pp. 126–127.

After Faisal's return to Damascus, the Syrians repudiated the agreement, announced in an historic resolution the independence of Syria (March 7, 1920), and proclaimed Faisal their king. His rule was hectic and short lived. On July 14 General Gouraud sent an ultimatum to Faisal pointing out that the peace conference had entrusted to France responsibility for guaranteeing to Syria the benefits of independence under the mandate system and demanding among other things Faisal's acceptance on behalf of Syria of a French mandate. Faisal's acceptance under threat of war arrived only after Gouraud's troops were advancing on Damascus. On July 28, after protesting vigorously and bitterly, Faisal departed for Damascus for Der'ā. After a French airplane had dropped leaflets in Der'ā giving the inhabitants ten hours in which to persuade him to leave their country on threat of aerial bombardment, Faisal left Syria for Haifa twenty-two months after his triumphal entry with T. E. Lawrence into Damascus on October 1, 1918.

Thus did the French mandate over Syria become a living fact.

Unrest in Iraq

Meanwhile events in Iraq had been moving with similarly disconcerting rapidity. A widespread unrest, nourished by nationalist leaders, culminated in open rebellion against the British in the summer of 1920. While the immediate contributing influences were varied and complex, its basic character is well expressed by H. St. J. Philby, who perhaps better than any other Anglo-Saxon understood the Arab ways and aspirations. In an address to the Central Asian Society on June 23, 1920, he stated,

What they [the Iraqis] want, and want because they are Arabs, is complete independence, nothing more and nothing less, and that is exactly what the British Government promised them in the most unequivocal terms by joining with the French in the momentous proclamation of November 1918. . . . Great as is the diversity of sentiment in various sections of the Arab world in regard to many matters—religion, politics and the like—there is the completest unanimity in regard to one point, the passionate love of liberty which is characteristic of the Bedouin stock to which not only the people of Syria but the bulk of those of Mesopotamia belong by origin.[19]

19. Quoted by Philip Willard Ireland, *Iraq, A Study in Political Development* (London: Jonathan Cape, 1937), pp. 243–244.

Announcement on May 3 that Great Britain was to assume the mandate for Iraq fanned the smoldering rebellion.[20] The British assertion that the mandate was an obligation laid on Great Britain by the League of Nations did not allay the hostility of Arab nationalists. Arabs viewed the assignment of mandates very much as did Lord Curzon who, as secretary of state for foreign affairs, stated frankly to the House of Lords on June 25, 1920,

It is quite a mistake to suppose . . . that under the Covenant of the League, or any other instrument, the gift of the mandate rests with the League of Nations. It does not do so. It rests with the Powers who have conquered the territories, which it then falls to them to distribute, and it was in these circumstances that the mandate for Palestine and Mesopotamia was conferred upon and accepted by us, and that the mandate for Syria was conferred upon and accepted by France.[21]

Treaty of Alliance Replaces British Mandate

The insurrection, while having little effect on basic British policy, indicated the strength of the nationalist movement and gave impetus to a move already under way to substitute for a mandate a treaty of alliance with an Arab government yet to be established.

Steps for establishing it involved setting up a provisional government, selecting a ruler, promulgating an organic law, and providing the necessary institutional paraphernalia for administering it. In all of these the British exercised an advisory and frequently a decisive role.[22]

About the administrative working of British control over Iraq during this period the High Commissioner had this to say:

It has been shown that during the period from November 1920 to August 1921, when the British officials received their orders direct from the High Commissioner, the resolutions of the Council of State were submitted to him for his approval, and legislation was enacted by him, the entire credit for the progress achieved can fairly be given to the British authorities, and that from

20. The British High Commissioner said of the early years of his administration "Public opposition in Iraq to any form of external control was rapidly becoming the most urgent national question of the time." *Special Report on Progress of Iraq,* p. 14.

21. Parl. Debates, H. of L. 5th S. XL (1920), 877. Quoted in Ireland, p. 263.

22. For a detailed discussion of Great Britain's role, consult Ireland, particularly Chapters XV–XVIII. For the official account, see *Report of the High Commissioner on Iraq Administration, October, 1920–March, 1924* (London: His Majesty's Stationery Office).

August 1921 to December 1924, the High Commissioner was still in the background as juridically the supreme authority, though King Faisal, as titular head of the administration, was regarded in Iraq as fully responsible. . . . It was not until November 1925, when Iraqi Ministers became responsible to an Iraq Parliament, and the British officials became the servants of the Iraqi Government in fact as well as in name, that Iraqi effort can be said to have made more than a subsidiary contribution to the building up of the Iraqi State.[23]

In drafting the treaty, British representatives consulted Faisal, who after a national referendum was proclaimed king of Iraq by the British high commissioner and recognized as such by "His Majesty's Government," but the British played the major role. Theirs was a delicate task. Their object was to create a document which permitted Great Britain to play its allotted role of mandatory power while offending as little as possible Arab national aspiration for complete independence. In the words of the high commissioner, "It was their intention not that the proposed treaty should replace the Mandate, but rather that the Mandate should be defined and implemented in the form of a treaty."[24]

Before the king ratified the treaty, protests against it became so vigorous as to occasion the resignation of the king's cabinet, the assumption by the British high commissioner of dictatorial powers, the deportation of the treaty's most vocal opponents, and the appointment of a new cabinet whose "sole purpose," according to Ireland, was to ratify the treaty.[25] On October 10 the cabinet ratified the treaty with the proviso that final approval should rest with the as-yet-uncreated national assembly.

Constituent Assembly Ratifies Treaty

Elections for assembly members were not completed until February 1925. The assembly, with eighty-four representatives present, met on March 27 of that year. Nationalist agitation against approval of the treaty culminated in demonstrations outside the Assembly on May 29 and May 31. Not until the British high commissioner had delivered to the king a draft law empowering him to dissolve the assembly was the assembly persuaded to approve the treaty. Of the sixty-eight deputies whom the prime minister had persuaded to meet in an emergency session on the night of

23. *Special Report on Progress of Iraq,* pp. 28–29.
24. *Ibid.,* p. 14.
25. Ireland, p. 36.

June 10, thirty-seven voted to ratify the treaty, twenty-eight voted against ratification, and eight abstained.

Significant sections of the treaty limiting the sovereignty of Iraq and guarding the British interests included[26]

Article 1, which gave Great Britain the right to advise and assist Iraq in any way "required" without prejudice to her national sovereignty and to be represented in Iraq by a high commissioner and consul general with the necessary staff.

Article 2, which provided for the appointment of British officials by the Iraq government and prohibited the appointment of officials of other nationalities except with the consent of the British government.

Article 4, under which the king agreed to be guided by the advice of the high commissioner "on all matters affecting the international and financial obligations and interests of his Britannic Majesty."

Article 5, which limited Iraqi rights to independent representation in foreign capitals.

Article 8, which prohibited the Iraq government from ceding to or placing any territory in Iraq under the control of any foreign power.

TPC Obtains New Concession

The above digression into Iraq's internal politics indicates Britain's determination to insure that the nationalistic aspirations of Iraqis should not defeat the political and material ambitions of the British. The TPC, which under the terms of the San Remo Agreement was to remain permanently under British control, was negotiating with a government which was obligated always to seek British advice and in emergencies to acknowledge British control. After the United States State Department challenged the validity of its prewar claims to Mesopotamian oil, TPC was impatient to have them confirmed through a new concession. It did not await the completion of the Iraq political structure to obtain approval of its concession. The king and his cabinet ratified the new concession on March 14, 1925, one week before King Faisal promulgated Iraq's organic law and approximately seven months before the Iraqi parliament ratified the treaty defining Iraq's obligations under the mandate principle; in short before an autonomous Iraq government had anything to say about it.

26. Hurewitz, pp. 111–114.

Clash of British and American Interests in Mesopotamian Oil

While Lord Curzon's picturesque statement that "the Allies floated to victory on a wave of oil" may have exaggerated oil's role in World War I, it reflects not only the significant part that oil plays in modern military operations but the postwar concern that the victorious powers manifested in getting control of potential oil-producing regions.

In this concern American oilmen shared and in promoting their interests, they had the help of the State Department.

In a vigorous and at times acrimonious correspondence with the London foreign office extending over several years, the state department displayed a determined solicitude for American public policy and private business interests—a solicitude that at times seemed to outweigh the department's devotion to the democratic principle of the right of the people to determine their own political destiny.

Shortly after France and Great Britain reached agreement at San Remo on their respective spheres of influence in the Arab world and their ownership rights in Mesopotamian oil, the State Department through the American ambassador in London set forth in a note to Lord Curzon the principle that ought to be applied in mandated regions.[27] Pointing out that the government of the United States had consistently taken the position in peace negotiations that any alien territory acquired under the Versailles Treaty must be held and governed in such a way as to assure equal treatment "in law and in fact to the commerce of all nations," the State Department called attention to the "unfortunate impression" that the "authorities of His Majesty's Government" had created in the minds of the American public—an impression that they "had given advantage to British oil interests" not accorded to American companies and that Great Britain had been preparing quietly for exclusive control of the oil resources in Mesopotamia.

The state department insisted that the rules properly applicable to any mandated territory forebade the granting of any exclusive economic concession or monopolistic privilege covering the region's natural resources which would place American citizens at a disadvantage compared with the nationals of the mandate nations.

27. Ambassador Davis to Lord Curzon, May 12, 1920. *For. Rel.,* II (1920), 651–653.

Later, Secretary Colby indicated the difficulty that the state department had in reconciling the specific provisions of the San Remo Agreement with the British foreign office's avowed intention of preserving to a future Arab state complete freedom of action in regard to Mesopotamian oil resources.[28]

Denying that the San Remo Agreement aimed at a monopoly of Mesopotamian oil or exclusive rights thereto and insisting that it could only become effective if it "conformed to the desires and laws of the countries concerned,"[29] the British foreign office presented a spirited defense of the validity of the concession "granted before the war by the Turkish Government to the Turkish Petroleum Company." The San Remo Agreement, which allots to the French government the former German interests in the Turkish Petroleum Company, Lord Curzon contended, represents merely the

adaptation of prewar arrangements to existing conditions, and in this respect His Majesty's Government far from acting in any selfish or monopolistic spirit, may reasonably claim to have sought the best interests of the future Arab state. Neither the rights of the Turkish Petroleum Company nor the provisions of the San Remo Agreement will preclude the Arab state from enjoying the full benefit of ownership or from prescribing the conditions on which the oil fields shall be developed.[30]

The American State Department challenged the validity of TPC's concession and insisted on the obligation of the British government to apply meticulously the principle of the open door and nondiscrimination in a mandate that it assumed. Submerging principles to realities, however, it proposed that the issue of the validity of TPC's concession be arbitrated.[31]

Standard Oil Company Negotiates Directly with Anglo-Persian Oil Company

Meanwhile the Standard Oil Company of New Jersey, more concerned with the practical problem of participating in the exploitation of Mesopotamian oil than with the abstract principle of the open door, had begun

28. Secretary Colby to British Foreign Secretary, November 20, 1920. *Ibid.*, pp. 669–673.
29. *For. Rel.*, II (1921), 81.
30. *Ibid.*, p. 83.
31. Ambassador Harvey to Curzon. *Ibid.*, p. 89.

negotiations directly with the Anglo-Persian Oil Company. It did so with the approval of the Department of State.

On June 24, 1922, Acting Secretary of State Harrison advised Ambassador Harvey in London that the state department did not want

to make difficulties or to prolong needlessly a diplomatic dispute or to so disregard the practical aspects of the situation as to prevent American enterprise from availing itself of the very opportunities which our diplomatic representatives have striven to obtain.[32]

The department made clear, however, that any private arrangement that might be worked out should recognize the right of "any responsible" American oil company to participate in it if it wished to do so, and that it should not recognize the legal validity of the claims of TPC except after "impartial and appropriate determination of the matter."[33]

Armed with these instructions, A. C. Bedford, chairman of the board of directors of Standard Oil Company of New Jersey, cabled Sir Charles Greenway, chairman of the Anglo-Persian Oil Company, that he had obtained the consent of the State Department "to discuss a practical basis of American participation" provided that "the principle of the open door already acquiesced in for mandated territories by the Allied Powers be maintained," that the plan so worked out "should be ratified or adopted by the Government ruling Mesopotamia which should possess sovereignty," and that "any arrangement of practical questions involved should be tentative and subject to acceptance by the State Department after they have been advised as to its details."[34]

Bedford also advised Sir Charles that the seven interested American companies regarded the percentage participation that Sir Charles had suggested for American participation as inadequate.

Later, Bedford submitted to the state department a memorandum dated July 21, 1922, which he had drafted and discussed with Ambassador Harvey in London, and to which the partners in TPC had informally agreed, outlining a plan for the development of Mesopotamian oil resources.[35]

Thereafter executives of the Standard Oil Company of New Jersey continued negotiations with representatives of TPC in an effort to reach an

32. Acting Secretary of State to Ambassador Harvey. *Ibid.*, II (1922), 337.
33. *Ibid.*, p. 338.
34. *Ibid.*, p. 339.
35. *Ibid.*, pp. 340–342.

agreement that would satisfy the political aims of the department of state and the business goals of the American oil companies. Negotiations extended over a period of about five years during which time Standard Oil kept the State Department informed of progress and problems and occasionally sought its advice and invoked its assistance.

In these negotiations, Standard spoke for the seven American oil companies that were definitely interested in sharing in the development of Mesopotamian oil.

Bedford's Plan Affords Basis for Opening the Door— and Then Closing It

Bedford's plan afforded the basis for an "open door" policy acceptable to the State Department and eventually to TPC participants. It was an ingenious plan. Apparently submitted originally as a device for opening TPC's concession to outsiders on a competitive basis, as finally adopted and modified by a working agreement among TPC participants it compromised the principle of the open door and made more certain that the concession would be exploited only by TPC.

The original plan provided that TPC, in which the American companies were to participate, would develop directly only twenty-four areas of eight square miles each. For the development of the remaining areas of the concession the company would offer similar areas periodically at public auction. Any responsible corporation, firm, or individual might bid for the exploration rights, and the highest bidder would then be granted a lease of the area acquired. Each lessee would pay to the company the royalty due from the company to the Iraq government.[36] Although the plan forbade TPC as such to bid in the auction, it did not forbid TPC's owners. This conceivably opened the way for machinations designed to keep exploitation within the family of TPC participants. In truth, Gulbenkian, who owned a 5 percent interest in prewar TPC but whose rights under the postwar plan of reorganization had not been definitely determined, initially opposed the plan because of this possibility. The American ambassador to Great Britain expressed Gulbenkian's fear in these words:

According to Gulbenkian the only possible purchasers are the four groups which are interested in the Turkish Petroleum Company. In the four groups is

36. *For. Rel.,* II (1923), 243–245.

included the American Group. He is of the opinion that unless he is protected the four groups will enter into an agreement to sell to each other these yearly parcels of presumably oil-bearing territory at a nominal price. He maintains that in this way his share in the actual value of the territory to be disposed of would be reduced practically to nothing.[37]

Apparently the American group was not averse to seeing Gulbenkian's share undergo such shrinkage. At any rate, Ambassador Houghton, in commenting on Gulbenkian's position, stated,

The American Group in basing its decision on the premise that Gulbenkian ought to be eliminated from sharing in the returns from all territory except the original 24 parcels is . . . within its rights. When, however, it requests the Department to support it in this position which it has assumed not as a question of right but as one of bargaining, then in my opinion it is requesting the Department to bring pressure on the Foreign Office to secure the elimination of Gulbenkian without adequate compensation or it is giving Gulbenkian an excuse for not participating equally with the other groups.[38]

American Group with State Department Approval Accepts the "Self-Denying" Ordinance

The negotiators did not work out an arrangement satisfactory to Gulbenkian and the American group until the spring of 1927. Then by confirming informal prewar agreements reached between the original owners of TPC, not only was Gulbenkian's interest adequately protected but the joint interests of the postwar owners were solidified.

It will be recalled that the British Foreign Office had reconciled and fused the interests of the various prewar claimants to Mesopotamian oil rights in a memorandum of agreement on March 19, 1914, signed by representatives of the German government, the British government, the National Bank of Turkey, the Anglo-Saxon Petroleum Company, Ltd., the Deutsche Bank, and the D'Arcy group.[39] By this agreement the several participants in TPC pledged themselves "not to be interested directly or indirectly in the production or manufacture of oil in the Ottoman Empire in Europe and Asia" excluding certain areas which they delimited on a map by a red line—otherwise than through TPC. The obvious object of this agreement, known subsequently as "the Red Line Agreement," was to

37. *For. Rel.,* II (1926), 367.
38. *Idem.*
39. The Agreement as reproduced in *For. Rel.,* II (1927), 821–822.

eliminate competition among the owners of TPC in developing Middle East oil resources. How to reconcile this objective with the interests of the American oil companies and with the State Department's open-door policy was a problem.

The collusive machinations that Gulbenkian had feared constituted a threat both to Gulbenkian's interests and to the open-door principle to which the State Department was dedicated. Final arrangements eliminated the threat to Gulbenkian but not to the open-door policy. The State Department, more trusting than Gulbenkian, approved not only the open-door plan as devised by private business but gave its approval to the associated action of TPC participants in future oil developments provided for under the Red Line Agreement. Concerning this, the assistant secretary of state wrote Guy Wellman, the associate counsel of the Standard Oil Company, on April 9, 1927, in part as follows:

> The Department notes . . . your statement that the effect of . . . [the self-denying] proviso, if adopted, would be to associate the American Group and its constituent members with the three foreign groups and Mr. Gulbenkian in the joint operation of all areas available for operation by the Iraq Petroleum Company (successor to TPC) under this convention, and also such areas under the Iraq Convention which the American Group and its constituted members, and the other groups of the Iraq Petroleum Company as well, might acquire through public offering.

The department, however, accepted the Standard Oil Company's conclusion that

> There would not result any modification of the so-called open-door plan adopted by the Iraq Petroleum Company . . . because the plan would still be operative so far as all nationals, including American nationals, are concerned in respect to their right to submit bids for tracts or areas offered for sublease.[40]

The department concluded its letter with an expression of its

> appreciation of your [Mr. Wellman's] courtesy in thus bringing to its attention the most recent developments in this matter and [its desire] to inform you that in the light of the information submitted it perceives no objection on the grounds of policy to the American Group taking up the preferred share participation in the Iraq Petroleum Company, Ltd., on the basis of the understanding recited.[41]

40. *Ibid.*, pp. 823–833.
41. *Idem.*

A year later Wellman submitted to the State Department the proposed agreement between TPC and its respective shareholders and a certificate of incorporation and by-laws of the Near East Development Company, under which the American interests were to be held. After noting the provisions of the two documents, the department gave them its blessing, stating that "the Department considers the arrangement contemplated . . . consistent with the principles underlying the open-door policy of the Government of the United States."[42]

Great Britain's Role in TPC Negotiations

While the American group were negotiating with TPC's original owners as to whether and on what terms they might be permitted to participate in the development of Mesopotamia's oil resources, TPC met the State Department's challenge of the validity of its concession by negotiating a new concession with a state as yet unborn and in whose birth Great Britain was playing the role of midwife. Britain's role necessarily contributed to, if it did not guarantee, insuring that a concession would be granted to TPC, and to TPC only, on terms satisfactory to Britain. As previously noted, King Faisal, who owed his position to the maneuvering of the British government, and his cabinet approved the concession before Iraq's organic law was promulgated and before its chief governing body had been duly assembled. Although the high commissioner reported that "The grant of the concession met with general public approval," the ministers of justice and education resigned in protest.[43]

That the State Department was well aware of the preferred position that TPC held in its efforts to obtain a new concession that in effect validated its prewar claims the secretary of state made clear in a communication to Ambassador Kellogg in London when he expressed fear that the American companies might not receive fair and equal treatment. In cabling the ambassador on September 24, 1924, he stated that

insomuch as the British interests hold a special position in negotiating with Iraq, since the British Government is the mandatory there and possesses special prerogatives in Iraq under treaties concluded with the Government of Iraq, it is . . . possible . . . that the British interests may try to obtain concessionary

42. *Ibid.,* p. 824.
43. Report on the *Administration of Iraq, 1925,* p. 11.

rights in Mesopotamia without according a fair share in them to the American companies ready and willing to participate.[44]

Iraqi Minister of Finance Guards Iraqi Interests

Although there is no doubt that British interests had a "special position" in their negotiations with the Iraq government, that position was not so secure that they could ignore entirely Iraqi interests. Iraqi Minister of Finance Sassoon Effendi, who conducted negotiations on behalf of his government, did not hesitate to express his views and insist on proper regard for the sovereign rights of his government. Anticipating possible objections from Effendi, TPC representatives at times found it discreet to conceal from him the details of their negotiations with the American group. Any arrangements worked out with them were to be embodied in the "working agreement" among TPC participants—a document separate and apart from the oil concession. The negotiators considered the open-door issue particularly delicate. Their problem was to satisfy the American commitment to the open-door principle without contravening Iraqi sovereignty. Article 34 of the proposed concession contract as drafted by TPC provided that the "Company shall have the right from time to time to underlet or transfer any part or parts of its rights and obligations hereunder with respect to portions of the defined area on such terms as it may think fit."[45] It was under such authority as this article provided that TPC would recognize the open-door principle as insisted on by the American group. This satisfied the Standard Oil Company whose president, W. C. Teagle, wrote the state department on October 25, 1923, in part as follows:

It is proposed that the Open Door Plan in the form already submitted to the Department shall have the sanction of an agreement between the Turkish Petroleum Company, Limited, and its voting shareholders, and it is our view that Article 34 of the draft convention enclosed will permit the Turkish Petroleum Co., Limited, to carry that plan into effect.[46]

Mr. Effendi, knowing nothing of the arrangement being worked out with Standard Oil and jealous of Iraq's sovereignty, proposed that Article 34 be modified to require approval by the Iraq government of every

44. *For. Rel.*, II (1924), 234.
45. Draft Convention of September 1923 between Government of Iraq and Turkish Petroleum Company, Ltd. Reproduced in *For. Rel.* (1923), pp. 247–257.
46. *Ibid.*, p. 246.

sublease that TPC might wish to make.[47] Teagle recognized that such a qualification would vitiate the principle of the open door and so wired Nichols of TPC, stating that the proposed change was "absolutely unacceptable."[48] At the same time he suggested phraseology for Article 34 which he thought would give adequate protection to the open-door plan. He also suggested in a letter to Nichols that careful consideration should be given to the question of making a specific reference in the draft concession to the plan of subleasing privately agreed on by Standard Oil and TPC representatives. Nichols later notified Teagle that the phraseology Teagle had suggested was acceptable to him but that it was not acceptable to the Iraq government, and he advised that if the proposed amendment were not withdrawn "the prospects of reaching a favorable agreement" would be "seriously prejudicial."[49]

TPC representatives at first demurred at disclosing to the Iraq government the arrangement it was privately working out on the open-door principle. Whether there was merit in the State Department's suspicion that the Iraq government's opposition to the unqualified right to sublease was shared by the Turkish Petroleum Company is not clear,[50] but eventually this issue was solved by incorporating in Article 6 of the convention the broad outlines of the subleasing plan as worked out by TPC and Standard Oil. Article 6 provided that the government within four years of the date of the convention and annually thereafter should offer plots for competitive bidding "by sealed tenders" by all responsible corporations, firms, or individuals without distinction as to nationality, the proceeds of which would be turned over to TPC.

The Government shall grant to the highest bidder for each plot, unless he shall be disapproved by the Government on reasonable grounds . . . a lease . . . conferring all the rights and imposing all the obligations . . . that initially fell to TPC.[51]

47. *Ibid.,* p. 255.
48. *Ibid.,* p. 247.
49. *Ibid.,* p. 260.
50. See Memorandum by Chief of Bureau of Near Eastern Affairs, Department of State, January 22, 1924. *For. Rel.,* II (1924), 224–225. The British High Commissioner in his *Special Report on Progress of Iraq* (1920–1931), p. 218, states that "The Iraqi Government at first resisted the plot system, principally because they were fearful that it would lead to difficulties when the time came for other companies to enter the lists, and they were prepared, so far as they were concerned, to grant the company exclusive exploitation rights over both Vilayets. In the end, however, they agreed to accept the plot system."
51. Hurewitz, pp. 131–142, reproduces the Turkish Petroleum Concession, granted on March 14, 1925.

Thus did the convention recognize the principle of the open door without rendering invalid the working agreement privately consummated between TPC and Standard Oil circumscribing this principle. The working agreement as finally approved granted the American companies the right to exploit independently any plot on which they might be the successful bidders. But this right promised little practical effect because the American companies were required to notify TPC in writing thirty days before the date fixed for opening the tenders, in return for which they were entitled to receive within ten days similar notice from such of the other groups as intended to bid. Wellman realized that the arrangements made it easy for the other participants as a group to outbid the Americans since "they would lose only one-fourth of the amount paid for a plot and the balance would be returned to them by TPC."[52]

All American Firms Drop Out except Standard of New Jersey and Socony

TPC participants signed the working agreement with the American companies on July 31, 1928, more than three years after TPC had obtained its concession from the Iraq government. The agreement allotted 23.75 percent of TPC's ownership to each of four groups: the D'Arcy Exploration Co., Ltd. (Anglo-Persian Oil Company group); Anglo-Saxon Petroleum Company (the Royal Dutch-Shell group); Compagnie Française des Pétroles (French group); and the Near East Development Company (the American group). It allotted the remaining 5 percent to Gulbenkian under the corporate title "Participation and Investment Co." Meanwhile the seven American companies had shrunk to five: Standard Oil Company of New Jersey, Standard Oil Company of New York, Gulf Refining Company, Atlantic Refining Company, and Pan American Petroleum and Transport Company (Standard of Indiana). Sinclair and the Texas Company had dropped out.[53]

Subsequent ownership developments included the merger of Standard

---

52. *The International Petroleum Cartel,* p. 60.

53. *Ibid.,* p. 54. Of the agreement allotting 23.75 percent interest in IPC to the Near East Development Co., the state department said, "This was secured entirely as a result of the insistence of the State Department that American firms be given equal opportunity in obtaining concessions." See *American Petroleum Interests in Foreign Countries,* Hearings before Special Committee Investigating Petroleum Resources, U.S. Senate, 69th Cong., 1st Sess., June 27–28, 1945, p. 23. TPC changed its name in June 1929 to Iraq Petroleum Company, hereafter referred to as IPC.

Oil of New York with the Vacuum Oil Corporation and the purchase by Socony-Vacuum and Standard Oil of New Jersey of Pan American's and Atlantic's interests in the Near East Development Company. Of the seven companies initially interested in Mesopotamian oil there remained only two: Standard of New Jersey and Socony-Vacuum (now Mobil Oil).

IPC Obtains a New Concession Abandoning Open Door

By 1931 all trace of the open-door policy had disappeared from the working agreement among TPC participants and from IPC's concession. Adopted as a device to gain the support of the state department and at the same time to afford the interested American companies an opportunity to share in the exploitation of Mesopotamian oil resources, it soon outlived its usefulness.

The Iraq government had accepted it in response to IPC's insistence. Standard Oil had devised the plan in response to the State Department's insistence on the open door. By the summer of 1928 it had served the purposes of diplomacy, and IPC owners could give consideration to more mundane factors. Apparently none of IPC's owners liked the leasing plan. If executed with integrity, it had three possible shortcomings. It might interfere with the efficient exploitation of underground oil-bearing structures which seldom conform to surface property lines; it constituted a threat to IPC's monopoly in exploiting Mesopotamian oil; and it interfered with the prompt development of discovered oil-bearing structures.

It will be recalled that the concession obliged IPC to choose twenty-four plots within four years which it would itself develop and thereafter to offer plots for competitive bidding. To determine probable oil-bearing plots required careful geological survey; this took time. Delay in the final delimitation of Iraq's frontiers impeded the work. By the end of 1926 IPC had chosen ten well locations, and in 1927 it completed a well with an initial daily production of more than 50,000 barrels. This indicated the potential richness of IPC's concession and stressed the importance of further exploration before releasing any acreage. In 1927 IPC complained to the government that restrictions in the movement of its geologists were interfering with the selection of plots and requested more time in selecting them.[54]

54. *Special Report on Iraq Progress 1920–31*, p. 219.

In 1927 and again in 1928 the government granted an extension. On its refusal to grant a third extension, IPC selected its plots under protest. Meanwhile IPC had been formulating a proposal for the abandonment of the leasing system. Accepting the government's suggestion that the company consider a Mediterranean railway project, proceed with the construction of a Kirkuk-Mediterranean pipeline, and make a substantial interim annual payment to the Iraq treasury, the company proposed that it be granted a new concession with exclusive rights. To get it IPC was willing to accept a reduced area. After protracted bargaining, the parties reached an accord. On March 24, 1931, the government granted a concession, and parliament later ratified it. The new concession, covering an area of approximately 32,000 square miles in the Mosul and Baghdad vilayets east of the Tigris, abandoned the subleasing system and placed control firmly in IPC's hands.[55]

Iraq Grants BOD a Concession

While the new agreement abandoned the subleasing system and left IPC firmly in control of Iraq's most promising oil lands, it left Iraq free to make such arrangements as it could for the exploration and development of the remainder of the country. It soon had a concession candidate. In May 1932 it granted a concession covering all of Iraq west of the Tigris and north of the 33-degree line (approximately 46,000 square miles) to the British Oil Development Company, Ltd. (BOD), a company formed by an English group in 1925 which was later extended through the instrumentality of Mosul Oil Fields, Ltd. (MOF) a holding company, to embrace Italian, French-Swiss, German, and Dutch capital. BOD's development program, which was instituted promptly, met with mediocre success. BOD, becoming discouraged in the face of the Great Depression, a world surplus of oil, and internal financial difficulties, looked to IPC for assistance. After granting MOF a loan, IPC organized a new concern. Mosul Oil Holdings, Ltd., to acquire the shares of MOF. By 1937 IPC had acquired a controlling interest and in 1941 MOF, renamed Mosul Petroleum Company, took over BOD's concession. It then liquidated BOD and MOF.

IPC's monopoly of Iraq's oil rights was not yet complete. With recurrent rumors of oil in the Basrah area, the significance of which the

55. Stephen H. Longrigg, *Oil in the Middle East,* (London: Oxford University Press, 1961), pp. 73–78.

discoveries of oil in Bahrain and al-Hasa accentuated, IPC began negotiations with the Iraq government for an additional concession covering the Basrah area. On July 29, 1938, the negotiators reached an agreement under which the Basrah Petroleum Company, a wholly owned IPC subsidiary, received a concession on terms similar to those of the IPC and BOD concessions.[56] By these several moves IPC acquired exclusive oil rights over the whole of Iraq, excepting only the small Khanaquian concession on the Persian border—rights that were not to expire for three quarters of a century.

56. *The International Petroleum Cartel,* pp. 84–85; Longrigg, pp. 79–83.

Saudi Arabia
and Aramco

SAUDI ARABIA is a vast country. It occupies most of the Arabian peninsula, which at its greatest length extends from the Gulf of Aqaba on the northwest 1,400 miles to the Arabian Sea on the southeast. The peninsula extends eastward from the Gulf of Aqaba to the Persian Gulf for about 750 miles. Saudi Arabia occupies about 865,000 square miles, of the peninsula's one million square miles, about one third of the area of the United States. No one knows how many people live within its ill-defined borders, but those familiar with the country estimate from three to seven million.

Mountain ranges 1,000 miles long and occasionally 9,000 feet high shut off most of Saudi Arabia from the Red Sea. These with a narrow coastal plain constitute the provinces of al-Hijaz (the barrier) and Asir (the difficult area). The province of Nejd (highlands) constitutes the central core of Saudi Arabia. It consists of a plateau about 5,000 feet high on the west, gently sloping to an elevation of about 2,000 feet on the east. It embraces a long reddish sand belt known as the Dahana; the Tuwaiq Mountains, generally less than 3,000 feet high but occasionally rising to 3,500 feet; and a portion of the Great Syrian Desert to the north. It extends southwest to the Rub al-Khali (empty quarter), a bleak and sandy waste stretching to the Trucial coast on the east and extending northward like a finger into al-Hasa (sandy ground with water close to the surface). Al-

Hasa, or the Eastern Province as it is now generally known, constitutes a coastal strip along the Persian Gulf about 200 miles long and on the average about 100 miles wide. It is here that the Standard Oil Company of California first found oil.

Rub al-Khali is a region more or less apart. About 750 miles long and 400 miles wide, it covers approximately 300,000 square miles, an area nearly as large as Texas. A relatively unexplored area consisting of slow moving sand domes and rolling sand ridges and sand hills, it supports little vegetation, little animal life, and virtually no human habitation. It is said to be the largest continuous body of sand in the world.

Saudi Arabia's vastness is matched by its barrenness. Sharp variations in elevation are accompanied by similar climatic variations. In the plateau region summer temperatures frequently reach 100 degrees Fahrenheit and occasionally soar to 130. In winter the thermometer may fall to 30 degrees. In the coastal region summer temperatures range from 100 to more than 120. Night temperatures may fall to 40 degrees. With a maximum annual rainfall in the mountain area and southern Hijaz of about ten or twelve inches, average annual rainfall for the entire country is estimated to be about four inches. Rub al-Khali may go ten years without a shower. Farming is confined to the southwestern area and in the neighborhood of oases scattered throughout central Arabia, most notable of which are those of Hofuf in the Eastern Province and al-Kharj near the modern Saudi Arabian capital of Riyad.

Abdul Aziz Unites a People by Conquest and Diplomacy

Although every oasis has its village or town with a sedentary population, until the discovery of oil most of the people made their living largely off camels and goats. Tribal and family groups were continuously on the move searching for better grazing. They were a fiercely independent people given to constant warring with each other or raiding the settled communities. Living on the fringe of the Turkish Empire, the people were untouched by Western civilization and lived very much as they had since the time of Christ.

Having given a common religion and a language to the world that united loosely a hundred million Arabic-speaking Moslems between the Persian Gulf and the Atlantic Ocean, the Arab tribes of the Arabian Peninsula had themselves remained disunited and isolated from their spiritual kin. It remained for Abdul Aziz ibn Abdur Rahman al-Faisal al Saud, or as he is better known, Abdul al-Aziz ibn Saud (member or son of the House of Saud) to weld the warring tribes of Arabia into a national unit and give to the newly created nation the name of Saudi Arabia.

Abdul Aziz ibn Saud is regarded by those who knew him best as one of the great men of the twentieth century. With courage and cunning, with military acumen and political finesse, with missionary zeal and religious fervor he established himself as supreme temporal ruler and spiritual leader of a people whose life was adjusted to the requirements of inhospitable desert land. Tempering punishment with mercy and ruthlessness with diplomatic magnanimity, and utilizing romantic affection and personal generosity to achieve the political aims of state, he obtained Bedouin loyalty. He systematically sought wives among the tribes whose loyalty he wished to consolidate and distributed alms and largess to retain it. Philby in his *Arabian Jubilee* in describing a day in King ibn Saud's court has the following to say about his amorous life:

> The king . . . confessed to having married no fewer than 135 virgins, to say nothing of 'about a hundred' others, during his life, though he had come to a decision to limit himself in future to two new wives a year, which of course meant discarding two of his existing team at any time to make room for them.[1]

Frequent marriages resulted in frequent offspring and the enlargement of a royal family with kinship ties to the nomadic tribes. The king's wives gave

1. H. St. John Philby, *Arabian Jubilee* (London: Robert Hale, Ltd., 1952), p. 111. In describing the physical, institutional, and personal background out of which the Saudi Arabian oil concession emerged, I have drawn heavily on the works of Philby, who knew the king and his country better than any other Westerner. Saudi Arabia became Philby's adopted country, its ways became his ways, its religion, his religion. A keen observer, with boundless energy, he wrote prodigiously, sympathetically, and brilliantly about his adopted country and life in it. Particularly helpful to me have been his *Arabian Jubilee,* his *Saudi Arabia,* and his *Arabian Oil Ventures.* The last, published posthumously, presents the most detailed account available of the negotiations that culminated in the California Standard Oil concession. Philby wrote from intimate personal knowledge, relying on his notes and his memory. Because his account may err occasionally in minutiae, students are fortunate in now having access to T. E. Ward's book, *Negotiations for Oil Concessions in Bahrain, El Hasa, The Neutral Zone, Qatar and Kuwait,* printed privately (1969). Ward, a pioneer in the oil fields of Trinidad, with wide experience in other oil-producing areas, in 1926 undertook negotiations for the sale of Eastern and General Syndicate's Bahrain concession. His book is based largely on personal experience and documentary evidence.

birth to some three dozen sons and an undisclosed number of daughters who counted for little in the Arabian scheme of things. The sons and daughters multiplied until the royal family is said today to embrace several thousand members.

As a result of his military exploits Abdul Aziz ibn Saud proclaimed himself Sultan of Nejd in 1921 with the approval of the tribal chiefs and religious leaders. In 1926 he became King of al-Hijaz and in 1927 King of Nejd. In 1932 he assumed the title of King of Saudi Arabia and no one disputed his sovereignty.

The King Needs Money but Is Skeptical of the West

Ruling over a barren country inhabited for the most part by a nomadic people loosely united by a fanatical devotion to Allah and his prophet Muhammad, a people remote from Western society and unfamiliar with its ways, King ibn Saud, responsible for both the material and spiritual welfare of his subjects, was confronted with the problem of how to finance the country's and his royal family's needs.

Although the future was to reveal petroleum resources unmatched by those of any other country, to exploit them he would need Western capital, Western business ability, and Western technical know-how. About these his skepticism was great and the prejudices of his people strong. His country had long been closed to Westerners. Religious taboos not only prevented their free travel within the country but shut out, as instruments of evil if not of the devil, the machines that were both a reflection of and a stimulus to Western industrial progress. Philby relates that the people of the fanatical village of Hauta burned in the market place the first automobile to enter the town and threatened the driver with a similar fate.[2] He describes how the king overcame opposition to the introduction of the telephone by asking the religious leaders to try the instrument for themselves. To their surprise and delight what they heard was the "familiar voice of an invisible friend reciting a passage from the Koran."[3]

Even before the conquest of al-Hijaz the king was in sore need of additional revenue. To maintain the Ikhwan colonies, which he had estab-

2. Philby, *Saudi Arabia* (New York: Frederick H. Praeger, 1955), p. 304.
3. *Ibid.*, p. 305.

lished in an effort to unite the people and to prevent mutual raiding by rival tribes, he needed funds for food, clothing, and religious instruction. According to Philby, his annual revenue from all internal sources was no more than £150,000. A British subsidy provided an additional £60,000.[4]

Major Holmes Seeks A Concession

The king had two possible ways to augment his revenues: increase Mecca pilgrimage fees or open the door to Western exploitation. The British frowned on the former; internal opposition to the latter was great and benefits from it were apt to be slow. Fortunately the Eastern and General Syndicate, organized in London by a New Zealand adventurer and promoter, Major Frank Holmes, offered the king modest financial relief without seeming to endanger his country greatly by foreign exploitation. Because ibn Saud had little faith in anyone's finding oil, he was willing to accept the best offer to demonstrate the futility of looking for it.

Ibn Saud's skepticism was not unique. Although the Anglo-Persian Oil Company was conducting a profitable business in Persia, producing 2½ million tons of oil in 1921–22, paying dividends of 20 percent on its common shares, and yielding the government £593,000 in revenues, the company did not believe Saudi Arabia likely to yield oil in commercial quantities. Sir Arnold Wilson, then General Manager of the Anglo-Persian Oil Company, wrote Philby in 1923,

> I personally cannot believe that oil will be found in his [ibn Saud's] country. As far as I know, there are no superficial oil-shows, and the geological formation does not appear to be particularly favorable from what little we know of it; but in any case no company can afford to put down wells into a formation in these parts (however favourable) unless there is some superficial indication of oil.[5]

Major Holmes, with less to lose and more to gain, was more venturesome. He was also persistent. He was undeterred by the king's reluctance to open the door to foreign capital and by the opposition of the British high commissioner of Iraq, Sir Percy Cox, who had advised Holmes to go slow about the concession. "The time is not ripe for it," he said. "The British

4. Philby, *Arabian Jubilee*, p. 68.
5. Philby, *Arabian Oil Ventures* (Washington, D.C.: Middle East Institute, 1964) p. 68.

Government cannot afford your company any protection."[6] Not only did Sir Percy try to discourage Holmes but he drafted a letter to Holmes for the king's signature informing Holmes that he (the king) was not free to make a decision without first consulting the British government.[7]

Abdul Aziz Grants Concession to Eastern and General Syndicate

Acting on the advice of Ameen Rihani, an American citizen of Lebanese extraction who had gained the king's confidence, and spurred on by his financial needs, the king ignored Sir Percy's advice and decided to grant a concession to the Eastern and General Syndicate. In doing so, the king's financial gain was small but the danger from foreign exploitation seemed to be equally small. Concerning the king's decision, Philby has this to say: "As always he was in dire need of hard cash, which had an astonishing way of slipping through his ever-generous hands."[8] For an annual rental, which according to Philby was only £2,000, payable in advance, Holmes got his concession and "everybody seemed satisfied with the deal." As for the king, he "was doubly so when he received the first year's rent; and there is reason to believe that he was optimistic about the prospect of receiving a similar sum for a number of years while the experts went about the dreary task of discovering that Saudi Arabia had no oil."[9]

Although in the light of the subsequent developments Philby characterized the terms of the concession as "ludicrous," if Major Holmes's optimism had triumphed over the king's pessimism, the king would have been more handsomely rewarded. The concession covered approximately 36,000 square miles along the eastern coast of Saudi Arabia, embracing what has today become the world's richest oil field. Clause 17 of the concession provided that ibn Saud "shall have the right to have allotted to him, as fully paid shares, twenty percent of all and every class of shares issued by any company which the concession may form or float for the exploitation of the concession. . . ."[10]

6. Ameen Rihani, *Makers of Modern Arabia* (Boston: Houghton Mifflin, 1928), p. 83; see also Philby, *Arabian Oil Ventures*, p. 57.

7. Rihani, pp. 84–85.

8. Philby, *Arabian Oil Ventures*, p. 65.

9. *Idem.* Ward states that the annual rental was £3,000, pp. 17, 34.

10. Philby, *Arabian Oil Ventures*, p. 64. See also, Ward, p. 34. Ward gives the area as approximately 40,000 square miles and states that the concession provided for customs duties of one percent on all oil exported. See also, p. 38.

The Concession Lapses

After signing the concession agreement (May 1923) and paying the first year's rental, the syndicate organized a team of Swiss geologists who in the winter of 1923–24 began a search for structures favorable to the accumulation of oil. Their search was unsuccessful. Nevertheless the syndicate paid a second year's rental and the geologists continued their hunt for another year. Meanwhile the syndicate tried to interest established oil companies in its venture. Finding no takers, the syndicate discontinued exploration and defaulted in its annual payments. When Holmes later tried to reinstate the concession by paying up arrears, the king had become more interested in other bidders. About the syndicate's permitting its concession to lapse, Philby states,

the weight of expert opinion seemed to be against the pioneers, though one is left with the feeling that the principal oil companies operating in the Middle East at that time [IPC and Anglo-Persian] were more concerned to preserve their existing monopolies than to encourage the search for and development of competitive fields.[11]

T.E. Ward, who was active in trying to find a buyer for the Eastern and General Syndicate's concession, supports Philby's judgment about the nature of Anglo-Persian's interest in obtaining additional concessions in the Persian Gulf area. He states,

It was recorded in a minute of the Management Committee meeting of Anglo-Persian on September 9, 1924, that Charles Greenway (later to become Lord Greenway) stated "Although the geological information we possess at present does not indicate there is much hope of finding oil in Bahrain or Kuwait, we are, I take it, all agreed that even if the chance be 100 to 1 we should pursue it, rather than let others come into the Persian Gulf and cause difficulties of one kind and another for us."[12]

Between the cancellation of the syndicate's concession in 1928 and the granting of the Standard Oil Company of California concession, two significant developments took place: oil companies became more interested in obtaining a concession and ibn Saud more interested in obtaining a concessionaire.

11. Philby, *Arabian Oil Ventures*, p. 67.
12. Ward, pp. 19–20.

The King's Mood Changes

Remote though it was from the epicenter of the economic disturbances which inaugurated the Great Depression, Saudi Arabia felt its tremors. Depression brought a sharp decline in pilgrimages to Mecca. As pilgrimages decreased, ibn Saud's spirits fell. He needed money.

Philby describes the king's mood on a day in the autumn of 1932 in these words:

He was so evidently in a despondent mood that I asked him straight out the reason for his gloominess. He sighed wearily, and admitted that the financial situation of the country was seriously worrying him. With a meager prospect of visitors for the coming pilgrimage, the exchequer was all but empty. And the situation was so bad that the Government could see no way of making both ends meet, and was at the same time confronted by difficulties with which it could not hope to cope. I replied, as cheerfully as possible in the circumstances, that he and his government were like folk asleep on the site of buried treasure, but too lazy or too frightened to dig in search of it. Challenged to make my meaning clearer, I said I had no doubt whatever that his enormous country contained rich mineral resources, though they were of little use to him or anyone else in the bowels of the earth. Their existence could only be proved by expert prospecting, while their ultimate exploitation for the benefit of the country necessarily involved the co-operation of foreign technique and capital. Yet the government seemed to have set its face against the development of its potential wealth by foreign agencies; and was content to sit back bemoaning its poverty. . . .

"Oh, Philby," the King exclaimed almost beseechingly, "if anyone would offer me a million pounds, I would give him all the concessions he wanted."

This was certainly a startling confession. A short experience of unaccustomed riches during the previous few years had made a continuance of wealth, even as a flash in the pan, an imperative necessity.[13]

The king's fear of poverty had apparently overcome his fear of the infidel.

Holmes Obtains Bahrain Concession and Bahrain Oil Company Discovers Oil

As the king's mood changed, so also did that of the oil hunters. In truth, Holmes had never lost faith in the Persian Gulf area. The year after

13. Philby, *Arabian Jubilee*, p. 176. About the King's habitual need of funds, Philby states, "At neither extreme of his life did ibn Saud's income suffice for his generous conception of the functions and obligations of a ruler." *Saudi Arabia*, p. 333.

he got the al-Hasa concession Holmes contracted with the sheik of Bahrain (summer of 1924) to drill twelve to sixteen wells on Bahrain Island. This contract led to the sheik's granting on December 2, 1925, a two-year oil exploration license to the Eastern and General Syndicate subject to a two-year prospecting license with the possibility of extension.[14] On a basis of a geologic report made by George Madgwich, professor of petroleum engineering at Birmingham University, the Eastern Syndicate tried without success to interest international oil companies in the concession. Anglo-Persian, Royal Dutch-Shell, Burmah Oil Company, and Standard Oil of New Jersey all passed up opportunities to buy the concession.[15] Eventually T. E. Ward, in whose hands Holmes had decided to leave negotiations, roused the interest of the Gulf Oil Corporation. Under a contract signed on November 3, 1927, Eastern and General Syndicate transferred its Bahrain option to the Eastern Gulf Company. At the same time the Eastern Syndicate transferred to Gulf such rights as it had under its 1923 al-Hasa concession in which it was in arrears; its right to a concession covering the Neutral Zone (owned jointly by Saudi Arabia and Kuwait), rights which had never been validated; and its rights to a Kuwait concession which Holmes had been trying to obtain.[16]

Following this transaction Gulf dispatched Ralph O. Rhoades to make a geological survey of the island.[17] On the basis of Rhoades's report, Gulf decided to exercise its option under the syndicate's concession. Membership in the Red Line Agreement blocked its decision to do so. Under the Red Line Agreement, it will be recalled, signatories obliged themselves not to develop any oil concessions within the territory which the Red Line delimited except through or with the consent of the Turkish Petroleum Company. TPC refused both to buy Gulf's rights and to permit Gulf to develop them independently without sharing oil which it developed with TPC.[18] Blocked in an independent program, Gulf found a ready buyer of its rights in the Standard Oil Company of California. But Standard Oil also encountered difficulties in exercising them. A Bahrain treaty with Great Britain forbade exploitation by a non-British company. Standard overcame

14. Ward, pp. 25, 35, 39.
15. *Ibid.*, pp. 27, 33–34.
16. Ward reproduces the agreement between Gulf and the Eastern Syndicate at *ibid.*, pp. 268–294.
17. *Ibid.*, p. 43.
18. *Federal Trade Commission, The International Petroleum Cartel,* Select Committee on Small Business, U.S. Senate, 82nd Cong., 2nd Sess., (Washington, D.C.: Government Printing Office, 1952), p. 72.

this obstacle by organizing the Bahrain Petroleum Company (Bapco), a Canadian incorporated subsidiary. With its political difficulties overcome, Bapco began drilling in October 1931, and on May 31, 1932, at a depth of 2,000 to 2,500 feet, struck oil in commercial quantities.

Standard Decides to Obtain a Saudi Arab Concession

Bahrain lies only twenty miles off the coast of Saudi Arabia. Standard's geologists, encouraged by their correct diagnosis of Bahrain's oil possibilities, looked to the mainland. They requested M. E. Lombardi, vice-president in charge of Standard Oil's foreign operations, to assist them in obtaining a permit to make a geological survey of the Eastern Province of Saudi Arabia. Lombardi in turn sought the help of Karl Twitchell, who through the generosity of Charles Crane, successful American businessman turned diplomat and philanthropist, had been conducting field research on economic development problems of Yemen and Saudi Arabia. As a result of these studies, Twitchell at the request of the king endeavored to interest American oil companies in Saudi Arabia. Among them he contacted Standard of California. Lombardi, who was already interested, recognized in Twitchell a useful contact with the king. Accordingly, he authorized Twitchell to negotiate on behalf of Standard of California and teamed him with Lloyd Hamilton of Standard's legal staff. Hamilton's major responsibilities were with the terms of a concession. Twitchell was to advise on political problems.[19]

Before recruiting Twitchell to aid in negotiating a concession, Standard executives had sought the advice and help of Philby. A few days before the completion of Bapco's well in Bahrain the American consul general in London, Albert Halsted, wrote Philby (May 26, 1932) introducing Francis B. Loomis, formerly undersecretary of state but then an executive of Standard Oil. About six weeks later while having lunch with Philby in London, Loomis told him of Standard's desire to obtain a concession and sought Philby's advice. Philby, while warning Loomis that "obtaining . . . a concession would involve a satisfactory arrangement regarding the price

19. About his and Crane's more remote role in the enterprise Twitchell states, "Saudi Arabia is presumably the only country in the world whose development of oil and mining resulted from purely philanthropic sentiment." Karl S. Twitchell, *Saudi Arabia*, (Princeton: Princeton University Press, 1947), p. 211.

to be paid" assured Loomis that he "would be glad to help in any scheme which would contribute to the prosperity of Arabia."[20]

Philby's Role in Negotiations

From this point on Philby played an increasingly important role in the negotiations, which culminated a year later in the king's granting a concession to the Standard Oil Company of California covering approximately 320,000 square miles.

In his *Arabian Oil Ventures* Philby has given a detailed and charmingly forthright account of the progress of negotiations and the part he played in them. To appreciate Philby's role one must bear in mind that he was an intimate friend and trusted counsellor of the king. The king expressed this confidence in a letter to Philby on February 25, 1933, in which he said,

I am confident that you will protect our interests, both economic and political, just as you would protect your own personal interests. So I shall expect your assistance in this matter and also I shall expect you to give me the benefit of your personal advice, which will be treated as confidential and with all consideration.[21]

Realizing that the king's interests would be better served if Standard Oil had an active competitor for Saudi Arabian petroleum rights, Philby encouraged Anglo-Persian to enter the lists, and when it had done so through the Turkish Petroleum Company, Philby kept both Standard Oil and Anglo-Persian informed of the progress of negotiations, shrewdly playing one against the other. He made it clear to both that for a consideration he would be glad to represent either in the negotiations. Pleased at the prospect of a "windfall" that would ease "the problem of bills for a son at Cambridge and three daughters at first-class schools," Philby eventually accepted Hamilton's invitation to work in behalf of Standard Oil for what he regarded as "generous enough" terms—£1000 a month for a minimum of six months with substantial bonuses on the signature of a concession and on the discovery of oil in commercial quantities.[22]

After Philby had become a paid representative of the Standard Oil Company of California in its negotiations with the Saudi Arab government,

20. Philby, *Arabian Oil Ventures,* p. 78.
21. *Ibid.,* p. 89.
22. *Ibid.,* p. 88.

he conferred with Hamilton and they agreed that Hamilton should be the active negotiator, with Twitchell in attendance and with Philby remaining in the background providing information and advice as needed.

Philby Apprises Bidders of the King's Needs but Insists on Quid Pro Quo

Before becoming Standard's paid negotiator, Philby had made clear to both Standard Oil and Anglo-Persian that the king because of his economic distress was in the market for a concessionaire. On December 22, 1932, in writing Loomis he referred to the government's inability to pay its debts and its urgent need of funds.[23] At about the same time he wrote Dr. G. Martin Lees, a geologist with the Persian Oil Company, that the government "is right up against a serious economic situation, and would be bound to accept a reasonable offer."[24]

However, he never allowed bidders to believe that they could take advantage of the king's plight to obtain a concession for a song. He continually emphasized the need of a *quid pro quo,* but with the progress of negotiations he revised downward his conception of what a concession was worth. Initially he suggested to Loomis an annual rental of £5,000, a government loan of £100,000, and 30 percent of the net profits of the enterprise. He suggested the same terms to Dr. Lees of Anglo-Persian Oil Company. Loomis thought the terms "quite burdensome."[25] As bargaining progressed, it became evident that the king would settle for the best cash advance he could get to meet his current financial needs, with the promise of continuing royalties to meet future requirements.

Standard of California Gets the Concession

It was the promise of future payments that eventually eliminated Stephen H. Longrigg, whom the Iraq Petroleum Company had sent to Saudi Arabia as its representative in the bidding. IPC was apparently more interested in keeping potential oil lands out of the hands of a competitor

23. *Ibid.,* p. 80.
24. *Ibid.,* p. 81.
25. *Ibid.,* p. 80.

than in developing them. In the course of negotiations, Longrigg revealed
to Philby that IPC needed no more oil: it had more in Iraq than it knew
what to do with. But he confessed that IPC "is vitally interested to keep
out all competitors."[26] What IPC apparently hoped to get was not a
concession but exclusive exploration rights by paying a modest annual
rental with preferential rights to a concession should oil be found.

At the government's request, both Hamilton and Longrigg eventually
submitted their offers in writing. When they had done so and Philby
learned that Longrigg was authorized to pay no more than £5,000 for a
concession, Philby told him he might as well pack up and go home. He
went. As Philby expressed it, "he faded out of the picture leaving the
Americans alone in the lists to fight it out with the champions of Arabia."[27]

Hamilton's offer included a cash advance of $50,000 against future
royalties. The king wanted £100,000. Largely through Philby's skill as a
backroom negotiator, Hamilton raised his offer to £50,000 and Finance
Minister Abdullah Sulamain, acting for the king, accepted. Hamilton and
the finance minister thereafter soon reached an agreement on other details.
On May 29, 1933, the finance minister, acting for the king, and Hamilton
for the company signed a concession agreement.[28] Before they signed, the
king and his privy council met to consider the agreement's details. The
king's lack of concern for these or his confidence in his advisors is revealed

26. *Ibid.*, p. 106. Anglo-Persian as a principal owner of IPC and sole owner of the
Persian concession was apparently more concerned than other IPC owners about the
operations of California Standard and the threat they constituted to its position in
world markets. G. S. Walden, Standard Oil of New Jersey representative on IPC's
board, in a memorandum to H. G. Seidel dated September 15, 1932, stated, "There is
no doubt that the Standard Oil Co. of California's well is giving real concern to the
Anglo-Persian." F.T.C., *The International Petroleum Cartel*, 73n15.

27. Philby, *Arabian Oil Ventures*, p. 20. About his negotiations on the behalf of
TPC, Longrigg in his *Oil in the Middle East* (Second Edition; London: Oxford
University Press, 1961), says only this, "The Iraq Petroleum Company had this time
decided to contest the issue. Their representative (the present writer) arrived at Jedda
to find negotiations in progress between Hamilton and the Sa'udi ministers and was
invited to make his own offers. Both negotiators interviewed the King, both advanced
their proposals, each was assured that his Company and nationality would, all things
being equal, be the more acceptable to the Sa'udi King. But the I.P.C. directors were
slow and cautious in their offers and would speak only of rupees when gold was
demanded. Their negotiator, so handicapped, could do little." p. 107. On May 5,
1933, IPC directors "decided that it was not desirable that the IPC should apply for
an oil concession over El Hasa." Minutes of meeting of IPC, May 5, 1933, cited in
F.T.C., *The International Petroleum Cartel*, 74n24.

28. Philby incorrectly gives the date of signing as May 11, 1933, *Arabian Oil
Ventures*, p. 125.

in Philby's account of this meeting. He relates that the king dozed off at intervals as the finance minister read clause after dreary clause until the king could take no more. A resumption of reading on the morrow found the king equally weary or indifferent. During the reading he went fast asleep and awoke with a start at its end. Fixing his one good eye on Philby, he asked his opinion. When Philby expressed satisfaction, the king replied, "Very well!" and turning to his finance minister, he said, "Put your trust in God, and sign."[29]

On July 7 the king signed Royal Decree No. 1135 granting the concession; on July 10 the official newspaper, *Uumm al-Qura,* published the decree and on July 14 published the text of the concession. In November 1933, Standard Oil assigned the concession to its wholly owned subsidiary, California Arabian Standard Oil Company (Casoc).

Standard Regards Concession as a Confidential Document

Standard Oil officials treated the concession more circumspectly than did Saudi Arabia. They not only refrained from making it public in the United States, but fourteen years after its publication in Saudi Arabia, when James A. Moffett—in a suit in the Federal District Court of Southern New York against the Arabian American Oil Company (Aramco), successor to Casoc—obtained a subpoena from the court ordering the defendant to produce the original concession, Aramco's counsel, Charles Evans Hughes, Jr., requested the court to quash the subpoena. In doing so, he declared that to make the concession public would annoy and embarrass the government of Saudi Arabia, the government of the United States, and Aramco. James Terry Duce, vice-president of Aramco, supported Hughes's request with an affidavit in which he said,

I am informed and believe that the full texts of said agreements have not been made public by the Saudi Arabian Government in Saudi Arabia or elsewhere. Defendant has not made public the full text of said agreement within the United States or elsewhere. It has been the understanding of defendant that the Government of Saudi Arabia does not desire such publication.

Duce further avowed,

The question of the production of said agreements has a significance transcending the private interests of either of the parties herein. The Saudi Arabian

29. *Ibid.,* p. 124.

Government is a sovereign state. It has oil resources upon which the United States is increasingly dependent. It is an area in which the international situation at the moment is delicate. Its views with respect to agreements which constitute a major source of its income are entitled to respect, and the Government of the United States itself has an interest in whether any action should be taken which would be contrary to the wishes of the Saudi Arabian Government.

He continued,

Accordingly, I suggest that, since I am advised the relevance of the text of such agreements is at best dubious, the Court should not resolve the doubt in favor of their production, or at least should not so resolve without seeking the advice and recommendation of the Department of State of the United States, which might well wish to consult the Saudi Arabian Government to ascertain the views of said Government.[30]

The court was convinced and on July 22, 1947, in response to Aramco's request, quashed the subpoena. Duce's apparent ignorance of the publication of the concession agreement fourteen years before is surprising. So also is his concern about making public the details of the agreement.[31]

The Concession's Principal Provisions

In its basic provisions the agreement is patterned after the Turkish Petroleum Company's Iraq concession and Anglo-Persian's Persian concession. It grants the company the exclusive right for sixty years to explore for crude petroleum and exploit its production in an area covering most of Eastern Saudi Arabia (Article 1) and a preferential right to a concession covering the balance of the Eastern area.

As previously indicated, it obliged the company to advance the government a loan of £50,000 against future royalties (Articles 4 and 6) and to pay an annual rental of £5,000 until the discovery of oil in commercial quantities (Article 5). It obliged the company to begin geological explora-

30. United States District Court of Southern New York, File Civ. 39–770. Quoted in Benjamin Shwadran, *The Middle East, Oil and the Great Powers* (New York: Council for Middle Eastern Affairs Press, 1959), p. 294.

31. The Saudi Arabian government reproduced the original and later concession agreements in its *First Memorial in Arbitration Proceedings between the Government of Saudi Arabia and Aramco*, Geneva, 1956. Terry Duce himself deposited bound copies of the proceedings, briefs, memoranda, and other relevant documents in the Hoover Institution at Stanford University, Palo Alto, California. Although the Institution keeps them in its locked vaults, it makes them available to scholars.

tion promptly (not later than September 1, 1933) and to continue to explore until it found a suitable structure on which to drill (Articles 8 and 10). On discovering oil in commercial quantities, the company obliged itself to advance an additional £50,000 against future royalties and a year later another £50,000 (Article 11). The company agreed to continue drilling diligently (with at least two strings of tools) until it had drilled the whole area in accordance with first-class oil field practices until it terminated the contract (Article 13). The agreement authorized the company to terminate the contract on thirty days' notice with the obligation to transfer to the government without charge all immovable facilities it may have constructed (Article 28).

Aside from the company's cash loans to the government, its most important financial obligation was to pay the government a royalty of four shillings gold, or its equivalent, on each (net) ton of crude oil it produced (Article 14). The government's most important concession to the company is its agreement to forgo for all time its right to impose any taxes on the company (Article 21).

The agreement provides that Americans shall direct the enterprise but shall employ "as far as practicable" Saudi Arab nationals (Article 23). The company pledges itself not to interfere with "administrative, political or religious affairs within Saudi Arabia" (Article 36).

The contract is drawn up in English and Arabic, but if a divergence of interpretation should arise, the English text shall prevail (Article 35).

Relative Gains of the Two Parties

Philby's account of the negotiations and the terms of the contract make it clear that both parties got what they most wanted. The government got a substantial downpayment against future royalties, prompt exploration of the concession, and the promise, on the discovery of oil, of continuous production on which it would receive royalties. The agreement seemed to provide for both the king's immediate and his continuing fiscal needs. The company got an exclusive right to explore and exploit a vast area which seemed to promise fruitful development. Neither party suspected the bonanza that the concession proved to be.

That each was satisfied with the immediate results is a tribute to Philby's skill in serving both his conscience and his purse. F. A. Davies, former chairman of the board of directors of the Arabian American Oil

Company, in a foreword to Philby's *Arabian Oil Ventures,* after recounting some of the major benefits that the development of the concession has brought to both parties, continues,

> Other benefits too numerous to mention here have stemmed from the Agreement which Philby helped so ably to bring into being. By the part he played, he has helped his adopted country and people. He served well his friend and master King 'Abd al-'Aziz, and the Company also. His honesty, frankness, and sense of balance enabled him to act as liaison or go-between without either side then or since having the feeling that he was unduly favoring the other. A truly remarkable performance![32]

California Arabian Standard Discovers Oil and Signs Supplemental Agreement

Soon after signing the contract, Casoc began its explorations. It promptly mapped Dammam Dome, whose demographic features "caught the geologists' eyes." It spudded in the first well on April 30, 1935, and despite discouraging results drilled six other wells to a depth of about 3,200 feet, tapping the same horizon that had proven so prolific in Bahrain. Not finding oil in commercial quantity, it decided to drill deeper. On March 5, 1938, in Well No. 7, tapping what became known as the Arab Zone at 4,700 feet, it found oil in large quantities. Davis epitomized the significance of Well No. 7 in this succinct statement: "The lid was off!"[33]

With its hopes thus rewarded the company initiated steps to fulfill its other obligations and to exercise the options that the concession provided. On October 16, 1938, it declared that it had discovered oil in commercial quantities. On May 31, 1939, it signed a supplemental agreement extending the area of the concession and prolonging the life of the concession.[34]

The Contract and the Bargaining Process

George Ray, counsel for Aramco, has described the concession contract as an instrument which united the skills, capital, and venture spirit of

---

32. Philby, *Arabian Oil Ventures,* foreword, p. xiii.
33. *Ibid.,* p. xii.
34. The Supplementary Agreement appears in Exhibit 2 in the government's *First Memorial.*

private enterprise with the national ambitions of a farsighted ruler of an undeveloped country anxious to promote its material progress—a union arrived at rationally by informed bargainers each knowing what he wanted and hopeful of what he was getting.[35] The foregoing account may raise some doubt about such a judgment.

Neither the king nor his people were familiar with Western industrial technology and business methods, and they feared both. Saudi Arabia was a young nation, largely desert, with limited resources and a loosely united people living close to a subsistence level under primitive conditions. The king ruled by right of conquest. As absolute monarch he had complete authority over any minerals that lay under Saudi Arabia's barren surface. The government—which was the king and his council—made no distinction between the privy and the public purse. Both were empty. Neither the king nor his council had any realistic notion of the value of the rights they were bargaining away. The king's minister of finance was apparently a shrewd trader advised by a faithful friend who was interested in getting the best bargain available, but who personally profited by doing so.

The company, with adequate capital, long experience in the technology of oil-finding, accustomed to risk-taking, but with a confidence born of its successful Bahrain venture, brought to the bargaining table a legal sophistication and technological understanding unmatched by the king's representatives. To recognize this is not to suggest that the company drove a hard bargain by unfair means. There is no evidence that it did. The bargaining took place in the midst of the Great Depression. The oil industry was suffering not only from a restricted demand but from a surplus of oil-producing capacity that it had developed during the prosperous 1920s. Prices were depressed. The leading American companies had for some time been laboring to establish a program of output restriction and price stabilization in the United States, and the leading international companies were similarly engaged on a world scale. As partners in the Iraq concession, they had obliged themselves under the Red Line Agreement to substitute group activity for independent action. Standard of California was not so handicapped. Its operations had been confined for the most part to the western United States. It was ambitious to establish itself in world markets. Its Bahrain venture had given it a foothold and valuable information.

The company had reason to be more anxious to obtain the concession

35. George Ray, in his opening statement, *Arbitration Proceedings*, especially pp. 7–12.

than any other oil company and it apparently got it for somewhat less than the maximum it might have paid had competition been more vigorous.[36] The king got less than he wanted but more than any competitor was willing to pay. Philby played an effective role in sealing a bargain that seemed to satisfy both parties. It was a bargain out of which both eventually received rewards far greater than anticipated but which eventually led to misunderstanding and dissatisfaction on the part of the government. But this is a later story.

Socal's Arabian Concession Threatens Market Stability

Although the well-established international oil companies were not sufficiently concerned that Standard Oil of California (Socal) obtain the Arabian concession to engage in effective bidding to prevent it, as the potentialities of Socal's Middle East production became apparent they became disturbed over the threat it constituted to the program of output control and price stabilization in which they were engaged. The United States' domestic oil companies initiated this program at the state level as an output and proration program in Oklahoma and Texas in the late 1920s. Later with the assistance of the federal government they broadened its domestic scope and tightened its application.[37]

The "As-Is Agreement"

In 1928 in the so-called "As-Is Agreement," formulated at an informal conference by the heads of the three most important international oil companies, Walter Teagle of New Jersey Standard Oil Company, Sir Henri Deterding of Royal Dutch-Shell, and Sir John Cadman of Anglo-Persian, the "Big Three," projected a plan designed to supplement the domestic control program and bring stability to world oil markets.

36. Philby states that "Hamilton accepted in principle the necessity of making a substantial loan to the Government, which was after all the chief hurdle to be negotiated: the sum he envisaged as an initial offer being £50,000 gold, which he was clearly prepared to improve on, if pressed by competition." *Arabian Oil Ventures,* p. 87.
37. See Chapter 18 for a more complete discussion of this.

About the As-Is Agreement Teagle was quoted in the *Oil and Gas Journal* (September 20, 1928) as having said,

Sir John Cadman, head of the Anglo-Persian Oil Co., and myself were guests of Sir Henri Deterding and Lady Deterding at Achnacarry for the grouse shooting, and while the game was a primary object of the visit, the problem of the world's petroleum industry naturally came in for a great deal of discussion. . . . Any attempt at regulation of overproduction of crude would obviously require cooperation of a vastly greater number and diversity of interests than were represented at Achnacarry Castle.

A document bearing the simple title, "Pool Association," under date of September 17, 1928, emerged from the Achnacarry grouse-hunting conference. The Federal Trade Commission in its study of *The International Petroleum Cartel* has set forth in detail its content and purpose.[38] Although the commission's study has engendered much controversy regarding the scope and effectiveness of the As-Is Agreement, it sets forth clearly the agreement's stated objectives and the principles regarded as necessary to achieve them.

Like the domestic program the international program is formulated within the framework of conservation. The As-Is Agreement laments that "certain politicians, with the support of a portion of the press, have endeavored to create in the public mind the opinion that the petroleum industry operates solely under a policy of greed and . . . methods of wanton extravagance."[39]

Denying this contention, the document alleges that the industry has been too diligent in trying to forfend against the day of oil shortage. "Excessive competition has resulted in the tremendous overproduction of today," with shut-in production equal to approximately 60 percent of consumption. Producers, trying independently to solve their problems by increasing sales at the expense of their rivals, have resorted to destructive competition. This, the document contends, has placed an intolerable burden of high costs on the industry. The remedy lies in co-operative action to effect economies, eliminate waste, and curtail expensive duplication. Each

38. F.T.C., *The International Petroleum Cartel,* pp. 199 *et seq.* The Achnacarry document is known as the "As-Is Agreement" because its objective seems to have been to preserve the existing market shares of the international oil companies. The Federal Trade Commission, which subpoenaed the document, quotes extensively from it.

39. *Ibid.,* pp. 199–200.

member of the industry by curbing his individual competitive zest by co-operating with his rivals can promote the interests of all.

IPC Participants Seek Co-operation with Socal

The As-Is Agreement, designed to achieve worldwide co-operation in the marketing of petroleum products, paralleled the Red Line Agreement, designed to insure co-operation in finding and developing new oil fields. Standard of California's go-it-alone policy threatened both. As this became increasingly evident, the leading members of the Red Line Agreement tried to eliminate Socal's threat in one of three ways: by modifying the Red-Line Agreement so as to exclude Bahrain and Saudi Arabia from its scope, thus permitting the signers of the agreement to negotiate independent agreements with Socal; by IPC's buying Socal's Bahrain and Arabian concessions; by the members of the Red Line Agreement buying the oil that Socal would produce from its concession.

They succeeded in none. The French and Gulbenkian interests, believing that modifying the Red Line Agreement was a scheme to exclude them from its benefits, blocked modification. Socal was not interested in selling its concession. A conflict of interest among IPC owners prevented them from making satisfactory arrangements with Socal to buy its oil.

About negotiations with Socal and with the French and Gulbenkian groups, H. S. Seidel, NEDC's representative on IPC's board, wrote W. C. Teagle, president of New Jersey Standard, on February 10, 1936, as follows:

it is our feeling that since the negotiations with Socal in America have not progressed materially, and since the proposals negotiated with the French and Gulbenkian groups were not acceptable . . . we should at least try to obtain the exclusion of Bahrain Island from any restriction of the group agreement, at the same time securing, if possible, the continued agreement of the French and Gulbenkian groups to negotiate and eventually conclude an agreement with the California Company for the El Hasa concession in Saudi Arabia. . . . We felt there was a reasonable chance of securing a settlement on this basis, and while this would not materially facilitate an agreement with the California Co., at least it would have excluded a considerable potential production from the present restrictions, and to this end would . . . help . . . the Standard Vacuum Co. in the protection of its eastern markets.[40]

40. *Ibid.,* p. 78.

While not successful in trying to buy Socal's Middle East concession, the IPC groups did not despair of bringing Saudi Arabia's underdeveloped area within its control. As late as March 1939, Seidel wrote W. S. Farish, the president of New Jersey Standard:

The groups in Iraq Petroleum Co. unanimously decided to try and negotiate a partnership agreement with the California Company or Caltex covering concessions remaining in Ibn Saud's territory and the neutral zone of Arabia.[41]

Again in April 1939, Seidel wrote Stuart Morgan:

All the groups . . . were of the opinion that in principle an attempt should be made to work out a deal with Caltex whereby each company would set aside and retain definitely explored areas or a mutually agreed to area and treat the balance of the Arabian Peninsula, including neutral zones, as areas to be posted and operated for joint account.[42]

The Trade Press Recognized Socal's Threat to Market Stability

What the Federal Trade Commission has revealed by painstaking investigation with access to confidential documents about the efforts of IPC participants to bring Socal's potential Middle East production under group control, the oil world in general and the trade press in particular were aware of. In its issue of May 2, 1936, the *Petroleum Times* of London had this to say of Socal's threat to world market stability:

Fight for markets has been under reasonable control . . . "gentlemen's agreements" are the order of the day, and what each large oil group fears more than all else is the entry of a powerful newcomer in the established order of world markets.
. . . the vast strides of international scope the Standard Oil Company of California has made in a relatively few years at least proves that romance is not yet dead in our industry, but it may lead to disturbing factors.

Commenting on the California company's proven field in Bahrain, the explorations in Arabia, and its ambitious plans with regard to Mexico and Sumatra, the *Petroleum Times* concludes that "All this implies an outlet and raises a question. Is this to be an orderly outlet or one regardless of the consequences?"[43]

41. *Ibid.*, p. 82.
42. *Idem.*
43. *The Petroleum Times*, May 2, 1936, p. 563.

Socal and Texas Company Form Partnership

While the IPC group were seeking a way for Socal's Middle East oil to get into world markets without greatly disturbing them, Socal proceeded independently to find its own way. As previously indicated, before acquiring its Middle East concession it had been primarily a domestic oil producer and marketer of oil products. As its Middle East production mounted, its problem was to find outlets. On the other hand, the Texas Company, which was already marketing oil products on a world scale, had no foreign production to feed markets that it had developed in Europe, Africa, Australia, China, and other areas in the Far East. A union of the Texas Company and Socal offered a happy solution to both companies' problems.

Accordingly, on July 1, 1936, Standard of California agreed with the Texas Company to acquire a one-half interest in the Texas Company's marketing facilities east of Suez. To accomplish this union, the Bahrain Petroleum Company, a wholly owned subsidiary of the Standard of California and owner of the Bahrain concession, doubled its outstanding capital stock and transferred the new issue to the Texas Company. The Texas Company in return transferred to the Bahrain Company the stock of its Eastern marketing subdivision. The Bahrain Company then transferred the stock of the Texas Company's marketing subsidiaries to a newly created and wholly owned subsidiary, California Texas Oil Company, Ltd. (Caltex). The net book worth of the marketing was approximately $27 million.

Concerning this union the *Petroleum Times* stated that it was

a natural outcome of the failure of the Standard of California to reach an agreement with the Royal Dutch-Shell, Jersey Standard, and Anglo-Iranian groups last year when a series of conferences were held.
the advantage of this merger [from the viewpoint of world oil circles] is that both companies being sound, stable and conservatively managed, it assures that Bahrein production, as well as any output that may eventually come from countries now being developed by Standard of California, will have assured and regulated outlets and will so lessen any possible danger of upsetting the equilibrium of international markets.[44]

About this transaction the Standard of California in its annual report for 1936 stated,

44. *Ibid.,* July 4, 1936, p. 8.

The management . . . considers the new arrangement most advantageous. . . . If the Bahrein Company itself had undertaken to market its own output, it would have been necessary to invest many millions of dollars in facilities and probably would have required many years to acquire a like market position.

In December 1936, Standard of California and the Texas Company broadened and tightened their union by the Texas Company's buying for $3 million in cash and $18 million in deferred obligations payable out of crude oil or oil products, a one-half interest in the California Arabian Standard Oil Company, owner of the Arabian concession, and a one-half interest in the N. V. Nederlandsche Pacific Petroleum Maatschappij, owner of a Sumatra concession. As a part of the same transaction the Texas Company granted Standard of California an option to buy a one-half interest in its European market facilities. Socal did not exercise its option, but ten years later Caltex paid the Texas corporation $28 million for its European facilities, thus uniting Socal's and the Texas Company's operations in Europe and throughout the Eastern hemisphere.

The United States Government Comes to the California Arabian Oil Company's Aid

When World War II broke out, the California Arabian Standard Oil Company (Casoc) controlled what has proved to be the world's richest oil concession. Casoc completed its first commercial well in 1937. Thereafter developments proceeded rapidly. By merging its foreign properties with those of the Texas Company, it obtained well-established channels for distributing its products in world markets. Its future was bright. Although it had not fully recognized the potentialities of Saudi Arabia's oil deposits, it had a good thing going and wanted to keep it.

So also did King ibn Saud. Unfortunately, the war slowed the company's developmental program with serious consequences to the king's revenues on which his prosperity and political security had come to depend. As previously indicated, before the company discovered oil, the king derived his income chiefly from Mecca pilgrims' fees. By 1939 oil royalties had come to represent a significant part of the king's revenue. In that year, while producing about four million barrels of oil, the company paid the king about $1,900,000 in oil royalties. Thereafter royalties declined sharply, averaging only about a million dollars during the war years.

Meanwhile the war had slowed the pilgrimage to a trickle. Poor harvests aggravated the country's poverty and increased the king's obligation to his people.[45] For relief he turned to his concessionaire. Regarding the California Arabian Standard Oil Company as the agency through which Arabia's oil deposits (then estimated at about one billion barrels) could be converted into current income and by now with unlimited confidence in the richness of the deposits and the wealth of the company, the king demanded that the company pay him advance royalties of $6 million in 1941 and indicated that he would need similar amounts until the emergency passed.[46]

American Owners Decide to Cut Costs

The American owners of the company demurred. At a conference early in 1941 company executives decided to curtail operations in Saudi Arabia, cut operating costs, and reduce the labor force to an essential minimum. At the same time the executives approved a commitment that the company representatives had made to lend the king $3 million to meet his emergency needs in 1941. But in doing so they instructed Davies to "intimate to King, as long as war continues and market for Arabian oil continues curtailed we do not want to increase amount advanced nor to lead him to feel that this is precedent for further advances in the future."[47]

They took this action knowing that company representatives in negotiating with the king had promised to try their best to obtain the full $6 million that the king had demanded. Finance Minister Abdulla Sulaiman, in a letter to Davies written on January 18, 1941, acknowledged the company's authorization of the $3 million loan but reminded Davies that he had promised personally "to return to America with Mr. Hamilton and, jointly, to endeavor, with all earnestness and diligence, to increase the said loan and to raise it to the figure of six million dollars."[48] At the same time

45. Hearings Before a Special Committee Investigating National Defense Program, U.S. Senate, 80th Congress, 1st Session, *Petroleum Arrangements with Saudi Arabia,* (Washington, D.C., 1948), Pt. 41, Ex. 2538, p. 25360. Also Ex. 2556, Memorandum introduced in Executive Session of Committee by W. S. S. Rodgers, Chairman of Board, Texas Company, p. 25380.

46. *Ibid.,* Ex. 2580, Letter from Abdulla Sulaiman, Minister of Finance, Saudi Arabia, January 18, 1941, p. 25410.

47. *Ibid.,* Ex. 2563, Confidential letter from President F. A. Davies, Aramco, p. 25389.

48. *Ibid.,* Ex. 2580, p. 25410.

the finance minister expressed his complete confidence in Davies's ability to arrange the larger loan. He said, "We have no doubt at all that your efforts will be crowned with success . . . and, accordingly, we are figuring our budget . . . on six million dollars as explained to you on several occasions."

And he reminded Davies of certain compensating concessions that his government had granted the company. It had agreed to extend the term of the concession by two years and to allow the company "freedom of action" in its operation "during the abnormal conditions occasioned by the present international emergency, provided that this freedom of action does not affect the government's revenue from the export of oil . . ." and provided that "after the present war is over the Company will resume its former activities and will do its utmost in all the work of the Concession."

Davies Acknowledges Company's Responsibility

To this statement of the minister's understanding of the mutual obligations that the company and the government had assumed Davies promptly replied, again acknowledging the "heavy responsibility" that he and Hamilton had undertaken and expressing the hope that their efforts would be "crowned with success." He continued,

In our many discussions with His Majesty and with Your Excellency we have come to a full realization of the Government's need for extraordinary assistance during the present international emergency. Should this emergency extend beyond the year 1941, the Company will, of course, continue to assist the Government as much as it can.

Aware of the decision that company executives had reached that they did not wish to increase the amount of the loan nor regard it as a precedent for future assistance, Davies warned that "in uncertain times such as these the amount of such assistance must of necessity depend upon the situation at the time the help is needed." He softened his warning, however, by emphasizing the identity of interest between the company and the king. "Your Excellency need have no doubt that the company realizes fully how closely its own interests are bound up with those of His Majesty's Government."[49]

49. *Idem.*

Company Turns to United States Government for Aid

The owners of the company, having spent approximately $27.5 million in developing its concession and having already advanced approximately $6.8 million against future royalties,[50] although not unsympathetic with the king's needs and not unmindful of the commitments that Davies and Hamilton had made, sought a less burdensome way of solving the king's financial difficulties. Recognizing the happy identity between their private corporate interests and their governments' political interests, they turned to the President of the United States. As an intermediary they utilized the services of James A. Moffett.

Moffett was not only chairman of the board of both the Bahrain Petroleum Company and its subsidiary, the California Texas Company, but had at various times in various ways served his country's government. Equally important, as a personal friend of President Roosevelt, he had easy access to the White House.

On April 9, 1941, Moffett interviewed the president, explained the plight of the king, emphasized the importance of the concession to the welfare of the United States, and explored the possibility of the Import-Export Bank, or the administrators of the recently enacted lend-lease program supplying the king with the money he needed. The president was sympathetic but indicated that no existing laws authorized the United States to extend financial aid directly to the Saudi Arabian government. However, he suggested that an arrangement might be worked out whereby through the sale of oil products to the navy the king might indirectly receive the aid he so sorely needed. At the president's request Moffett later submitted a specific proposal that the California Arabian Standard Oil Company deliver to the United States government from the account of King ibn Saud at favorable prices $6 million worth of oil products annually for the next five years. In a memorandum prepared for Moffett by executives of the company, Moffett again emphasized the plight of the king, his pro-allied sympathies, the importance of the concession, its ownership by American companies with 168,000 stockholders, the advances already made by the company, and the impossibility of a private corporation continuing to assume responsibility for financing an independent country. Specifically the memorandum stated:

50. *Ibid.*, Ex. 2554, p. 25379.

It has now come to a point where it is impossible for the company to continue the growing burden and responsibility of financing an independent country, particularly under present abnormal conditions. However, the King is desperate. He has told us that unless necessary financial assistance is immediately forthcoming he has grave fear for the stability of his country.[51]

Moffett not only outlined a naval oil purchase program to relieve the king,[52] but he suggested that

our State Department approach the British not only to increase the amount of money which the British have been advancing to the King, amounting to 400,000 pounds sterling per year, but also to request the British to continue to make such advances in sufficient amount, which, added to those made by the United States Government plus any other revenues received by the King, will total approximately $10,000,000 per year.

Moffett cautioned, however, that the British advances should be purely political and military and should not represent in any way a claim against the concession.[53]

At President Roosevelt's suggestion, Moffett discussed the company's proposal with Navy Secretary Knox. When it became apparent that "the Navy could not use petroleum products in such quantity in the Persian Gulf," Moffett suggested that the navy might buy the oil as a naval reserve. "But Secretary Knox said he had no appropriation which would cover anything of that kind.[54]

United States Government Enlists Aid of British Government

But the matter did not rest here. After discussing it with Secretary Knox and Harry Hopkins, President Roosevelt suggested that Jesse Jones, who as Federal Loan Administrator was arranging a loan of more than $400 million for Great Britain, should stipulate that the British government should take care of the king's financial requirements from the proceeds of the loan.[55]

Accordingly, Harry Hopkins wrote a "personal and confidential" letter to Jones (June 14, 1941) advising him that the president was "anxious to

51. *Ibid.*, Ex. 2539.
52. *Ibid.*, p. 25361.
53. *Idem.*
54. *Ibid.*, James Moffett, p. 24719.
55. *Ibid.*, p. 24747.

find a way to do something about this matter," and suggesting that the king's needs might be met in part by "the shipment of food direct under the Lend-Lease Bill." But with engaging frankness Hopkins expressed the judgment that "just how we could call this outfit a 'democracy' I don't know."[56]

Accordingly on August 11, 1941, Jones advised Moffett by letter that

Further and due consideration has been given to the matter of aid to the King of Saudi Arabia.

It is clearly the responsibility of the British to furnish the King with such aid as in their opinion he is entitled to and they feel would be helpful to their cause.[57]

Whether Jones was merely expressing an opinion or referring to an action he had taken was not entirely clear from this letter, and to clear up the matter he wrote Moffett again on October 9, 1941, stating,

At the instance of the President and the Secretary of State I suggested to the British Ambassador that Britain consider providing King Ibn Saud with such funds as in its opinion were necessary to meet his requirements.[58]

Jones wrote Moffett as quoted above at the request of Mr. W. S. S. Rodgers, chairman of the board of the Texas Company, who apparently wanted both more definite assurance that Jones had acted on the president's suggestion and evidence which might be shown to the king that the company had been faithful in its obligation to obtain for him the money he needed. As Rodgers put it, "We didn't want the British to run away with all of the credit on this thing."[59]

Company Representatives in United States and Saudi Arabia Keep Each Other Informed of Developments

Between the time that Moffett first spoke with President Roosevelt (April 1941) and Jones informed Moffett (October 1941) that he was soliciting British aid for King ibn Saud, company officials in Saudi Arabia and the United States kept each other informed by cable and letter about the negotiations. These throw light on the company's progress in trying to

56. *Ibid.*, Ex. 2583, p. 25415.
57. *Ibid.*, Ex. 2556, p. 25384.
58. *Ibid.*, W. S. S. Rodgers, p. 25384, and Ex. 2556.
59. *Ibid.*, Rodgers, p. 24828.

satisfy the king's demands while keeping down its own cost in doing so. More specifically they reveal the company's continued effort to obtain funds that the king needed either directly through the United States government or indirectly through the British government; the company's hope that for every dollar of aid offered through either of these channels their own obligation would be reduced by a corresponding dollar; the company's fear that the king might regard such aid as a supplement, not a substitute, for company aid; the importance of the king's being made to realize the key role the company and the state department played in enlisting British aid; and of convincing the king that the company had not reneged on its own promises; the king's ire at the company's delay in meeting its promised payments; the company's belief that the king was making unreasonable demands and accompanying them by "continued nagging and threats"; and the ignorance of the king in financial matters and his laxity in dealing with them.[60]

The skill with which company officials handled these problems is reflected in Roy Lebkicher's wire to Davies on November 3, 1941, stating, "Negotiations now under way regarding future British financial aid and feel our local difficulties over Government financing have disappeared except possibly with respect to immediate cash requirements if British action too slow."[61]

British Come to King's Aid but Endanger Concession

The company's success in substituting governmental aid to the king for its own is reflected in the following data: In 1940 the company advanced against future royalties and rentals, $2,980,988; in 1941, $2,433,222; in 1942, $2,307,023; in 1943, only $79,651. As the company decreased its advances, the British government increased its payments. In 1940 the British government advanced $403,000; in 1941, $5,285,500; in 1942, $12,090,000; in 1943, $16,618,280.[62]

Rodgers testified that the company made its advances in 1941 and 1942 to offset British influence.[63] The company officials were obviously

---

60. *Ibid.*, Ex. 2605, F. W. Ohleger to Davies, quoting from Roy Lebkicher, p. 25426.

61. *Ibid.*, Ex. 2616, p. 25431.

62. *Ibid.*, Ex. 2556, p. 25381.

63. *Ibid.*, Rodgers, p. 24829.

growing uneasy about the stake the British government was acquiring in Saudi Arabia's welfare and its possible effect on the political interests of the United States and the material interests of the California Arabian Oil Company. As Rodgers put it,

We didn't like it. We realized that he [the king] had to have the money, but we didn't like the British Government making these advances. And to offset their influences, we made these advances . . . in 1941 and in 1942 . . . And then at the beginning of 1943 it got so bad that we came down here to Washington and called on many people, trying to get the matter straightened out.[64]

In a memorandum introduced on April 27, 1944, in an executive session of the Special Senate Committee Investigating Petroleum Resources, Rodgers described British–Saudi Arabian relations as follows:

The situation in Saudi Arabia had now reached the point that the British were backing the Saudi Arab Government so far as its finances were concerned. It would appear that they were looking for an opportunity to remain as the financial advisor and backer of the Saudi Arab Government without having to advance any actual gold or silver. This crystalized in March 1943 when they proposed a plan for a Saudi Arabian note issue. This plan provided for the creation of a Saudi Arab Currency Control Board in London, composed of the Saudi Arabian Minister, Government of Great Britain Representatives, and Bank of England Representatives.[65]

The constantly increasing influence of the British government on the affairs of Saudi Arabia, according to Rodgers had reached "alarming proportions." He feared not merely that British prestige would be enhanced but that British influence might ultimately lead to a loss of a part of the California Arabian Oil Company's concession.[66]

Company Officials Again Appeal to Washington

To forestall the threat that British aid to Saudi Arabia constituted to his company's and his country's security, Rodgers arranged for conferences in Washington (February 1943) with Secretaries Ickes, Stimson, Knox, Welles, and Forrestal. Preliminary to these conferences he submitted a memorandum to Secretary Ickes in which he again outlined the importance

64. *Idem.*
65. *Ibid.*, Ex. 2556, p. 25385.
66. *Idem.*

of the company's concessions to the future petroleum needs of the United States. To emphasize what he regarded as an alarming situation, he pointed out that additions to United States reserves "during the past several years have been far short of those required to maintain either current or future production." He prophesied that "Maximum efficient production from all domestic wells will soon be insufficient to meet this country's expanding consumption and exports."[67]

To meet this situation he again proposed that the United States government extend direct aid to Saudi Arabia through lend-lease and that the government assume the indebtedness already incurred by King ibn Saud to Great Britain. To compensate the United States for assuming the king's indebtedness he proposed that the California Arabian Standard Oil Company set aside reserves in its Arabian concession from which it would supply the United States with its oil at prices well under world prices.

To assure the United States government oil for its future needs and to compensate it in part for additional funds supplied the king, it would set aside additional reserves.[68] By thus providing the help that the king needed, "United States' prestige would be enhanced and the United States armed forces assured of a one hundred percent owned reserve of great magnitude in a neutral country in a portion of the world where it does not have such a reserve, and at prices less than such oil could be purchased elsewhere even if available."[69]

Government Comes to Aid of the King—and the Company

Company officials evidently succeeded in communicating their sense of urgency to governmental officials. Within ten days (February 18, 1943) President Roosevelt wrote Stettinius as follows:

My dear Mr. Stettinius:
For purposes of implementing the authority conferred upon you as Lend-Lease Administrator by Executive Order No. 8926, dated October 28, 1941, and in order to enable you to arrange lend-lease aid to the Government of Saudi Arabia, I hereby find that the defense of Saudi Arabia is vital to the defense of the United States.
Sincerely yours,
Franklin D. Roosevelt[70]

67. *Ibid.*, p. 25386.
68. *Ibid.*, p. 24859.
69. *Ibid.*, p. 25387.
70. *Ibid.*, Ex. 2557, p. 24861.

Eventually Saudi Arabia got the help it needed; the company, the relief it sought; and the United States re-established its prestige without the company's finding it necessary to set aside the reserve Rodgers had proposed. About this proposal the following colloquy took place between Senator Brewster and Rodgers:

Senator Brewster: Within a week after you made the proposal, the Government went ahead to put 18 million dollars into exactly the proposition that you had made, but somewhere between you and the Government that proposal seemed to evaporate.

Mr. Rodgers: It did not evaporate with us. It evaporated with the Government.[71]

Ickes Proposes Buying out the Company

Actually it seemed not to have evaporated at all, but to have sunk deep in the fertile soil of Ickes's creative mind. There the roots of the idea soon burgeoned into an awesome plant. The essence of the Rodgers proposal was that the company set aside reserves from which to supply oil at favorable prices to the United States government as and when needed. Ickes countered with a proposal that the government buy out the company lock, stock, and barrel.

As a preliminary step on June 10, 1943, he wrote a letter to President Roosevelt recommending that the president instruct the secretary of commerce to organize a petroleum reserves corporation with authority to "acquire and participate in the development of foreign oil reserves," to be managed by a board of directors including the secretaries of war, navy and interior. The corporation's first order of business was to acquire a participating and management interest in the crude-oil concession of the California Arabian Standard Oil Company.[72]

In so recommending, Ickes warned that if steps were not taken to prevent it, the country would soon be confronted with an oil shortage dangerous to the armed services and crippling to the domestic economy. Such a move would not only forestall these developments, but it would also serve to "counteract certain known activities of a foreign power which presently are jeopardizing American interests in Arabian oil reserves."

71. *Ibid.*, Rodgers, p. 24869.
72. *Ibid.*, Ickes, Ex. 2633, pp. 25237–25238.

The president promptly created the Petroleum Reserve Corporation and made Ickes its president and chairman of its board. Thereafter Ickes began negotiations for the purchase of an interest in the California Arabian Standard Oil Company. He first proposed that the company sell its entire stock to the Petroleum Reserves Corporation. This, in the language of Ickes, "scared them" so much that "they nearly fell off their chairs."[73] Ickes gradually "whittled down" his proposal to 70 percent, then to 51 percent, then to 33⅓, and then suddenly called off the whole deal.

Concerning this Rodgers testified:

Well, the Petroleum Reserves Corporation . . . wanted to buy the whole company for what we had put in it and then pay us a small royalty. Naturally that did not appeal to us. Then they got down to a 33⅓ basis and we did not like it, or at least I did not personally, but we began to get a little bit closer. We did not get close enough, and one day, Mr. Ickes said, I don't know why, "The negotiations are off," and I was very much relieved.[74]

Mr. Ickes' account runs as follows:

my records show a good deal of talking back and forth and rapid retreat on the part of the companies. . . . In the meantime . . . they came up here to the Hill and built a fire under us on the theory that this was an attempt on the part of the Government to take over a private-business enterprise, which, of course, was against the American tradition, as they put it, and perhaps it was. But this was more than a business enterprise, this involved the defense and safety of the country.

Ickes discerned a change in the attitude of the owners of the Arabian concern with a change in the progress of the war. "They felt in the meanwhile that since Rommel had been chased out of North Africa they were secure in their concession and more disposed to thumb their nose at us."[75]

Not only were the owners reluctant to sell, but the oil industry at large vigorously protested the government's going into the oil business. In the language of Rodgers, "They were violent on this whole thing." Meanwhile Ickes had discovered what he regarded as a more promising way of providing for the government's petroleum needs.

73. *Ibid.*, Ickes, p. 25234.
74. *Ibid.*, Rodgers, p. 24845.
75. *Ibid.*, Ickes, pp. 25240–25241.

Ickes Recommends the Government Build a Middle East Pipeline

Blocked in his effort to obtain control of the California Arabian Standard Oil Company, Ickes adopted an idea first suggested by Admiral Andrew Carter, who had visited Saudi Arabia and conferred with company and local government representatives, that the United States government construct a pipeline from the Persian Gulf to the Mediterranean.[76] The pipeline, which was to cost from $100 million to $120 million, although designed to serve as a common carrier, would in fact have served as a Mediterranean outlet for the California Arabian Company's concession and for the oil of the Kuwait Oil Company, 50 percent owned by the Gulf Petroleum Company from its Kuwait concession. The pipeline would eliminate the long tanker haul from the Persian Gulf through the Indian Ocean, the Red Sea, and the Suez Canal. California Arabian heartily supported the project and in return agreed to supply oil to the United States government at a discount of 25 percent from commercial prices. In recommending to the president that he approve the plan, already approved by the Joint Chiefs of Staff, James Byrnes, then secretary of state and director of the Office of War Mobilization, explained its advantages as follows:

The plan has great advantage over all previous proposals. It does not involve the Government in the business of producing oil. It does not involve the Government in business as a partner of two oil companies. If the Government went into business by itself, it would be in violation of the policy that heretofore has been pursued in other countries. . . .

The contract now presented requires the Government to do only what it has done in this country—construct a pipeline. It will operate the pipeline practically as a common carrier. The pipeline is constructed at this time because we will need Middle East oil for military purposes. At the same time, it will conserve our domestic resources and assure us of an additional oil supply when the war is over.[77]

Ickes supported the pipeline proposal not only because of his belief that it would insure the government a continuous supply of cheap oil ("it was practically as good as an interest in the oil concession")[78] but because he thought that it would strengthen his hand in negotiating an Anglo-American oil agreement which he characterized as promising "American oil

76. *Ibid.*, Ex. 2588, p. 25387. See also, Ickes, p. 25243.
77. *Ibid.*, Ex. 2558, Memorandum dated January 25, 1940, by James F. Byrnes.
78. *Ibid.*, Ickes, p. 25244.

companies greater protection than they had ever had in that area of the world, or where they came in conflict with the British with respect to oil."[79] Ickes thought that it would "alert the British to the idea that we really meant business in the Middle East on oil."[80]

Pipeline Proposal Falls Through

Both the pipeline and the Anglo-American oil agreement died a-borning. California Arabian Oil Company officials favored both. The industry at large vigorously opposed both. In the language of Senator Brewster, "The pipeline . . . evaporated as a result of . . . formidable opposition. . . ." from the industry and from Congress. Or as Rodgers put it, "I would say that the opposition was about 98 percent of the oil industry, and it was pretty strong from up here on the Hill."[81]

Thus without surrendering any of its rights in its Arabian oil concession, or without any binding agreement to supply the United States with cheap oil, the California Arabian Company was able to substitute American government subsidies to King ibn Saud for the advances which it and the British government had previously been making. With the war's end, both the company's prestige and its property rights were intact. Its hold on its concession seemed secure.

Secretary Ickes did not abandon easily his project for the construction of a trans-Arabian pipeline, but in an article in *Colliers* (December 2, 1944) he was forced to acknowledge that it had been "done to death without benefit of clergy." Even as the Ickes plan was being defeated, the owners of California Arabian recognized its commercial feasibility—the most economical way to get Persian Gulf oil to European markets.

Problems Confronting Tapline: Obtaining Right of Way

Accordingly, in July 1945, they organized the Trans-Arabian Pipe Line Company (Tapline) to construct a pipeline extending from their Saudi Arabian fields on the shores of the Persian Gulf to the Mediterranean. Construction of the pipeline involved three major problems: obtaining

79. *Ibid.*, Ickes, p. 25247.
80. *Ibid.*, Ickes, p. 25244.
81. *Ibid.*, Rodgers, p. 24871.

right-of-way permits across the countries through which the pipeline would be laid; obtaining the necessary steel in a time of acute steel shortage; obtaining capital with which to execute the project. Soon after its organization Tapline began right-of-way negotiations with Palestine, Lebanon, and Syria. With Palestine it encountered no difficulties; Sir Alan S. Cunningham, British high commissioner, on June 7, 1946, granted the company free of charge a concession to construct the line across Palestine to the Mediterranean. Later the British colonial secretary, George Hall, explained to the House of Commons that while Palestine would "receive no direct financial benefit from the pipeline . . . its construction and maintenance would provide increased employment and indirect revenue."[82]

Negotiations with Lebanon and Syria soon became ensnarled in diplomatic and political difficulties. Both countries, jealous of their newly-won statehood, sought stiffer terms than Palestine had granted, but Lebanon, anxious to attract the pipeline's western terminal to its coast, soon reached an agreement with Tapline, granting it right of way for an annual payment of $180,000. Syria demanded more. Out of a diplomatic crisis between Syria and Lebanon on the one hand and between Syria and the United States on the other, followed by a two weeks' conference at Beirut attended by the American ambassador to Iraq, James Wadsworth, the American minister to Lebanon, Lowell Pinkerton; and Tapline's representative, W. J. Lenahan, emerged an agreement between Syrian Premier Jamil Mardam and Tapline.

The Israel-Arab War of 1948 disrupted proceedings, and the Syrian Parliament never ratified the agreement. After the armistice, Syria and Lebanon, fearing that Tapline might choose to use its Palestine right of way, reached an agreement on January 28, 1949, under which they were to share annual pipeline royalties based on the amount of oil transported, with a guaranteed minimum annual payment. The agreement exempted Tapline from import, export, and all other taxes. On March 30, Colonel Husni Zaim seized control of the Syrian government, and on May 16, under his dictatorship, the government, which the United States quickly recognized, signed an agreement in Damascus with a Tapline representative.

The Problem of Steel Shortage

Obtaining the necessary steel for the pipeline involved domestic, economic and political problems. Shortly after the first Tapline-Syrian agree-

82. P.D.C., p. 422, col. 1860, May 15, 1946.

ment was signed, the United States Department of Commerce approved an export license for 20,000 tons of steel. The department justified its action on the grounds that Arabian oil would serve American strategic, political, and economic interests. The Senate Small Business Committee viewed the matter differently. After hearings extending over approximately nine months, the committee found no such public interest.[83]

While investigating the matter further, the Department of Commerce held up steel shipments for more than a year, but shortly before Syria reached its final agreement with Tapline the department again authorized steel shipments. It did so on the recommendation of the departments of interior and state, the national military establishment, and the Economic Co-operation Administration. The defense department refrained from joining in the recommendation, holding that military considerations did not justify it.

The Independent Petroleum Association's dissent was more partisan and vigorous. It was also bitter. In its view the department's action was but a further step

in a decision made long ago. Through all the evasions, the dodgings and the squirming that have occurred over a period of eight years, there has run one consistent thread. The Government of the United States was committed by certain officials to a course of securing the position of the oil companies which hold the Arabian concession.[84]

New Jersey Standard Oil and Socony Form Partnership with Aramco

On January 31, 1944, California Arabian Standard Oil Company (Casoc) changed its name to Arabian American Oil Company (Aramco). Soon thereafter Aramco began negotiations with Standard Oil Company (New Jersey) and Socony Vacuum Oil Company whereby the New Jersey company would obtain a 30 percent participation in the Arabian oil venture and Socony-Vacuum a 10 percent interest. In December 1946, the parties agreed in principle on this division; in 1947 they signed contracts incorporating the principle; and on December 2, 1948, they consummated the arrangement. Pending its consummation, Standard of New Jersey and Socony-Vacuum in March 1947 guaranteed an Aramco bank loan of $102 million with the understanding that if and when Jersey and Socony received

83. Cong. Record, Senate, 95, part 2, p. 2222, March 11, 1946.
84. *Ibid.*, p. 2225.

their ownership share, Aramco would use the proceeds of the purchase to pay off its bank loans.

New Jersey Standard and Socony also guaranteed with Aramco's parents a separate loan to provide funds for the construction of a pipeline from the Persian Gulf to the Mediterranean, a 30-inch, 1,050-mile line.

Jersey Standard's and Socony's partnership in California Arabian's oil venture accomplished three things: it provided capital for the construction of the pipelines; it assured the Jersey Company and Socony a low-cost oil supply for their rapidly expanding world markets; and it expanded greatly California Arabian's channels to world markets without Caltex's engaging in a vigorous and perhaps ruinous competiton to obtain them.

About Caltex's need for capital, Standard of New Jersey said,

The requirements in men, money, and equipment were to grow to such proportions that, despite the large resources of these two corporations, [The Standard of California and The Texas Corporation] it later became necessary to accept the participation of other companies in the development of Saudi Arabian oil.[85]

Jersey and Socony, which had pooled production and marketing interests in the Far East and were together probably the world's largest marketer of petroleum products, found themselves handicapped in the race for positions in Middle East oil production, which was rapidly establishing itself as the world's most prolific and lowest-cost producing area. The British Petroleum Company was the sole exploiter of Iran's rich deposits. The Iraq Petroleum Company, in which the New Jersey Standard and Socony owned only a 23¾ percent interest, controlled virtually all of Iraq's output, and the Red Line Agreement prevented both companies from developing independent sources of supply in the promising Persian Gulf area.

About this the Jersey Company said that because of the growing demands for oil it

decided . . . to approach other participants in Iraq Petroleum Company and associated companies to determine whether some modification of the Group Agreement of 1928 could be worked out, so the Jersey Company could participate separately in the development of other sources of crude oil within the "Red Line."[86]

85. *Standard Oil Company* (N.J.) *and the Middle East Oil Production, A Background Memorandum on Company Policies,* Revised, 1954, p. 16.
86. *Ibid.,* p. 17.

While Standard of New Jersey and Socony were looking for additional low-cost oil supplies, Aramco's potential output had so expanded that Caltex's established markets were no longer adequate to handle it. The 1936 Calso-Texas union afforded only temporary relief. In its *1945 Annual Report* the Standard of California (Socal) could say about Aramco and Bapco: "These joint operations have now been developed to a point where they rank among the more important holdings in the world." The following year it could more precisely claim that Saudi Arabia was now the world's fifth largest producing country. By linking Aramco's operations with Standard of New Jersey and Socony on the one hand, and by Caltex's acquiring the European marketing subsidiaries of the Texas Company on the other, Aramco's owners again acquired outlets that would permit their oil's continuing to flow into world markets without disturbing them. Concerning Aramco's new partnership, Socal in its *1946 Annual Report* stated,

One of the advantageous results of participation in Arabian American by these two additional companies will be the opening of their extensive marketing outlets to the oil products of Saudi Arabia thus permitting increased production from the vast stores of oil in that country. Production has been limited by markets, not by supply.

Concerning Caltex's purchase of the Texas Company's European marketing subsidiaries it stated, "This move enables Caltex to expand immediately into markets of Northern and Western Europe and the countries bordering on the Mediterranean Sea."

Both unions were happy ones. They not only supplied Aramco with additional capital, Jersey Standard and Socony with additional crude, and Aramco easy access to world markets, but they created a pattern of international oil company relationships more nearly in keeping with the aims of the As-Is Agreement.

Red Line Agreement Modified

To consummate their partnership in Aramco, Jersey Standard and Socony had to free themselves from the restrictions to which they submitted in becoming parties to the 1928 Red Line Agreement. This proved to be an arduous task. In October 1946, on the advice of its English legal counsel (IPC was a British company), because of the German occupation of France where Compagnie Française des Pétroles (CFP) was incorporated and where Gulbenkian was residing at the time, the 1928 agreement was dissolved. Standard of New Jersey announced that it was no longer bound by it. Both Gulbenkian and the CFP protested, insisting that if Jersey Standard and Socony participated in Socal's Arabian venture other than through IPC they too would be entitled to share in it. In February 1947, CFP brought suit in the British courts to establish the continuing effectiveness of the 1928 agreement. Jersey Standard and Socony filed a counter suit contending that "any agreement contained in the said declaration [the 1928 agreement] was in restraint of trade and contrary to public policy and void and unenforceable in law."[87]

Explaining his company's having been a party for almost twenty years to a restrictive agreement contrary to public policy and void at law, the Jersey Standard's president had this to say: "There has been a substantial change in the attitude of the American public and Government toward restrictive agreements and under current conditions, the affirmation of the agreement seemed inadvisable."[88]

CFP was less charitable in explaining Standard's motives. In January 1947, its attorney wrote the counsel for NEDC (holders of Jersey's and Socony's interest in IPC) as follows:

Matters have now changed. It would appear that the Sherman Act no longer is the real course [cause?] of concern, but was merely an excuse upon which to bring a contemplated and now effected breach of the group agreement, namely, the negotiations leading up to your client's proposed acquisition of a large interest in Caltex.[89]

None of the parties wished to air matters in court, and after almost two years of negotiation they reached a new agreement leaving Standard and Socony free to complete their acquisition of an interest in Socal's Arabian venture.

It was in contemplation of a favorable outcome of their legal difficulties that Standard of New Jersey and Socony Vacuum had signed the contract in 1947 with Casoc's owners providing for their future acquisition of a 30 and a 10 percent interest. On December 2, 1948, they consummated the deal. Standard and Socony acquired a similar ownership interest in Tapline which had already begun constructing its Persian Gulf–Mediterranean pipeline. Completed in late 1950, Tapline eliminated an expensive 7,300-

87. F.T.C., *The International Petroleum Cartel*, p. 104.
88. *Idem.*
89. *Ibid.*, 103n75.

mile around trip from Ras Tanura to the Mediterranean to become one of the most strategically located pipe lines in the world.

Between 1938 and 1949 Aramco increased oil production from 65,618 long tons to 22,820,783 and its payment to the Saudi Arabian government from less than one-half a million dollars to $39 million.

Despite its phenomenal increase in oil revenues, the Saudi Arabian government was about to enter negotiations for a new deal. But that is a later story.

4

Kuwait and the Kuwait Oil Concession

On December 23, 1934, the ruler of Kuwait granted an exclusive concession to the Kuwait Oil Company, Ltd., to explore for, produce, and market Kuwait's oil. The concession covered the whole country and its territorial waters, an area approximately the size of New Jersey. Thirty-two years later the Anglo-Iranian Oil Company (Anglo-Persian's name was changed to Anglo-Iranian in 1935) and the Gulf Oil Corporation, joint owners of the Kuwait Oil Company, had produced more than nine billion barrels of oil from their concession, most of it from the Burgan field, the world's largest, containing an estimated reserve of more than 64 billion barrels, twice that of the entire United States. In developing Kuwait's oil resources, the concessionaire had paid the Kuwait government in royalties and taxes more than £2 billion sterling. In the fiscal year 1967–68 alone they paid the government more than £242 million sterling. These payments account for Kuwait's having had until recently the largest per capita income of any country in the world.[1]

Meanwhile oil reserves enabled the government to transform a little tribal state—most of the untutored residents made their living as best they

1. Statistics from Organization of Petroleum Exporting Countries, *Annual Statistical Bulletin, 1968,* p. 110. Since 1965 Abu Dhabi, with a population of about 16,000 and oil revenues ranging from £11.9 million in 1965 to £68.6 million in 1968, has had the largest per capita income of any country.

could from the desert and the sea by camel- and goat-raising, date-farming, fishing, pearl-diving, boat-building, and trading—into a welfare state which assumed responsibility for all its citizens from the cradle to the grave.

Great Britain Obtains Special Privileges in Kuwait

That the Anglo-Persian Oil Company, at the time sole owner of the Persian concession and joint owner of the Iraq Petroleum Company with a concession covering most of Iraq, should have been a prime mover in obtaining the Kuwait concession is a tribute to Great Britain's then far-flung colonial interests.

Fearing Turkish encroachment on their mutual interests, Great Britain and Kuwait had established cordial relations throughout the nineteenth century. But it was not until after Sheik Mubarak became ruler of Kuwait in 1896 that Britain formally enjoyed any exclusive privileges or assumed any responsibility for Kuwait's political welfare. On January 23, 1899, Mubarak signed an agreement with the British political resident on behalf of the British government pledging and binding himself and his heirs "not to receive the Agent or Representative of any Power or Government at Koweit . . . without the previous sanction of the British Government. . . ." He further bound himself and his heirs not "to cede, sell, lease, mortgage, or give for occupation or for any other purpose any portion of his territory to the Government or subjects of any Power without the previous consent of Her Majesty's Government. . . ."[2]

The formal agreement contains no *quid pro quo*, but Great Britain as beneficiary assumed the role of Kuwait's protector. However, it was not until the outbreak of hostilities between Turkey and Britain in 1914 that the British government, in requesting Sheik Mubarak to attack and occupy certain Turkish-controlled Arab towns, formally recognized the "Shaikhdom of Kuwait" as an "independent Government under British protection."[3]

In thus reassuring the sheik, the British political resident in a letter to Mubarak personally promised the ruler "that your gardens which are now in your possession, *viz.*, the date gardens situate between Fao and Qurnah,

2. C. U. Aitchison, *Treaties, Engagements and Sanads Relating to India and Neighboring Countries* (Delhi, 1933), 14 vols.; XI, No. 36, 262.
3. *Ibid.*, No. 42, p. 265.

shall remain in your possession and in possession of your descendants without being subject to the payment of revenues or taxes."[4]

Meanwhile, in anticipation of a visit from "His Excellency the Admiral" (Slade), Sheik Mubarak acknowledged a British interest in Kuwait's oil possibilities. In writing the political agent on October 2, 1913, he stated:

> We are agreeable to everything which you regard advantageous and if the Admiral honours our (side) country, we will associate with him one of our sons to be in his service, to show the place of bitumen in Burgan and elsewhere and if in their view there seems hope of obtaining oil therefrom, we shall never give a concession in this matter to anyone except a person appointed from the British Government.[5]

Major Holmes Becomes Interested in a Kuwait Concession

For more than a decade no British subject tried to capitalize on the sheik's promise. As previously noted, under the leadership of Major Frank Holmes, the Eastern and General Syndicate in 1925 obtained an exclusive right to hunt for oil in al-Hasa province of Saudi Arabia. Holmes's chief interest was in getting and selling concessions wherever oil might be found, and his hunting ground covered much of the Persian Gulf area. Although he allowed his al-Hasa option to lapse, he got his Bahrain concession in December 1925. In selling it to Gulf Oil on November 30, 1927, he also contracted to sell to Gulf a Kuwait concession should he succeed in getting one.[6]

As a negotiator in a desert environment, Holmes was a picturesque character. In his *A Golden Dream,* Ralph Hewins characterized him as an "apparition, monstrously out of keeping with the surroundings, not knowing a word of the language, unfit, and by no means young at nearly fifty years of age to tackle the scorching desert."[7] Colonel H. R. P. Dickson, at one time British political agent in Bahrain and later in Kuwait, who found Holmes "amazingly amusing," describes in these words his anachronistic

4. *Idem.*

5. *Ibid.,* No. 41, pp. 264–265.

6. F.T.C., *The International Petroleum Cartel,* Select Committee on Small Business, Senate, 82nd Cong., 2nd Sess. (Washington, D.C.: Government Printing Office, 1952), p. 130.

7. Ralph Hewins, *A Golden Dream: Miracle of Kuwait* (London: W. H. Allen, 1963), p. 211.

appearance when they first met in 1922 at the 'Uqair boundary conference:

He carried a large white umbrella lined green, wore a white helmet as issued to French troops in Africa, and over his face and helmet a green gauze veil—quite like pictures one has seen of the tourist about to visit the Pyramids.[8]

Despite his lack of familiarity with the Arabic language and his unconventional dress, his sense of humor and personal charm won a warm friendship with Arab sheiks and tradesmen. All of this stood him in good stead as a negotiator—a talent bulwarked by an understanding of Arab generosity both in giving and receiving gifts. Dickson relates that on Holmes's first visit to ibn Saud he brought an "amazing number of presents . . . over fifty cases, leather bags, boxes and guns."[9]

Ameen Rihani characterizes his penchant for finding oil in these picturesque words:

But he knows what's in the bosom of the land, this man; can see the invisible streams of water that flow from the Persian mountains under the Gulf through the veins of the Hasa soil; can track the bubbling oil and the sparkling minerals to their depths and beyond;—has the modern Argus eye of science and finance.[10]

Major Holmes Uses Money as Well as Charm in Bargaining

In gaining the friendship of Arabs and influencing sheiks, Major Holmes relied not solely on his personal charm and trading talents. He bought good will and truckled to avarice by a liberal use of his client's money. In the spring of 1928, after Holmes had traveled 90 miles across the desert for a visit with the Kuwait ruler in his palace, negotiations proceeded amicably and Holmes left to draw up a concession and to return in a month. About this visit Holmes remarked, "It is interesting to observe the greedy attitude the Sheikh of Kuwait exhibited when I hinted to him that the Company, I was representing, had an American tang about it."[11]

8. H. R. P. Dickson, *Kuwait and Her Neighbours* (London: George Allen & Unwin, 1956), p. 269.
9. *Idem.*
10. Ameen Rihani, *Maker of Modern Arabia* (Boston: Houghton Mifflin, 1928), p. 83.
11. T. E. Ward, *Negotiations for Oil Concessions in Bahrain, El Hasa, The Neutral Zone, Qatar and Kuwait,* printed privately (1969), p. 73.

In May of the previous year in presenting his expense account to the Eastern and General Syndicate which was "probably double that of any one of the Geologists," he explained that he had to keep open house, be ready to entertain Arabs at any moment, and distribute tips liberally.[12]

As negotiations proceeded, T. E. Ward, who had been granted power of attorney by Eastern and General Syndicate, kept in constant touch with William J. Wallace, representing the Gulf Corporation. In a communication to Wallace, dated June 8, 1928, Ward acknowledged receipt of $2,932.38 (£ 600 sterling) to be disbursed "to the proper parties" in Kuwait. In a communication dated July 27, 1928, from Ward to Wallace, Ward tells of making cash payments of 4,000 rupees, "divided among four members of the Council of State" and 2,000 rupees for lesser members of the government. On delivery of the signed concession he notes that an additional 4,000 rupees were to be paid to state advisors and 500 rupees for office people. The concession draft provided for an annual rental to the state of 30,000 rupees and for 45,000 rupees to be divided among the highest state officials, of which around 20,000 rupees were for the ruler's private purse. Ward explained that certain of these payments were in the nature of "baksheesh," which may be interpreted as "tips" or "gifts" or by the more sordid as "bribes."[13]

Meanwhile it became evident that Kuwait's commitments to the British government were complicating Holmes's negotiations. The British Colonial Office objected to any concession being granted to a non-British corporation. Anglo-Persian, which at the outset had manifested a dog-in-the-manger attitude, was becoming more positively interested and the sheik had seen fit to reject Holmes's 1928 offer. The Gulf Company, impatient with the rate of progress, suggested that talks with the Colonial Office be speeded up, authorized Holmes to liberalize his offer, and to present as a personal gift to the sheik a Sunbeam car costing £905/18/8 or $4,389.25 at the current rate of exchange.[14]

Gulf Enlists Department of State Aid

Equally important, Gulf enlisted the State Department's help in overcoming British opposition. Charles Rayner, petroleum advisor to the State Department, has summarized the State Department's role as follows:

12. *Ibid.*, pp. 80–81.
13. *Ibid.*, p. 95.
14. *Ibid.*, p. 162.

On November 27, 1931, the Eastern Gulf Co. formally called the Department's attention to the fact that the so-called British Colonial Office was insisting on the so-called "nationality clause" in the Kuwait concession. This clause in effect prevented anyone except a British subject or firm from obtaining a concession in Kuwait.

The Department's reaction to this information was prompt, and on December 3, 1931, the instructions were sent to our embassy in London [over which Andrew Mellon, with Gulf interests, presided as Ambassador] to make representations with a view to securing equal treatment for American firms.

These negotiations were long, and complicated at a later date by the fact that the British controlled Anglo-Persian, which had previously expressed its disinterest in Kuwait, suddenly endeavored to secure a concession from the sheikh of Kuwait. Here again the Department insisted on the "open door" policy, and our Embassy in London was assiduous in its endeavor to expedite a settlement, and continuously and frequently pressured the British authorities for action.[15]

Developments While the State Department Negotiated with British Government

Between the secretary of state's telegram of December 3, 1931, instructing the Ambassador's office to intervene on Gulf's behalf and December 23, 1934, when Kuwait's ruler granted a concession to the Kuwait Oil Company, jointly owned by Anglo-Persian and Gulf Oil, the official account of the "long and complicated negotiations" brings out the following facts:[16]

Through oral and written communication the department kept the Gulf Oil Corporation continually informed of the progress of its negotiations. Secretary Stimson made it clear that he was not asking for exclusive rights for Gulf but was interested in establishing the "principle of the open door." In doing so, he cited the pertinent section of the U.S. Mining and Leasing Act of 1920 extending to British subjects the same treatment that American citizens receive. While Eastern and General Syndicate was trying to persuade the British Colonial Office to waive the requirement that only a British company could receive a concession, Anglo-Persian in August 1931 formally renewed its efforts to obtain a concession. Anglo-Persian thereafter dispatched a geological mission to Kuwait armed with drilling equipment to explore for oil. Eastern and General Syndicate submitted a revised concession application. The British Colonial Office, exercising its preroga-

15. From a memorandum by Charles Rayner, petroleum advisor to the department of state, submitted in a hearing on *American Petroleum Interests in Foreign Countries* before a special committee of the U.S. Senate investigating petroleum resources, 79 Cong., 1st Sess., June 27 and 28, 1945, p. 24.

16. *Foreign Relations of the United States,* Department of State, II (1932), 1–28.

tive under its treaties with Kuwait, analysed both the Anglo-Persian and the Eastern Syndicate's concession drafts in order that the sheik, who "is naturally not well versed in such technical matters, may understand what in fact will be the effect of the main provisions of each offer (e.g., the financial side, conditions of working the oil, etc., etc.)."[17] Thereafter the Colonial Office submitted its report to the Sheik of Kuwait embodying certain "safeguards" as "a matter of discussion after the Sheik . . . has made his decision from the point of view of what is to the best advantage of his own State."[18]

Anglo-Persian and Gulf Decide to Join Forces

It became evident in the course of the diplomatic correspondence that both Gulf and Anglo-Persian would be satisfied with something less than an exclusive concession, and when the sheik rejected both offers they decided to join forces in further negotiations. That they did so may not have been entirely owing to diplomatic pressure, which at times was characterized by American representatives as reaching the point of "anxiety"[19] and "expectation."[20]

It apparently was in part a response to the current instability in international oil markets. Both Gulf and Anglo-Persian were parties to the so-called cartel agreement formulated at Achnacarry, Scotland. The Achnacarry agreement was designed to substitute co-operation for competition as a regulator of oil markets. Sir John Cadman in addressing the American Petroleum Institute in Houston, Texas, in November 1932 on "Petroleum Policy" expressed the spirit of Achnacarry when he said,

> Consumption everywhere has decreased, competition is too keen; and prices contain no margin for gain. One possible step toward rehabilitation of trade is evidently the readjustment of supply to demand and the prevention of excessive competition by allotting to each country a quota which it will undertake not to exceed.

About Cadman's address T. E. Ward, Eastern Syndicate's attorney and negotiator with Gulf said, "The sentiments so well expressed were not only

17. *Ibid.*, p. 26.
18. *Ibid.*, p. 25.
19. *Ibid.*, Chargé Atherton, p. 8.
20. *Ibid.*, Atherton for Ambassador Mellon, p. 20.

a keynote for the convention but also for the Anglo-Persian and the Gulf Company directors on the subject of equal partnership in Kuwait."[21]

Anglo-Persian was apparently willing to compromise on a jointly owned concession rather than run the risk of having its control of Middle East oil further weakened. Until the Bahrain Oil Co., Ltd., Standard of California's subsidiary, completed its first commercial well in Bahrain on May 31, 1932, Anglo-Persian as sole owner of the Persian concession could exercise effective control over shipments of oil from the Persian Gulf. Although the Iraq Petroleum Company, in which Anglo-Persian had a 23¾ percent interest, had completed a commercial well as early as 1827, until IPC had constructed a pipeline from the Kirkuk field to the Mediterranean, it had never produced as much as a million barrels of oil in any year.

The Standard of California's successful Bahrain venture constituted a more serious threat to Anglo-Persian control of Middle East oil. Confronted with this situation, Anglo-Persian apparently decided that in Kuwait half a loaf was better than none.

Anglo-Persian and Gulf Agree to Restrict Competition

As a guarantee that Gulf's participation in the development of Kuwait oil would not unduly disturb international oil markets, on December 14, 1933, Anglo-Persian made an agreement with Gulf designed to limit competition between them.[22] The agreement provided, among other things, (1) that Kuwait oil would not be used to "upset or injure" the marketing position of either "directly or indirectly at any time or any place;" (2) that the parties would "confer from time to time as either party may desire and mutually settle in accordance with such principles any question that may arise between them regarding the marketing of Kuwait oil and production therefrom"; (3) that either party would have the right to require the Kuwait Oil Company "to produce such quantity of crude oil as may be decided by the party making the request"; (4) that all of the oil produced at the request of both parties should be allocated "50–50 to Gulf and Anglo-Persian at cost"; (5) that oil produced at the request of only one party should be "allocated in full to the party making the request at cost for such oil."

21. Ward, p. 224.
22. F.T.C., *International Petroleum Cartel*, pp. 131–132.

Provisions 3, 4, and 5 were limited by the following significant agreement:

The parties have in mind that it might from time to time suit both parties for Anglo-Persian to supply Gulf's requirements from Persia and/or Iraq in lieu of Gulf requiring the company to produce oil or additional oil in Kuwait.

Provided Anglo-Persian is in position conveniently to furnish such alternative supply, *of which Anglo-Persian shall be the sole judge* [emphasis supplied] it will supply Gulf from such other sources with any quantity of crude thus required by Gulf provided the quantity demanded does not exceed the quantity which in the absence of such alternative supply Gulf might have required the company to produce in Kuwait—at a price and on conditions to be discussed and settled by mutual agreement from time to time as may be necessary—such price f.o.b., however, not to be more than the cost to Gulf of oil having a similar quantity produced in and put f.o.b. Kuwait.[23]

In addition to its contribution to the stability of international oil markets, joint action by Anglo-Persian and Gulf permitted the parties to get better concession terms than either might have gotten alone. With Archibald H. T. Chisholm of Anglo-Persian and Holmes of Eastern Syndicate no longer bidding against each other, they persuaded the sheik to reduce the royalty rate from 3¾ rupees to 3 rupees per ton[24] and they overcame his insistence that he be allowed to appoint a director to the board of the Kuwait Oil Company.[25]

The Ruler Grants Concession

After Anglo-Persian and Gulf joined forces, the negotiations between the Kuwait ruler and the British colonial office proceeded smoothly. On December 22, 1934, the British political agent in Kuwait, Colonel H. R. P. Dickson, notified the sheik that the British government formally approved of the concession. And a day later with pomp and ceremony "His Excellency Sir Shaikh Ahmad al Jabir as-Sabah," as "Knight Commander of the Most Eminent Order of the Indian Empire and Companion of the Most Exalted Order of the Star of India," and more realistically as ruler of Kuwait, signed the concession. Major Holmes representing Gulf Oil

23. *Ibid.,* p. 133.
24. Ward, p. 246.
25. *Ibid.,* p. 248.

and Chisholm representing Anglo-Persian signed on behalf of the Kuwait Oil Company. Colonel Dickson witnessed the signature.[26]

In its general pattern the concession conformed to Persia's as modified in 1933 and Iraq's as modified in 1931. It covered the whole of Kuwait, approximately six thousand square miles. It was to last for seventy-five years—later extended seventeen years. At the expiration date the Kuwait Oil Company was obliged to surrender all movable and immovable property to the government. In return for exemption from all taxes the concessionaires paid a bonus of 475,000 rupees (£35,625 sterling). They obliged themselves to pay an annual rental of 95,000 rupees (£7,125 sterling). Royalty payments, which were to be paid in Indian currency, not guaranteed by gold as were those of Persia, Iraq, and Saudi Arabia, were at the rate of three rupees a ton with a minimum guarantee of 250,000 rupees per year. Until the early 1950s when the 50-50 profit-sharing agreement was generally introduced throughout the Middle East, Kuwait's royalties averaged about ten cents a barrel less than those of Persia, Iraq, and Saudi Arabia.

26. *Ibid.*, pp. 251–252.

PART II

THE CONTROVERSIES

5

Conflicts Inherent in the Concessions

In the area south of the Taurus Mountains of Turkey, east of the mountains and highlands of the Arabian Peninsula and west of the western mountain ranges of Iran, Muscat, and Oman, lies a vast topographical basin embracing the Persian Gulf. In terms of regional geology this area constitutes a structural basin particularly favorable to the generation and accumulation of crude oil. In it lie the world's largest and most prolific oil reserves. By 1950 exploration and exploitation had established the pre-eminence of the Middle East oil fields. Geologists estimated the combined reserves of this area at more than 32 billion barrels. Middle East combined production in that year exceeded that of any single country except the United States. With the accumulation of data on the region, by 1965 geologists estimated the reserves to be approximately 215 billion barrels, almost six times the estimated reserves of the United States and 60 percent of the world's total reserves. Production was more than twice that of any other free-world area outside the United States.

Although the concessionaires had paid Iran, Iraq, Saudi Arabia, and Kuwait together more than $100 million in oil royalties in 1949 and were playing an increasingly important role in the economic progress of the countries in which they operated, evidence mounted of the unsatisfactory relationship between the concessionaires and the governments of the several countries. By 1952 this relationship had culminated in a hostility so great that Iran canceled the Anglo-Iranian concession and nationalized its

121

oil industry. By 1961 Qasim, dictator of Iraq, by unilateral action reclaimed for Iraq more than 99 percent of the lands covered by the Iraq Petroleum Company's concession. And at the Second Arab Petroleum Congress in Beirut in 1960 Abdullah Tariki, later Saudi Arabian minister of petroleum and mineral resources, in effect accused the oil companies of having defrauded the oil countries of hundreds of millions of dollars in revenue by their allegedly arbitrary pricing policies. Everywhere in the Middle East governments were giving evidence of a deep-rooted hostility toward their benefactors. Why was this?

Attitude of the Oil Companies

Oil company executives have made the question seem more perplexing by having acknowledged their responsibilities to the underdeveloped countries and having expressed satisfaction regarding the manner in which they have discharged their stewardship. American oil executives believe that they have served well the interests of the United States, with which they identify themselves, and the interests of the Middle East, where the oil companies are regarded as a symbol of the United States. In expressing satisfaction over Aramco stewardship, F. A. Davies, while chairman of Aramco's board, stated,

> I may say . . . that our contribution to the welfare of the United States and the free world is not inconsiderable. We represent a major American interest abroad. We have in the course of our operations made a major contribution to the development of an underdeveloped area.
>
> We have created an American interest in a vital resource and in a vital area in the present world struggle. We are proud of our accomplishment.
>
> . . . .
>
> We seek no special privilege but I submit as an American and with a conviction bred of some 25 years' experience in the oil industry of the Middle East that we serve in full measure the interests of both the United States and Saudi Arabia.[1]

The Texas Company in pleadings in a civil action in a United States District Court echoed a similar sentiment in these words:

> Despite all the difficulties and problems . . . the activities of the American oil companies abroad have constituted a private Point 4 program which has

1. U.S. Congress, Senate Subcommittee of the Committee on the Judiciary and Committee on Interior and Insular Affairs, 85th Cong., 1st Sess., 1957 Joint Hearings on *The Emergency Oil Lift Program and Related Oil Problems,* 4 Pts, Pt. 2, p. 1412.

brought to the countries and the people involved improved economic and social conditions and a knowledge and understanding of American business customs and ideals in a way that no amount of direct government aid could ever do.[2]

Why have the Middle East countries treated so lightly the benefits that the oil companies have brought them? Why have their loyalties been so easily undermined?

A brief review summary of the circumstances under which the oil companies obtained their concessions, a description of the corporate relationships of the concessionaires, and a detailed analysis of the terms of the concessions will aid in answering these questions.

The Persian Concession

As the preceding chapters have disclosed, when they granted foreign corporations the rights to exploit their oil resources the Middle East countries differed substantially in their historical backgrounds, their political maturity, their social institutions, and their role in international affairs. Persia had a long history as an independent state. But its glory and grandeur were of bygone days. Long the pawn in the international rivalries of Russia and Great Britain, Persia was ruled by an irresponsible monarch lacking the restraints that an articulate middle class might have exerted, with no responsible nobility to check his actions, with an insecure control over hostile tribal groups, heading a hierarchy of provincial and local governmental officials who derived their sustenance from a corrupt and inequitable tax system, and confronted with an impoverished peasant class tied to the land and exploited by landlords many of whom owned whole villages. To meet his own and his favorites' extravagant tastes, the shah found inadequate both the public and privy purses between which custom made little distinction. In debt to foreign treasuries, faced with social unrest, habituated to foreign intrigue, and heading a government in which corruption had become a way of life, "the Persian monarchy itself was," in the words of Sir Arthur Hardinge, "an old, long-mismanaged estate, ready to be knocked down at once to whatever foreign power bid highest, or threatened most loudly its degenerate and defenseless rulers."[3]

2. U.S. District Court Southern District of N.Y., Civil Action, No. 80–27, U.S.A. Plaintiff v. Standard Oil Co., N.J. *et al.* Defendants, Reproduced in *Emergency Oil Lift,* Pt. 3, pp. 2007–14 at p. 2013.

3. Sir Arthur H. Hardinge, *A Diplomatist in the East* (London: Jonathan Cape, 1928), p. 280.

It was against this background that D'Arcy and his associates, more concerned with business objectives than with the niceties of modern morals, obtained the original concession. Chapter One traced the circumstances by which a strictly commercial venture eventually became the property of a tightly held corporation more than 50 percent of whose stock was owned by the British government and related how under Churchill's guidance the oil resources of Persia were mobilized for the benefit of the British Navy.

The Iraq Concession

Iraq's independence as a sovereign state is a post–World War II phenomenon. After several hundred years under the suzerainty of the Turkish Empire it was made a British mandate after World War I in violation of the nationalistic ambitions of articulate Arabs and of British promises made during the war. As a mandated country it was regarded by its creators, the League of Nations and Britain, as lacking in the capacity for self-government and was forbidden the full rights of sovereignty. Great Britain, in acquiring the League mandate, reflected not so much its concern with the political welfare of a fledgling nation as with insuring its own political welfare while the fledgling under its tutelage grew to maturity. The Allies had "floated to victory on a wave of oil" and Great Britain, as a colonial power, was determined to utilize Middle East oil in maintaining control of the sea.

It was under British influence that Iraq validated the Iraq Petroleum Company's claim to a concession promised before World War I by the Turkish Grand Vizier to the Turkish Petroleum Company, jointly owned by the D'Arcy group (Anglo-Persian Oil Company, the Deutsche Bank, and the Anglo-Saxon Petroleum Company, forerunner of Royal Dutch-Shell). The San Remo Agreement transferred the former German rights to French interests and, as a result of long negotiations with the several owners, American companies acquired part ownership in IPC. In the process the American State Department had in effect delegated authority to the Standard Oil Company of New Jersey to negotiate an arrangement in harmony with the State Department's open-door policy.

As previously described, by a series of maneuvres extending over more than a decade IPC emerged as sole owner of the oil rights covering virtually the whole of Iraq.

IPC's initial bargaining for the Iraq concession represented basically negotiations between representatives of the mandated country and representatives of the holders of the mandate. So circumscribed, the outcome of the bargaining, no matter how fair it may have been, was inevitable. To whom the concession would be granted and the general terms under which it would be granted were foreordained. That the American companies eventually became partners in the contract represents a tribute to the State Department's insisting that the United States, having shared the burden of the war, was entitled to share in the fruits of victory, and to the skill of the Standard Oil Company, bargaining not with Iraq but with its concessionaires.

In the long process of bargaining the open door had been closed and the national ambitions of articulate Arabs had been blunted. This had not happened without leaving in Iraq a deep-rooted hostility toward IPC as a symbol of Western colonial exploitation.

The Saudi Arabian Concession

Saudi Arabia represented a loose consolidation of numerous nomadic, warring, desert tribes eking out a miserable existence largely through a camel and goat culture, and the more settled urban communities along the coastal areas of the Arabian Peninsula peopled by more stable but equally impoverished groups. Out of these disparate elements a bold and courageous warrior had constructed a medieval state. It was a twentieth century anomaly. Outside the settled areas, land ownership was a vague concept, as indeed it was throughout the Middle East, with a tribal rather than an individual significance. When the nation was formed, no Arab had any notion of the potential wealth that underlay its barren surface and no one except the king, who had established his spiritual and secular leadership by the sword, had any claim to it. An absolute monarch, unhampered by any of the paraphernalia of modern governments, restrained only by Moslem customs and the advice of his tribal and religious leaders, making no distinction between the public and the privy purse, badly in need of funds with which to meet the increasing cost of governing his loosely knit kingdom, unschooled in the ways of the West and ignorant of its technology, bargained as an Arab trader with sophisticated representatives of a modern oil company anxious to obtain a source of oil with which to supply its domestic needs and permit it to enter world markets.

The New Jersey Standard Oil Company in pointing out the impropriety of applying American antitrust laws to the joint corporate ventures of the Middle East describes, with particular reference to Saudi Arabia, the cultural gap between the modern corporation and the Middle East countries in these words:

> Most Western business methods and concepts are without meaning in the Middle East. The corporation was unknown until very recent years. Local business was conducted according to ancient Islamic law supplemented by the dictates of a monarch, like the late King Ibn Saud who has been described as a "desert prophet" not a modern or medieval man but the last of the great figures of the Old Testament.[4]

Bargaining between the "desert prophet" and the modern corporation eventuated in a complex legal document framed by attorneys skilled in corporate and contract law. Making the rate of material progress in Saudi Arabia dependent on the decisions of a foreign corporation, its terms led, not surprisingly, to misunderstanding and controversy.

Intercompany Relationships in Middle East Oil

The contemporary hostility of the Middle East countries toward their concessionaires is in part a heritage of the circumstances under which the concessions were obtained. The corporate relationships existing among the concessionaires, the united front which the oil companies have presented in their dealings with the Middle East countries, have exaggerated the cultural gap between the countries and the concessionaires and aggravated the hostility between them. Each of the countries came to believe that it was being victimized, not by a single gigantic corporation, but by a combination of industrial giants that together controlled the international oil market and set prices in a way to enrich their coffers at the expense of their hosts. The publication in 1952 of the Federal Trade Commission's study, *The International Petroleum Cartel,* with which a small group of Middle East technocrats, educated in the West, promptly familiarized themselves, exacerbated their suspicions.

As previously indicated, control of Middle East oil was highly concentrated. In the four major Middle East oil-producing countries a single company accounted for each country's entire output. In Iran, Anglo-Ira-

---

4. U.S. Congress, H. of R. Subcommittee on Antitrust, Committee on the Judiciary, *Hearings on Antitrust Problems,* 84th Cong., 1st Sess., 1955, p. 828.

nian was the sole producer; in Iraq it was the Iraq Petroleum Company and its two affiliated companies, the Mosul Petroleum Company and the Basrah Petroleum Company; in Saudi Arabia it was Aramco; in Kuwait, the Kuwait Oil Company. Although these companies were separate corporate entities, they were closely linked by ownership ties. It will be recalled that the Iraq Petroleum Company was jointly owned by the Royal Dutch-Shell Company (23.75 percent); Anglo-Iranian (23.75 percent), in which the British government owned a controlling interest; the Near East Development Corporation (23.75 percent), itself jointly owned by New Jersey Standard and Socony Vacuum; Compagnie Française des Pétroles (23.75 percent), in which the French government owned a substantial interest; and the Gulbenkian interests (5 percent). Aramco was jointly owned by Standard Oil Company of California (30 percent), Texas Company (30 percent), New Jersey Standard (30 percent), and Socony (10 percent). The Kuwait Oil Company was jointly owned by British Petroleum and the Gulf Oil Corporation (50 percent each).

The common interest of the four corporate entities that accounted for virtually the whole of Middle East production not only depended on their relations as joint owners of producing enterprises but was strengthened by various contractual or partnership arrangements between and among two or more of the four. Royal Dutch-Shell and Anglo-Iranian were closely associated not only as joint owners of IPC but as joint marketers throughout much of the Eastern hemisphere, including the United Kingdom and Ireland, much of Africa, the Near East, Ceylon, India and Pakistan.

Just as Shell and Anglo-Iranian operated jointly in many world markets, so did Socony and Jersey Standard. Here the relationship was perhaps even closer and of longer standing. After May 1933, Jersey and Socony conducted their production and refining operations in the Dutch East Indies and their marketing activities over a considerable extent of the Eastern hemisphere through the Standard-Vacuum Oil Company in which each had a 50 percent ownership. Until the Socony Vacuum and Jersey Standard dissolved their corporate ownership of Standard-Vacuum in 1960 in conformity with a court decree following antitrust proceedings, Standard-Vacuum owned refineries in Indonesia, Australia, South Africa, India, and Japan and marketed petroleum products in southern and eastern Africa, Australia, southern Asia and certain areas of the Far East.

Not only were Jersey Standard and Socony Vacuum themselves partners in Aramco and IPC and associated with Anglo-Iranian Petroleum as joint owners of IPC, but they had more closely identified their interest with

Anglo-Iranian through long-term crude oil–purchase contracts designed to meet their own market requirements and at the same time assure Anglo-Iranian a continuing market for its surplus crude oil output.

The arrangement between Anglo-Iranian and the Gulf Exploration Company under which they agreed to operate the Kuwait Oil Company for their joint benefit provided that neither party would use Kuwait oil to "upset or injure" the other's "trade or marketing position directly or indirectly at any time or place." As an aid in achieving this objective, Gulf, which lacked foreign markets, entered into a complicated contract with Shell, which was short of crude, covering transportation, refining, and marketing of Kuwait crude oil and its products. The contract not only contributed to an economical utilization of resources in conducting their operations but insured an outlet for Gulf's crude without disturbing Eastern hemisphere markets.[5]

Commission's Study Confirms Suspicions of the Middle East Governments

These and similar arrangements the Federal Trade Commission characterized as "The International Petroleum Cartel." Although some scholars and oil-company representatives generally had denied that they in fact functioned as a cartel, the Middle East technocrats readily accepted the commission's point of view. The commission's study confirmed their fears and suspicions. This is not surprizing. Our own lawmakers have been quick to find in the commission's allegations a strong fiber of truth. They have feared, as have other Americans, that the power of international corporations increased with the scope of their operations and that their privileges and authority transcended those of their creators. Senator O'Mahoney expressed his concern about Aramco's operations to its president in these words,

Aramco was not created by the Government of the United States. It was created by the State of Delaware which has no power to carry on foreign policy for the United States, which has no power to regulate foreign commerce, which has no power to regulate interstate commerce, and yet your company, created

5. *The International Petroleum Cartel*, staff report of the Federal Trade Commission, submitted to the Subcommittee on Monopoly of the Select Committee on Small Business, U.S. Senate (Washington, D.C.: Government Printing Office, 1952), pp. 131–134.

by the State of Delaware, is operating in the Middle East as a subsidiary of four giant corporations, and you are engaged in all of these things.[6]

The Middle East governments were not concerned, of course, with the relationship between the power of their concessionaires and the governments that had created them, but with the power which the concessionaires had over their own material welfare. They had come to believe that the concession contracts were being used to promote the private commercial interests of an international oil combine at the expense of the economic progress of Middle East countries. As they came to realize the full significance of the specific terms of the concession contracts, their conviction grew that the contracts had deprived them of the full measure of wealth that their oil resources promised.

The concessions follow a common pattern. The provisions which with the passing of time have proven particularly objectionable to the Middle East countries include those relating to the size of the grants, the duration of the concessions, the exclusive nature of the privileges granted, the surrender of the right of taxation, the surrender of all management functions, the utilization of natural gas, the settlement of disputes, the size of governmental royalties, and the procedures of the companies in cost accounting and the pricing of oil after the sharing-of-profits agreement of the 1950s.

As previously indicated, the grants covered vast areas. The original D'Arcy concession covered the whole of Persia except five northern provinces in which Russia had a particular interest. The grant covered about 600,000 square miles with exclusive rights over about 500,000 square miles. IPC's concession as modified by the 1931 agreement covered about 32,000 square miles. The British Oil Development Company's 1932 concession covered all the lands in the vilayets of Mosul and Baghdad north of the thirty-third parallel, totaling about 46,000 square miles. The Basrah Petroleum Company's 1938 concession covered the remainder of Iraq except a small territory on the Persian border, a total of about 93,000 square miles. After IPC acquired control of BOD and BPC its total holdings covered about 172,000 square miles, virtually the whole of Iraq. Aramco's original concession, including the area within which it had preferential rights, covered considerably more than 500,000 square miles of Saudi Arabia's total area of about 800,000 square miles. The Kuwait Oil Company's concession covered the whole of Kuwait, about 6,000 square miles.

6. *Emergency Oil Lift,* Pt. 2, p. 1489.

Duration of the Concessions

Never in modern times have governments granted so much to so few for so long. The original D'Arcy concession was to last sixty years. After Reza Khan canceled it and negotiated a new one with Anglo-Persian in 1933, the date of expiration was moved forward to 1993. Iran negotiated a new agreement with the consortium in 1954 which covers twenty-five years with the right of renewal for three additional periods of five years each. If it runs its full term, when it expires Iran will have surrendered its exploitation rights on its richest oil reserves for about a century.

The agreement between Saudi Arabia and the Standard Oil Company of California, signed May 29, 1933, was to run for sixty years. The supplemental agreement signed May 31, 1939, enlarging the areas over which the California Arabian Standard Oil Company had exclusive rights, extended the duration of the concession by six years, bringing the expiration date to 1999. The Iraq Petroleum Company agreement as signed on March 14, 1925, and revised on March 24, 1931, extends seventy-five years from the date of signing. The Basrah Petroleum Company's agreement, signed on July 29, 1938, and the BOD Company agreement, signed on April 20, 1932, had identical expiration dates. The Kuwait Oil Company's original concession, dated December 23, 1934, runs for seventy-five years. Article 4 of the Agreement of December 30, 1951, extended the grant seventeen years bringing the date of expiration to 2,026 A.D.

United States Policy in Leasing Public Lands

The prodigal nature of the Middle East concessions can be more readily appreciated by comparing them with the United States' policy in granting oil leases on public lands. Our federal policy has been designed to prevent monopoly exploitation of oil lands. Federal laws limit the area covered by exploration permits (not more than 246,080 acres in any one state except Alaska, where the limit is 300,000 acres). They limit the area a prospector may lease after the discovery of oil to one-fourth of the lands embraced in the prospecting permit with a minimum of 160 acres. They limit the terms of a lease to twenty years. They require that three-fourths of the acreage covered by an exploration permit, to which the original lessee has a preferential right, be leased by competitive bidding with a minimum royalty of 12½ percent. They require that lands located in a known geological structure of a producing field be leased by competitive bidding in units of not more than 640 acres. They limit the duration of a lease resulting from competitive bidding to five years or as long as oil or gas is produced in paying quantities.[7]

Relinquishment of Areas

The original D'Arcy concession contained no provisions for relinquishing undeveloped areas. The new concession granted Anglo-Persian Oil Company in 1933 after Reza Shah had canceled the old, reduced the total area of the concession to 100,000 square miles and gave the company six years to select the areas it wanted to develop.[8] The new agreement made no provision for relinquishing any areas. Nor did IPC's concession. Standard Oil of California's Saudi Arabian concession obliged the company to return within ninety days after it began drilling such areas as it might then decide not to explore further and thereafter to make periodic releases of those areas that it should decide not to explore further. Four years after it began drilling, the company had not released any acreage, and in May 1939 it obtained from the government a ten-year extension on its obligation to do so.

The Kuwait concession made no provision for relinquishment. As a result of negotiations conducted in the early part of 1962, the Kuwait Oil Company agreed to relinquish exploration rights on approximately 50 percent of the original concession. The land released lies mostly in the western half of Kuwait, and as yet no oil has been discovered on it.

Power to Tax

Under the fiscal provisions of the original concessions, the governments of Iran, Iraq, Saudi Arabia, and Kuwait bartered away their right to tax the concession throughout the life of the agreement.

Article 11 of the renegotiated Anglo-Iranian concession (1933) exempted the company from all taxation by the state or any local authority.

7. U.S. Code, Vol. 7, 1964 ed., Table 30, Chap. 3, Secs. 184 (b); 223; 226 (b); and 226 (e).
8. Anglo-Persian changed its name to Anglo-Iranian in 1935.

For this exemption the company agreed to pay ninepence for each of the first six million tons of oil produced and sixpence for each of the second six million tons produced. It guaranteed that these payments would amount to at least £225,000 sterling a year. After fifteen years it agreed to pay one shilling for each of the first six million tons produced and ninepence for each of the next six million tons, but not less than £300,000 sterling a year.

Article 21 of the Standard of California original Saudi Arabian concession reads in part as follows:

> In return for the obligation assumed by the Company under this contract, and for the payments required from the Company hereunder, the Company and enterprise shall be exempt from all direct and indirect taxes, imports, charges, fees and duties (including, of course, import and export duties), it being understood that the privilege shall not extend to the sale of products within the country, nor shall it extend to the personal requirements of the individual employees of the Company.

Article 27 of IPC's Iraq concession reads as follows:

> No other or higher taxes, impositions, duties, fees or charges, whether Government or municipal or port, shall be imposed upon the Company, or upon its property or privileges or employees within Iraq, than those ordinarily imposed from time to time upon other industrial undertakings, or upon their property or privileges or employees. No taxes, impositions, duties, fees or charges, whether Government or municipal, shall be imposed upon the borings of the Company, or upon the substances comprised in Article 1 hereof before their removal from the ground, or upon substances comprised in Article 1 hereof and used by the Company for the purpose of its operations hereunder.

The language of Article 27 of IPC's concession later raised doubts in the minds of company executives about its tax liabilities. An exchange of letters between Prime Minister Nuri Said and the Iraq Petroleum Company on March 24, 1931, laid to rest all doubts. In return for the company's obliging itself to pay £9,000 (gold) on January 1, 1932, and annually thereafter until exports began, and £60,000 annually thereafter for the first four million tons exported and pro rata, and £20,000 for each subsequent million tons and pro rata, the government made more explicit IPC's exemption from taxation in the following language:

> The Company shall be exempt from all taxation falling due on or after 1st of April, 1931, of whatever nature whether State or Municipal on their capital, borings, plant, machinery, buildings (other than houses and offices within Municipal limits) and profits (other than those accruing from the transport of

oil not produced from the defined area) and on the substances comprised in Article 1 of the Convention before or after their removal from the ground and upon the technical processes utilized in connection with the said substances.

Article 7 of the Kuwait Oil Company's concession provided that aside from 475,000 rupees payable within thirty days from the signature of the concession grant (Article 3) and aside from the annual royalty payment thereafter (3 rupees was changed to 12½ percent of posted price on January 1, 1955 in Article 10)

the Company, its operations, income, profits, and property including petroleum shall be exempt and free during the period of this Agreement from all present or future harbour duties, import duties, export duties, taxes, imports, and charges of any kind whether state or local tolls, and land surface rent of whatever nature; and in consideration thereof the Company shall in addition to the payments provided for in Art. 3 pay to the Sheik on each anniversary of the date of signature of this Agreement, four annas per ton of petroleum on which royalty is payable (Art. 7).

Exclusive Rights

The rights which the oil companies acquired over these vast areas are exclusive and broad in their terms. Article 1 of the IPC-Iraq concession grants IPC the exclusive rights "to explore, prospect, drill for, extract and render suitable for trade petroleum, naptha, natural gases, ozocerite, and the right to carry away and sell the same and the derivatives thereof." The other agreements have similar terms. While all of the concessions contain provisions designed to insure that the concessionaires proceed with reasonable dispatch to explore and exploit the concession, all leave to the concessionaire's discretion how rapidly the areas will be exploited once oil is discovered in commercial quantities, and what areas will be developed and for how long areas may lie in idleness.

Managerial Functions

It is, of course, not to be expected that any concessionaire would be willing to entrust the management of its enterprise to the citizens of the concession-granting country. The capital investments were large. The risks were great. Middle East countries even today lack the managerial talent essential to the operation of an enterprise using a complex technology and comprising intricate physical and chemical processes. Accordingly, each concession recognizes the right of the concessionaire, with some restrictions, to employ managers, engineers, and technical staff of its own choosing.

Article 23 of the Saudi Arabia–Standard of California agreement provides that

> The enterprise under this contract shall be directed and supervised by Americans who shall employ Saudi Arab nationals *as far as practicable,* and insofar as the Company can find suitable Saudi Arab employees, it will not employ other nationals. [Emphasis supplied.]

Article 29 of the Iraq-IPC agreement of March 14, 1925, as revised by the agreement of March 24, 1931, provides that

> The employees of the Company within Iraq shall, in so far as possible, be subjects of the Government, but managers, engineers, chemists, drillers, foremen, mechanics, other skilled workmen and clerks may be brought from outside Iraq if qualified persons of these descriptions cannot be found in Iraq, and provided that the Company will, *as far as reasonably practicable,* and as early as possible, train Iraqis in these capacities. [Emphasis supplied.]

Article 8 of the Kuwait concession provides that the company shall employ subjects of the sheik in so "far as possible for all work for which they are suited." If the local labor supply is inadequate or unsuitable, the agreement permits the company with the sheik's approval to import labor, giving preference to labor from neighboring countries. It also permits the company to import skilled and technical employees.

Article 12 of the original D'Arcy-Iranian concession provided that the workmen employed in the services of the company were to be "subjects of his Imperial Majesty the Shah, except the technical staff such as managers, engineers, borers and foremen."

Article 16 of the 1933 concession agreement between AIOC and the Iranian government provided that the company recruit its artisans, as well as its *technical and commercial staff* from among Persian nationals to the extent that it can "find any Persian persons" who possess the requisite competence and experience.

The agreement obliges the company to recruit its unskilled workers exclusively from subjects of the shah and in co-operation with the government to prepare a plan for an annual reduction in the company's non-Persian employees and their replacement "in the shortest possible time and progressively by Persian nationals." The company also agreed to provide a yearly grant of £10,000 to provide Persian students with the professional education in Great Britain essential in the operation of the oil industry.

The Middle East governments have been jealous of their rights under these provisions relating to employment. Despite the training programs that the concessionaires have inaugurated, they have alleged that the companies have deliberately disregarded the spirit of the agreements, if not their letter. The companies, they profess to believe, have been slow to train their nationals for positions of responsibility. The modifications in the employment provisions of the Iran-AIOC agreement are indicative of the distrust of the oil companies by government.

The companies deny that there is merit in the criticisms of their employment policies and present statistics to support their position. In its 1965 annual report the Kuwait Oil Company indicates that its percentage of Arabs in staff positions increased from 36 percent to 42 percent in a single year. The company is pursuing an active training program "designed to develop the potential of its Kuwaiti and other Arab national employees at all levels. In particular, departmental training programmes were expanded to fit increasing numbers of other Arab employees for higher-grade posts."

The annual review of operations of IPC, Basrah, and Mosul oil companies shows that their Iraqi technical and administrative personnel increased between 1958 and 1965 from 486 to 618 persons or from 30 percent of their total staff employees to 85 percent. The review states,

Training of the Companies' personnel in Iraq, Syria and Lebanon has been maintained at a high level, and 100 administration and specialist staff underwent development courses in their respective areas of employment. On-the-job training continued throughout the year. Courses were arranged for skill improvement and development in the countries concerned and some one hundred Company personnel attended training courses abroad. A six weeks advanced development course for senior technical and administrative staff from all areas of the Companies' operations was held in London.

Review of Operations by the Arabian American Oil Co. (Aramco, 1965) describes Aramco's training program as follows:

Training of employees was a prime concern. Two of every five Saudi Arab employees participated in training and study programs ranging from lectures and short courses to post-graduate work.
Enrollment in the three Industrial Training Centers and three Industrial Training Shops amounted to nearly one-third of the Saudi Arab employees.

About 2,500 employees attended classes during working hours and more than 1,000 attended during leisure hours.

Four hundred and sixty-three Saudi Arab supervisors and supervisor candidates completed one or more of twenty different courses in managerial techniques.

Three hundred and sixty-six employees were enrolled in another program designed to develop technical and supervisory personnel.

. . . .

One hundred and seventy-seven Saudi Arab employees were on study and training assignments in the Middle East and the United States. At the university level fifty-seven were undergraduates and three were post-graduate students. Seventeen were enrolled in secondary schools preparing to meet entrainee requirements for universities, technical schools or other educational institutions. Eighty-eight employees took technical, craft and business administration courses and twelve trained in job situations.

In addition to citing impressive statistics to support their contention that they are meeting effectively their obligations to develop competent employees at all levels, company executives point out that it is in the interest of economy to do so. It is cheaper to hire nationals than to import and support employees at relatively great costs to fill jobs at any level. And a former employee of long years at the managerial level alleges that his company's efficiency has deteriorated because of the speed with which it has placed nationals in important technical, skilled, or supervisory positions.[9]

But it is not the merits of this issue that concern this study so much as attitudes toward them. Government officials still resent that their own nationals have little part in important managerial decisions. They want to control their own destiny.

The Utilization of Natural Gas

Oil is found in a great variety of underground reservoirs generally in close association with natural gas and salt water. The three substances customarily are distributed on a basis of their specific gravity: the gas, which is lightest, generally is found in a free state above the oil, and the water, which is heavier, underlies the oil deposits. Efficient utilization of the gas and water pressures may play a significant role in the rate of oil recovery and in the amount of oil ultimately produced. Frequently in oil production oil companies have found it economical or expedient to flare

9. Personal communication.

the gas as it is separated from the oil on recovery. In recent years they have with increasing frequency forced the free gas back into the underground reservoir with significant results in the ultimate recovery. Whether to repressure by utilizing the gas or with a water drive depends on both engineering and pecuniary considerations. For the most part Middle East concessionaires have flared the gas. The concessions have customarily left decisions on such matters to the discretion of the concessionaires.

Article 15 of the Saudi Arabia–Standard of California agreement provides that

If the Company should produce, save and sell any natural gas, it will pay the Government a royalty equal to one-eighth of the proceeds of the sale of such natural gas, it being understood, however, that the Company shall be under no obligation to produce, save, sell, or otherwise dispose of any natural gas.

The Iraq-IPC agreement places no obligation on the company to save, sell, or utilize natural gas it may produce but provides that

The Company shall pay a royalty of twopence per thousand cubic feet of all natural gas it sells calculated at an absolute pressure of one atmosphere and at a temperature of 60 degrees Fahrenheit [Art. 10].

The flares have been a constant reminder to the host countries that the gas may not be used in their best interests and they have sought to change such provisions.

Settlement of Disputes

D'Arcy's original Persian concession was written in French and translated into Persian with the stipulation that should any dispute arise "in relation to such meaning the French text shall alone prevail" (Article 18). The Iraq, Saudi Arabia, and Kuwait concessions are written in English and Arabic with the proviso that should any discrepancy in meaning between the English and Arabic arise, the English version shall prevail.

Aramco's Saudi Arabian concession (Article 35) offers a plausible explanation of why the English version should prevail:

This contract has been drawn up in English and in Arabic. Inasmuch as most of the obligations hereunder are imposed upon the Company and inasmuch as the interpretation of the English text, especially as regards technical obligations and requirements relating to the oil industry, has been fairly well established through long practice and experience in contracts such as the present one, it is agreed that while both texts have equal validity, nevertheless in case of any divergence of interpretation as to the Company's obligations hereunder, the English text shall prevail.

Under contracts of great length, couched in legal language or the technical language of the oil industry, with a foreign language taking precedence over the native language of the concession-granting country, it was inevitable that disputes would arise. All the contracts anticipate them and provide that they be settled by arbitration. Article 40 of the IPC concession reads in part as follows:

If at any time during or after the currency of this convention any doubt, difference or dispute shall arise between the Government and the Company concerning the interpretation or execution hereof, or anything herein contained, or in connection herewith, or the rights and liabilities of either party hereunder, the same shall, failing any agreement to settle it in another way, be referred to two arbitrators, one of whom shall be chosen by each party, and a referee who shall be chosen by the arbitrators before proceeding to arbitration. Each party shall nominate its arbitrator within 30 days of being requested in writing by the other party to do so. In the event of the arbitrators failing to agree upon a referee, the Government and the Company shall, in agreement, appoint a referee, and in the event of their failing to agree, they shall request the President of the Permanent Court of International Justice to appoint a referee. The decision of the arbitrators, or in the case of a difference of opinion between them, the decision of the referee, shall be final. The place of arbitration shall be such as may be agreed by the parties, and in default of agreement shall be in Baghdad.

Article 31 of the Saudi Arabia–Standard of California contract provides an almost identical procedure for settling disputes. The only significant variation is the provision for place of arbitration. If the parties do not agree on the place of arbitration, it shall be The Hague, Holland.

The arbitration clause of the Kuwait Oil Company agreement contains a unique provision indicative of the British political influence over Kuwait when the concession was granted. Before referring a dispute to arbitration, the parties shall consult with the British political agent in Kuwait or the British political resident in the Persian Gulf (Article 18).

Article 17 of the D'Arcy concession provided for an arbitration board consisting of one arbitrator chosen by each of the parties and a neutral arbitrator chosen by both. The board's decision was to be binding. The 1933 renegotiated concession set up a similar procedure but provided that if the two arbitrators could not agree on a third, the president of the Permanent Court of International Justice should nominate a third arbitrator, provided that the president is neither a British or Persian subject. If he should be, the vice-president shall nominate a neutral arbitrator. The board of arbitration would sit in Teheran.

In view of the character of the native courts, their position in the political and religious life of the nation, their composition, their ignorance of Western institutions, and the sort of disputes they customarily handled, it is readily understandable that the oil companies would prefer to rely on neutral arbitration boards rather than entrust disputes to the judicial system within the concession-granting countries. It could scarcely have been otherwise. Yet the Middle East governments, and more particularly the propagandists who frequently speak for them, have come to resent the arbitration provisions of the concession contracts and have been reluctant to use them.

While acknowledging that disputes between the governments of two countries or between private parties within a country may be appropriate matters for conciliation and arbitration, they contend that disputes between a government and a private corporation should be settled within the system of laws within the country. Hence, the governments have generally chosen to rely on negotiation (rather than arbitration) in the conduct of which their governments through the threat of unilateral action can exert great pressure on their concessionaires.

The Clash of Cultural Concepts

Although differences have been inevitable between the concessionaires and their hosts regarding the specific obligations and responsibilities that the two parties have assumed and the way in which they have acquitted them, the fundamental cause of the more critical controversies has been the basic difference in the attitude of the companies and the countries toward the meaning and significance of concession grants. Oil company executives regard them as contracts, freely negotiated between two informed parties, each of which knew what he wanted and what he was getting. This concept is the product of legal and industrial developments extending over centuries. It is, of course, a phenomenon of Western civilization. Business men and the philosophers of a competitive society regard the concept as essential to the operation of a free-enterprise economy. Without it, stability and progress will disappear. Davies, when chairman of the board of Aramco, in testifying before a committee of Congress, articulated in nontechnical language the Western point of view when he said,

> I think it is one thing to nationalize a business within a country purely for the welfare of the people of that country, and I think it is entirely different when an industry is nationalized, which is operating in the country through a *sacred* concession agreement or a *contract.* I think that that is a pure violation of the *sanctity of contract,* and I don't believe it should be justified or condoned by the United States Government or any other government.
> I think when a Government has entered into a contract knowingly and of its own accord, that that contract must be respected.
> I don't think that the world is going to advance very rapidly until we can be sure that a *contract like that means what it says.* [Emphasis supplied.][10]

Middle East governments reject this point of view. More acceptable to them are ideas articulated by Frank Hendryx, the legal adviser on oil affairs to the Saudi Arabian government, before the First Arab Petroleum Congress in Cairo in 1959. He contended that it was not merely the right but the obligation of a government to modify unilaterally or even revoke a concession when it believed it to be in the interest of its people to do so. In articulating this idea, Hendryx was giving philosophic and legal expression to a common-sense notion generally held by Middle East governmental officials and more recently developed by scholars whose interests are not at stake.

Neither Iraqians, when Qasim reclaimed for the state by unilateral action more than 99 percent of the lands covered by IPC's concession, nor Iranians, when their government nationalized Iran's oil industry, felt bound by agreements made a generation earlier by governments long repudiated. Nor do the Arabs of Kuwait or Saudi Arabia feel bound by the agreements made by their tribal chiefs under radically different conditions.

Not only is the gulf between the contemporary notions of Middle Eastern governments and the traditional notions of corporate executives on the sanctity of a concession contract wide, but so is the divergence between what the oil companies hoped to get and what they have gotten. Middle East oil reserves can be described only in superlatives. They are the world's largest and they are continuing to increase. As exploration and development proceed, yesterday's estimates are superseded by today's, and today's give way to tomorrow's. In 1961 an IPC geologist, long familiar with

10. *Emergency Oil Lift,* Pt. 2, p. 1462.

Middle East oil deposits and with a penchant for forecasting, estimated that the peak of production in the Middle East would not be reached until the second decade of the twenty-first century.[11] While this was nothing more than a sophisticated guess, it reflects the promise of Middle East oil. Middle East oil is not only abundant but it is the world's lowest-cost oil. Some notion of the relative advantage of Middle East oil is reflected in a comparison between the productivity of wells in the United States and those of the Middle East. In the United States, wells, most of which are on the pump, produce an average of about fourteen barrels daily. In the Middle East, wells under natural flow average 7,500 barrels daily.

From wells whose daily production is intentionally restricted or from standby wells, oil companies at any particular time can increase Middle East production by the turn of a valve. Whether they turn it depends on how turning it will affect their profits.

More important to the countries is how such decisions affect their prosperity. Iraq's former minister of economics, Dr. Nadim Pachachi, expressed his government's attitude towards the private exercise of power in these words:

> Our whole economic development is geared to the revenues from the flow . . . of oil. Our whole political stability depends on the flow . . . of oil.
> It is not just a private matter involving a private company's profits. It is a matter that affects the whole state of Iraq.[12]

As they have developed a sense of responsibility for the economic progress of their impoverished countries, as they have come to realize how much their countries have contributed to the prosperity of the oil industry and how little in terms of need the oil industry has done for their own people, Iraqian and Iranian political leaders have demanded a new deal in oil revenues. The Saudi Arabian royal family, first because of their own improvidence and avarice and later realizing that to perpetuate their rule they can't ignore the needs of their millions of impoverished people, have joined in the demand for a new deal. So also have the rulers of Kuwait, aspiring to become the bankers to their Arabic neighbors, parceling out their income through loans and grants with an eye alike to their own political welfare and their neighbors' needs.

11. Personal communication.
12. "Petroleum Comments" by Wanda M. Jablonski, *Petroleum Week,* April 19, 1957, reproduced in *Emergency Oil Lift,* Pt. 4, p. 2699.

A Western Industrial Complex in a Medieval Social Structure

That the oil companies have pride in the manner in which they have exploited Middle East oil resources is readily understandable. They have combined Western capital, technology, enterprise, and administrative ability in transforming inert oil deposits of unknown quantity into a dynamic energy resource of great value. In so doing, they have not only enriched themselves but have contributed substantially to the industrial progress and material welfare of their host countries. Through intensive training programs they have converted camel drivers into industrial mechanics. Through construction and maintenance activities they have transformed Arab tradesmen into modern contractors. They have improved communication and housing facilities and have provided educational opportunities and health services never before dreamed of. They have paid hundreds of millions of dollars in oil royalties to the governments and thereby created opportunities for governments to improve the standard and extend the scope of governmental administration and to plan and execute domestic development programs.

Their success in doing so has created resentment, not appreciation. As the Middle East governments have been confronted with more complex administrative problems, they have drawn into their service an increasing number of educated and dedicated public servants disturbed by vast poverty in the midst of plenty, restless at the rate of social and economic progress, impatient with the failure of their governments to utilize more equitably and efficiently their oil revenues, aware of the inadequacy of the revenues no matter how fairly and efficiently used to meet the rising expectations of an increasingly articulate population long habituated to ignorance, poverty, and disease. They view the oil companies as having created in their midst a vast industrial complex controlled by foreign corporations which together have presented a united front in adjusting the rate of oil production to their international commercial needs. They identify the oil companies with monopoly, and monopoly with nineteenth-century imperialism. They look forward to the ultimate nationalization of their respective oil industries, detained only by expediency. It is the timing of the outcome, not its certainty, that is their primary concern. Meanwhile the terms of the concession agreements and the manner in which they have been administered have been in constant dispute. Negotiations over their provisions have frequently been conducted on the principle of brinkmanship, government negotiators employing as their ultimate weapon the expropriation of the companies' properties and the nationalization of the industry.

6

From Tonnage Royalties to Profit Sharing

THE most significant changes in the terms of the Middle East oil concessions have been in their provisions relating to governmental revenues. As previously indicated, the original concessions provided for royalty payments based on the tonnage of oil produced. As the postwar output of crude oil increased, governmental revenues, geared to production, showed a corresponding increase. Aramco paid the Saudi government $4,300,000 in 1945; in 1949 it paid $39,100,000. In 1946 KOC paid the Kuwaiti ruler $800,000; in 1950 it paid him $12,400,000. In 1945 IPC and associated companies paid Iraq £2,600,000; in 1950 they paid £6,700,000. AIOC's annual payments to the Iranian government were remarkably stable during the war at about £4,000,000. In 1945 they increased to £5,600,000; by 1950, to £16,000,00.

Although the postwar years witnessed a tremendous increase in Iran's oil revenues, they brought an even greater increase in AIOC's profits and its tax payments to the British government. AIOC's after-tax net profits increased from £5,792,000 in 1945 to £33,103,000 in 1950 and its tax payments to the British government increased from £10,381,000 to £50,707,000.[1] AIOC's profits were derived from its entire international operations, and its tax payments to the British government were internal

1. *Chairman's Report to the 42nd Meeting of the Shareholders, December 20, 1951.*

144

fiscal matters related to the entire British system of providing revenues to meet governmental requirements. Hence most students of fiscal problems would regard AIOC's total income and its payment of British taxes as irrelevant to Iran's problem of deriving greater revenue from AIOC's operations in Iran. But Iranian government officials viewed the matter differently. They did not approach these issues as students of fiscal problems but as Iranians who had long felt that their greatest source of wealth was being exploited by a foreign corporation for the benefit of Britain. They resented it. The occupation of their country by foreign troops during World War II and the hardships occasioned by the war and the postwar readjustments accentuated their resentment. By 1950 the Iranian government had begun negotiations with AIOC for a new deal.

Saudi Arabia Negotiates a 50–50 Profit-Sharing Deal

It was Saudi Arabia, however, that first negotiated the transition from fixed royalty payments to profit sharing. In doing so the king was impelled not so much by the poverty of his people as by his own and his royal family's prodigal and conspicuous consumption. This had quickly adapted itself to expanding revenues. Moreover, to retain the loyalty of his tribal leaders he found it necessary to enlarge continuously his gifts to them. The king needed additional funds and he recognized in Aramco's increasing profits and the United States' corporate tax policy an opportunity to get them. In 1938 Aramco produced only 495,000 barrels of crude and paid the Saudi government only $100,000. Its production first reached 7.8 million barrels in 1944, when it paid the government $1,700,000. In the next five years Aramco's production showed a phenomenal increase. In 1949 it produced 174 million barrels. During this period its profits increased from $2,800,000 to $115,062,120. Government officials were not unaware of these developments and wanted a greater share in what they regarded as an Aramco bonanza.

In February 1949, the Pacific Western Oil Corporation, controlled by rugged individualist J. Paul Getty and later renamed the Getty Oil Company, anxious to obtain a concession in what promised to be the world's richest oil-producing area, signed a concession agreement with Saudi Arabia covering its part of the Neutral Zone lying between Saudi Arabia and Kuwait. For this Getty paid a cash bonus of $9,500,000 and obliged Pacific

Western to pay annual rentals of $1 million until the discovery of oil and a royalty of $.55 on each barrel of oil produced thereafter.

Meanwhile Venezuela had pointed the way to higher revenues for the Middle East countries in levying a tax of 50 percent effective as of 1948 on profits of its concessionaires from their Venezuelan production. Middle East countries knew about this and were jealous. In truth, Venezuela's tax policy had played an important role in Mossadegh's demand for a new deal from AIOC. Under increasing pressure from Mossadegh, AIOC had offered to share profits on a 50–50 basis on its Iranian crude production, but the government allegedly refused the offer and demanded a profit split on AIOC's integrated operations throughout the world. The critical status of the Mossadegh negotiations flashed a danger signal to all Middle East concessionaires. Their long-run interests seemed to require a conciliatory approach to the demands of their host countries.

Moreover Saudi Arabia knew about the income taxes that the United States government had imposed for a long time on corporations and resented a foreign government's taxing profits that Aramco derived from developing its oil resources. Aramco had paid the United States government more than $100,000,000 on its net earnings before 1950. In 1949 alone it paid about $43,000,000, five million more than it paid the Saudi Arabian government. In testifying before a congressional committee, Aramco's chairman stated with engaging informality, "They [the Saudi government officials] weren't a darn bit happy about that." And he alleged that "one of the chief causes of the blowup in Iran was that the British Government was obtaining more [in taxes] from the venture out there than the Iranian government was."[2] Summing up the influence of the several factors that led to the Saudi Arab-Aramco negotiations, he stated,

The Saudi Arabian Government had just entered into the contract with the company [Getty] on the Saudi Arabia's half of the Neutral Zone, . . . and the terms they obtained were much more than ours. Venezuela had gone to its tax arrangement. The Saudi Arabian Government knew of all those things. Our concession had greatly increased in value. We had developed a big reserve out there. They wanted more. They asked as early as 1948, "Isn't there some way in which we can get a greater take?" and a little later than that they said, "Isn't there some way in which the income tax you pay to the United States can be diverted to us in whole or in part?"[3]

2. U.S. Congress, Senate Subcommittee of the Committee on the Judiciary and Committee on Interior and Insular Affairs, 85th Cong., 1st Sess., 1957, *Joint Hearings on The Emergency Oil Lift Program and Related Oil Problems,* 4 Pts., Pt. 2, p. 1431.
3. *Ibid.,* p. 1429.

Saudi Arabia began negotiations with Aramco late in 1949 and concluded them in December 1950, when the parties signed an agreement bringing Aramco under the jurisdiction of the Royal Decree of December 30, 1950. This is the so-called hydrocarbon income-tax law which inaugurated the so-called 50–50 profit-sharing agreement.

50–50 Profit Sharing a Misnomer

To describe the agreement as a 50–50 profit-sharing arrangement is somewhat misleading. To allege, as some Middle East governmental spokesmen have done, that Saudi Arabia in the exercise of its sovereignty imposed a 50 percent corporate tax on Aramco's net earnings is equally misleading. The agreement does not provide for the equal sharing of what in customary oil production accounting are regarded as profits. Under the original agreement Aramco paid the Saudi Arabian government annually a royalty of five shillings on each ton of oil it produced (except for what it used in its own operations). It appropriately treated this as a cost of operation in calculating such income taxes as it was required to pay under United States corporate income tax laws, and it continued to do so after the 1950 profit-sharing agreement. The 1950 agreement provided that Aramco pay the Saudi Arabian government one half of net profits after meeting its foreign tax obligations, and it permitted Aramco to treat all royalties, rentals, and other sums paid or payable to the government as credits toward the government's share in its net earnings.[4] The payments to the government as a tax accordingly represented somewhat less than 50 percent of net earnings as customarily calculated.

Although this profit-sharing plan was inaugurated within the framework of an income tax law, it did not represent an assertion of Saudi Arabia's sovereign power to tax. This the government had surrendered in 1933 when it granted Standard of California its concession. The profit-sharing plan was a negotiated plan between the government and Aramco. While negotiations were under way, the king issued two royal decrees, the first on November 20, 1950, and the second on December 26, 1950. The November decree imposed a 5 percent tax on the personal income of individuals derived either as wages or salaries and a 10 percent tax on profits from investment of capital. Personal incomes of less than 20,000

4. Par. 1 of the Agreement of December 30, 1950.

Saudi riyals (at the then current rate of exchange, $5,700) were exempted as were the first 20,000 riyals earned from capital investments. It also levied a tax of 20 percent on net profits of companies. The decree did not apply to members of the royal family, officers and men of the armed forces, police, and those holding religious posts in mosques.

The decree of December 26, 1950, known as the hydrocarbon tax law, applied only to Aramco and provided, as explained above, for the so-called 50–50 profit sharing. Paragraph 1 (a) of the December agreement reads as follows:

Anything in Article 21 in Aramco's Concession Agreement notwithstanding, Aramco submits to the income taxes provided in Royal decrees No. 17/2/28/3321 and No. 17/2/28/7634 hereto attached for reference, it being understood
(a) that in no case shall the total of such taxes and all other taxes, royalties, rentals, and exactions of the Government for any year exceed fifty percent (50%) of the gross income of Aramco, after such gross income has been reduced by Aramco's cost of operation, including losses and depreciation, and by income taxes, if any, payable to any foreign country, but not reduced by any taxes, royalties, rentals, or exactions of the government for such years.

Although the December agreement provided that royalties and rentals be credited toward Aramco's tax obligation, it also provided that there be a guaranteed payment should the company earn no profit in any year.[5]

Zakat—The Chief Internal Source of Saudi Revenue

These two decrees represented innovations in Saudi Arabia's tax laws. Before the 1950 decrees the government's principal internal source of revenue aside from pilgrim fees was the ages-old Zakat tax. Only Muslims paid the Zakat, which was traditionally fixed at one-fortieth of a person's money and goods. The king as religious leader of his people (Imam) was the caretaker of the tax. It represented something in the nature of a religious tithe. A royal decree of June 12, 1951, established a uniform payment to the king and fixed the rate at one-eightieth of the payee's money and goods. The decree did not free good Muslims from their

5. Letter of confirmation on December 30, 1950, from F. A. Davies, acting for Aramco, to Sheik Abdulla Sulaiman, Saudi Arabian Minister of Finance, reproduced in *Middle East Basic Oil Laws and Concession Contracts, Petroleum Legislation*, P.O. Box 1591, Grand Central Station, New York 17, New York, Vol. I, 1959.

religious obligations with respect to the other eightieth. It left to the individual Muslim, however, the responsibility for distributing it "to those in whose favor God has imposed Zakat"—poor relations and destitute persons—and held him accountable to God for having fulfilled his duty.[6]

Income Tax Laws Relatively Burdenless to Aramco

Apparently the income tax decrees of 1950 were designed to provide more revenue to Saudi Arabia without unduly burdening Aramco. E. H. Brown, a Harvard-educated lawyer and a member of Aramco's staff at the time, states that "An attorney from Washington, D.C., Mr. John F. Greaney, advised the Saudi Arab government on this matter because of the American citizens and companies concerned."[7]

While Aramco was conducting negotiations with the Saudi government, it was also negotiating with the United States Treasury and State Departments with regard to the propriety of treating payments to the Saudi Arabian government as income tax credits under American corporation tax laws. After it had won approval for this procedure, Aramco agreed with the Saudi Arabian government to delete the phrase "and by income taxes, if any, payable to any foreign country" from Paragraph 1 (a) of the December 26, 1950, agreement. Since the modification Aramco has calculated its tax obligations to the Saudi Arabian government before paying any corporate income tax to the United States and has received credit for these payments against its tax obligations to the United States.

F. A. Davies, while president of Aramco, explained this shift from a tonnage basis to a tax basis as owing in part to the Saudi Arabian government's solicitude for Aramco's welfare. While admitting that to have increased royalty payments to Saudi Arabia to bring its revenue up to that which it had received under its 1950 agreement "would have made our position out there pretty bad competitively," he alleged that the Saudi Arabian government was insisting on an income tax which would result in no additional burden on the company.[8]

Fifty-fifty profit sharing enriched Saudi Arabia without impoverishing Aramco. Under the last year of payments based on tonnage of crude oil

6. Edward Hoagland Brown, *The Saudi Arabia-Kuwait Neutral Zone,* (Beirut, Lebanon: Middle East Research and Publishing Center, 1963), p. 14.
7. *Ibid.,* p. 13.
8. *The Emergency Oil Lift Program,* Pt. 2, p. 1430.

produced Aramco paid the Saudi government $56,700,000. Under the first year of 50–50 profit sharing it paid the government $110,000,000. Saudi Arabian revenues increased by about 94 percent while Aramco's production increased less than 40 percent.

Company Opinion on 50–50 Profit Sharing

Aramco owners have viewed the shift from royalty payments to profit sharing as an appropriate adjustment to the greatly changed conditions since the Middle East concessions were first granted. H. W. Page, Standard of New Jersey vice-president in charge of Middle East affairs, writing in the *Economic Journal* (London) in September 1960, said of it,

The oil companies operating in the Middle East recognized that the fixed royalty payment no longer gave the equitable division originally intended, and that additional royalty payments were neither economically practical nor a permanent method of maintaining equity between the parties. The "rapid spread" of 50/50 was the result of this recognition of the need to restore the equity which had been frustrated by drastically changed conditions, beyond the control of either party and unforeseen at the time the agreements were negotiated.

M. J. Rathbone, while president of Standard of New Jersey, characterized the principle in the following words:[9]

Fifty-fifty relieves the country of any financial risk, and places that risk upon the company which is in a position to evaluate it and spread the risks of the search for oil over many areas. Any basis other than 50–50 could create an imbalance of interests which would reduce the attractiveness of the venture for one or the other of the parties. The 50–50 represents a tested principle for maintaining an equality of interest through all the aspects of an inevitably complex relationship intended to endure for many years.

Other Middle East Countries Follow Saudi Arabia's Lead

Having found a relatively burdenless method of sharing the risk and the revenues, other Middle East countries were quick to follow Saudi Arabia's lead. On December 29, 1951, the sheik of Kuwait issued Decree No. 5, providing for an income tax applicable to corporations conducting trade or

9. The *Financial Times* (London), March 6, 1958, p. 6.

business in Kuwait. The tax rate was designed primarily for the Kuwait Oil Company's operations. The maximum rate of 50 percent provided for credits for "royalties, rentals, duties, imports and other exactions of like nature." (Article 1 of Decree No. 5.)

On February 3, 1952, the IPC group of companies signed an agreement with the Iraq government adopting the 50–50 profit-sharing principle as a basis for sharing revenues with the Iraq government. In its preamble the agreement set forth its purposes in these words: "Whereas it is the desire of the Government and the Companies to enter into an agreement for the equal sharing of the profits resulting from their operations in Iraq. . . ."

The Consortium Agreement, under which the controversy engendered by Iran's having annulled the AIOC concession contract was eventually settled, adopted the principle of 50–50 profit sharing between the consortium members and the Iranian government. To provide for this, Iran modified its Income Tax Act of 1949 by the addition of Article 35. This addition in effect requires the trading companies, created by consortium members to export the crude oil produced by the Iranian Oil Exploitation and Producing Company and the products refined by the Iranian Refining Company, to pay the government 50 percent of their profits as calculated on a basis of their posted prices after reducing the government's share by the 12½ percent royalty paid to NIOC. In 1954 the British government granted tax relief to the Anglo-Iranian Company which in effect relieved it of the burden of increased taxes paid to Kuwait and Iran.

Iran Nationalizes and the Consortium Takes Over

THE most significant feature of the 1933 Anglo-Persian concession agreement was its royalty provisions. Under the D'Arcy concession Anglo-Persian paid revenues to the government on a basis of its annual net profits. This led to disagreement over the way the company calculated profits and with the Great Depression seriously curtailed government revenues. The 1933 concession retained the profit-sharing principle but added a straight royalty payment of four shillings for each ton of oil consumed in Persia or exported. It resulted in a substantial increase in revenues, interrupted only by the depression of 1937–38. Except for disagreement with the company over minor issues, the government appeared satisfied with the arrangement. The depression brought minor discontent; World War II brought major difficulties.

World Wars Intensify Search for Oil

The period between the two world wars witnessed an intensified search for oil and brought American, British, and Russian interests in Persia's northern provinces into open conflict. Concession-hunters utilized both commercial and political strategies in trying to get what they wanted. In trying to get concessions, Standard of New Jersey, Anglo-Persian, and Sinclair Consolidated succeeded in blocking each other and were in turn blocked by Russia. World War II, which brought British and American troops into the south of Iran and Russian troops into the north, accentuated national rivalries and intensified each country's concern with Iranian oil. The war had taxed the capacity of the American oil industry to meet the needs of the United States and aroused the government's concern over the adequacy of American reserves. It forced Britain to a new recognition of the problem of protecting her oil and political interests in the Persian Gulf area and it intensified Russia's determination to keep northern Iran within the sphere of its influence and its resources under Russian control. At the Teheran Conference of 1943 the three powers are said to have discussed their mutual interest in Iran's oil. While governments were discussing their joint concerns, private companies were asserting their individual interests.

Russia Obtains and Loses a Concession

Meanwhile Russia temporarily outwitted them all and by overplaying her hand defeated herself. As a price for withdrawing her troops from the Northern Province, Russia exacted a 50-year concession from Iranian Premier Ahmed Ghavam es Sultaneh, who headed an Iranian delegation to Moscow to negotiate a settlement. The agreement provided that a joint Irano-Russian oil company exploit the oil of the Northern Province over a 50-year period. Russia was to own 51 percent of the company's shares during the first 25 years of the contract; Iran to own a 50 percent interest during the second 25 years. The company was to split profits in proportion to ownership.

An agreement having been reached, Russia withdrew her troops. With Russian troops out of the country, submerged nationalist sentiment erupted and Ghavam never submitted the agreement for ratification. Some believe that he never intended to do so.

The occupation of the country by foreign troops accentuated resentment against foreign influence. With British, American, and Russian interests so anxious to obtain rights to *hunt* for Iranian oil, it is not surprising that Iran should re-examine the terms under which AIOC,[1] formerly

1. In 1954 Anglo-Iranian Oil Company again changed its name, becoming the British Petroleum Company, Ltd. For convenience the text will refer hereafter to Anglo-Iranian Oil Company (AIOC) throughout the entire life of the second concession and to the British Petroleum Company, Ltd. (BP) during the period of the Consortium Agreement.

APOC, (50 percent British government owned) was exploiting her *known* reserves. In the years 1944–50, AIOC's profits had increased more than tenfold; Iranian oil revenues, only fourfold. As the company had grown richer, the war hardships had compounded Iran's traditional poverty. Stirred by the agitation of the Tudeh party, in October 1947, the government enacted a brief but significant oil law.[2] In a single article the law declared the Russian concession null and void, provided for a technical, scientific inquiry into oil possibilities throughout Iran, prohibited the granting of concessions to foreigners, and, where the rights of the nation in its oil resources had already been impaired, as it was believed to have been in the 1933 concession, required the government to begin negotiations to recapture its national rights. Obligated by this statute and impelled by the sentiments that prompted it, the government began negotiations with AIOC for a better deal. What began in early 1948 as negotiations to improve the terms of the concession ended three years later with Iran's canceling the concession and nationalizing the industry.

The Nature of the Controversy

It will serve no useful purpose to pass judgment on the fairness of the position of AIOC and the British government on the one hand and the Iranian government on the other. What is fair depends on the preconceptions of the participants. As Pope expressed it, " 'Tis with men's judgment as their watches; none/Go just alike, yet each believes his own." The British government and AIOC in particular viewed the concession agreement as a business contract under which the parties to it had assumed certain legal obligations. AIOC executives were aware, of course, of the magnificent returns with which their risk-taking, their capital, and their management and technical skills had been rewarded. Whether impelled by the exigencies of the situation or responding to their sense of fair play in business matters, they were willing to make concessions that would permit Iran to share more liberally in the fruits of their endeavors. But at no time were they willing to surrender their concession or permit it to be taken from them. To prevent nationalization they brought into play all of the diplomatic, business, and economic strategies at their command. When

2. George Kirk, *The Middle East 1945–50: Survey of International Affairs* (London, 1954), p. 88.

negotiations failed, they sought the support of their governments, which resorted both to economic pressures and to military moves, of the International Court of Justice, which ruled against them, and of the United Nations Security Council, which refused action, and they accepted, with some reluctance at first, the aid of the United States government in conciliation.

Against all of these the Iranian government under the leadership of their dedicated, some say fanatical, premier, Dr. Muhammad Mossadegh, stood steadfast on the principle of national sovereignty, Iran's right to reclaim control over its principal source of wealth which others had alienated. Nasrollah Entezam, the leader of his country's delegation, in addressing the General Assembly of the United Nations, of which body he had once served as president, stated simply the position of the more sophisticated and articulate leaders of his country when he said: "As you know, oil is the main source of our national wealth. It is therefore proper that we can countenance its exploitation only as a way of ensuring the general welfare of our people."[3]

The Controversy Reaches a Climax

The rank and file of Iranian citizens—living in squalor, mostly illiterate, and many diseased, exploited by their more fortunate brothers, and governed by leaders to whom corruption was a way of life—on every opportunity supported Mossadegh, whom they regarded as their saviour. As he grew more intransigent, eventually alienating the shah himself, they grew more loyal. On July 29, 1951, the day on which the International Court decided it was without jurisdiction, when the shah called on Mossadegh to form a new government, Mossadegh received a unanimous vote of confidence in the Majlis. When the senate hesitated to grant him the dictatorial powers he demanded, the Majlis, more responsive to the electorate, voted the senate out of office. After Mossadegh ordered the dissolution of the supreme court and demanded a new election law, the Majlis gave him a unanimous vote of confidence. When the Majlis showed some reluctance to approve his demand that it extend for a full year his right to govern by decree, a wave of demonstrations swept the country and the Majlis voted

---

3. U.N. General Assembly, Sixth Session, *Official Record,* November 14, 1951, 128; A/PV, 344.

59 to 1 to grant him the dictatorial power he sought. When he challenged the authority of the shah and a majority of the Majlis turned against him in August 1953, he ordered a plebiscite to dissolve parliament and won more than 99 percent of the votes cast and counted.[4] When the shah fled the country after failing to oust Mossadegh by a decree naming General Fazlollah Zahedi premier and another ordering Mossadegh's arrest, Mossadegh's supporters demonstrated in the streets and demolished the statues of the shah and his father. Only after hundreds of persons had been killed or wounded was the military able to restore order and to enable General Zahedi to control the government, force Mossadegh out of office, and guarantee the safety of the shah. The C.I.A. generally receives credit for the coup that restored the shah to power.[5]

More than two years had elapsed since the parliament passed the Nationalization Law and the Enabling Act. Meanwhile oil production had come to a standstill and the country had been thrown into political and economic turmoil.

Negotiations Reopened

These unhappy events cleared the way for the reopening of negotiations and an eventual settlement by the Consortium Agreement. In the resettlement negotiations the United States State Department, acting as conciliator, but not unmindful of American interests, sent Herbert Hoover, Jr. as petroleum advisor to American Ambassador Loy Henderson in Teheran. There in October 1953 they opened negotiations with representatives of the Iranian government. Teheran remained the center of negotiations but for weeks Hoover traveled between Teheran, London, and Washington, working out the details of a settlement. It was a delicate task. The revolution that had overthrown the intransigent Mossadegh had placed compromisers in control of the government but had not changed the basic character of the problem: how to reconcile the national aspirations of Iran with the commercial and political interests of AIOC. Even Zahedi with his military backing could not ignore the sentiments that Mossadegh had so effectively mobilized in support of his national aspirations. The revolution that laid

4. Benjamin Shwadran, *The Middle East, Oil and the Great Powers* (New York: Council for Middle Eastern Affairs Press, 1959), pp. 136–140.

5. David Wise and Thomas Ross, *The Invisible Government* (New York: Random House, 1964), pp. 110–114.

him low had submerged but not eradicated them. They were too deep-seated.

The Consortium Compromise

On the other hand, AIOC and the British government, as winners in the struggle, could not now be expected to grant more than they had previously offered. They were willing to recognize the principle of nationalization while stubbornly refusing to surrender their commercial rights and privileges. The Consortium Agreement salvaged both. It gave to the Iranians the shadow of what they sought while retaining for the British the substance of what they had. The Consortium Agreement is shot through with provisions that recognize that the oil properties belong to Iran, but they as clearly reserve the right of foreign companies to exploit them.

The new concession holder is the consortium. Anglo-Iranian Oil Company, Ltd., sole owner of the older concession, retained a 40 percent interest in the new. It surrendered for a substantial consideration a 14 percent interest to N.V. De Bataafsche Petrolemi Maatschappij (Royal Dutch-Shell), a 6 percent interest to Compagnie Française des Pétroles (both AIOC partners in the Iraq concession) and an 8 percent interest to each of five American oil companies (Standard Oil of New Jersey, Socony-Vacuum Oil, Standard Oil of California, Gulf Oil Corporation, and the Texas Company), all previously associated with AIOC or with each other in other Middle East concessions.

American Participation

The American companies, because of the problem of bringing Iranian oil back into world markets without disturbing their existing marketing interests, are said to have been reluctant to join the consortium. It is not clear why they should have been. Iranian oil could not re-enter world markets without causing readjustments in oil commerce, and it would seem more compatible with the interests of existing marketers that they share in control of the process. Moreover, they were not a group that had hitherto evidenced an unwillingness to share in virtually riskless but highly profitable oil ventures. At any rate, assurance from the United States Department of Justice that their participating in the consortium would not consistitute an illegal restraint of trade overcame their seeming reluctance. They joined, and subsequent events indicate that it paid them well to have done so.

This ownership arrangement, by placing exploitation of Iran's major oil reserves in several hands, overcame one of Iran's objections to the old AIOC concession—the monopolizing of her oil resources by a single company. It left, however, the seven major companies' control of Middle East oil intact. But with this arrangement the state department intervened at the insistence of a group of aggressive independent companies and worked out another whereby each of the five major American companies surrendered one percent of its participation to a group of eight independent companies. This brought the Richfield Oil Company, the Hancock Oil Company, the Signal Oil and Gas Company (which later absorbed Hancock), the Atlantic Oil Company, the Pacific Western Oil Corporation, the San Jacinto Petroleum Corporation, the Standard Oil Company (Ohio), and the Tidewater Associated Oil Company into the venture.

Concerning this, P. H. Frankel, an experienced, sophisticated, "dyed-in-the-wool oil man," but not always a reverent student of international oil affairs, has since stated,

For the additional members of the Iranian Consortium, their minor share proved an exceedingly profitable investment and the prospectus which had been drawn up by a large firm of chartered accountants showed clearly for all concerned to see that even after the compensation payable to Anglo-Iranian any stake in that venture was like getting a "license to print money."[6]

The Consortium Agreement and NIOC's Role

The area covered by the new agreement is only slightly less than that covered by the old Anglo-Iranian Oil Company agreement. It includes approximately 100,000 square miles extending from a point about the middle of Iran's western border in a southeasterly direction to the north end of the Persian Gulf, an area approximately 750 miles long and varying from 50 to 200 miles in width. It embraces all the fields that Anglo-Iranian had brought into exploitation except one small one in the extreme northwest corner of the area.

The parties to the Consortium Agreement are Iran and the National

6. P. H. Frankel, *Mattei: Oil and Power Politics* (London: Faber and Faber, 1966), pp. 95–96.

Iranian Oil Company (parties of the first part) and the several consortium members (parties of the second part). The National Iranian Oil Company (NIOC) is the corporate instrument that the government created to own and operate the oil industry under Mossadegh's nationalization program. Under the new agreement NIOC became the owner of all producing, refining, and auxiliary installations owned by AIOC before nationalization and of such other properties as the operating companies(created under the agreement) may subsequently install.

Although NIOC thus became the owner of the physical properties used in producing and refining oil, the agreement provides that two companies organized by the consortium members under the laws of the Netherlands should operate the properties "on behalf of Iran and NIOC" (Article 4). The operating companies are the Iraanse Aardolie Exploratie en Productie Maatschappij (Iranian Oil Exploration and Producing Company) and the Iraanse Aardolie Raffinage Maatschappij (Iranian Oil Refining Company). The operating companies have substantially the same responsibilities and rights in operating the concession that AIOC previously had, and they perform the same basic functions. Although Netherlands corporations, they are registered in Iran, and NIOC has the right to nominate two of the seven members of each of the operating companies' board of directors. It also has the right to subject all accounts of the operating companies to independent audit by qualified accountants of its own choosing and to inspect through its technical experts all technical activities of the operating companies. To facilitate inspection, NIOC may request and the companies must furnish NIOC with accurate copies of all topographical, geological, drilling, refining, and other relevant data (Article 4, Paragraphs G and E). These provisions permit NIOC to exercise some influence in the management of companies that operate "on its behalf" and give it access to a great deal of technical and practical information that may contribute to more efficient conduct of its independent operations that have greatly expanded since the 1954 agreement.

The Role of the Trading Companies

The operating companies do not buy or sell oil. They produce it for NIOC. Each member of the consortium has set up a trading company that individually and independently of the other buys oil at the wellhead from NIOC at a "stated price" of 12½ percent of the price that the trading company posts at the point of export. The Iranian Oil Exploration and Producing Company then delivers oil for the account of the trading company either to the crude oil loading ports at the upper end of the Persian Gulf or to the Abadan refinery. At the loading port the trading company sells its crude oil for export usually, but not always, to an affiliate. The Iranian Refining Company processes the oil that is delivered to the Abadan refinery and delivers the products to the trading company for sale to its affiliate or to others. The trading companies pay the costs which the producing and refining companies incur in their operations and in addition pay them a fee of one shilling for each cubic meter of oil that they receive either for export or for refining.

The trading companies not only meet the costs of producing and refining oil but they earn the profits that grow out of buying and selling it.

Source of Iranian Government's Revenues

The Iranian government's oil revenues are derived from taxing the trading companies' earnings at the rate of 50 percent. Under the 1933 concession, the government derived its oil revenues from a royalty on each barrel of oil produced and a modest share in the earnings of AIOC. This change to a tax basis is the most significant difference between the old and the new concession. The large gap between AIOC's cost of producing oil and the selling price, the large profits this brought to the company, and the relatively meager returns it yielded the government created the basic issue which led eventually to nationalization.

To arrive at their taxable income, the trading companies value the oil they sell at the applicable posted price of each crude shipment. From this valuation they are allowed under the agreement to deduct their share of production costs, the fee of one shilling per cubic meter paid to the producing company, and an allowance of approximately one percent of the posted price of crude oil.[7]

In the sale of oil products from the Abadan refinery the trading companies calculate their taxable profits by reference to the value of crude at the applicable posted price plus five percent. This is to permit Iran to share in the enhanced value of the oil occasioned by refining. As with crude oil, so with refined products, the trading companies deduct the cost of produc-

7. This has since been reduced to ½ cent.

tion, the production fee, the expense allowance, and the refining fee of one shilling per cubic meter.

In settling their tax liability the government under the consortium agreement permits each trading company to credit against its tax liability the 12½ percent of posted price customarily referred to as a royalty that they pay NIOC for their oil.[8] The only taxable income the producing and refining companies have is derived from the fee of one shilling per cubic meter of oil produced or refined. On this they pay income taxes at the standard rate.

The effect of these somewhat complicated tax arrangements is, in the language of a consortium official, "to bring about an equal sharing with Iran of the profits arising from the operations of the oil Consortium."[9]

While this may represent a somewhat euphemistic characterization of Iran's oil revenues, the Consortium Agreement brings its tax provisions into harmony with the so-called 50–50 profit-sharing agreements that the other Middle East countries had meanwhile negotiated with their concessionaires. Equally important, it more than tripled Iran's oil revenues from 1950 to 1956 despite a decrease in average daily production from 660,000 to 540,000 barrels.

This brief summary of the more important provisions of the Consortium Agreement indicates the way in which the consortium pays homage to Iran's aspirations to bring exploitation of petroleum wealth under her control. NIOC owns the resources, the facilities used to exploit them, and the oil produced. NIOC representatives sit on the boards of directors of the producing and refining companies which make decisions on the way in which production is conducted. This is the shadow of control, not the substance. Decisions on how much oil the companies will produce, to whom and at what price they will sell it, how fast they will develop Iran's resources are left to the trading companies. In commercial decisions vital to prosperity and economic health, Iran has no voice. To say this is not to say that either the Iranian oil industry or the world industry of which it is a part would be more prosperous if Iran had the sole responsibility to develop the industry according to its discretion and wisdom. It is merely to emphasize the gap between what Iran wanted and what it got.

But in other ways the Consortium Agreement has enhanced greatly NIOC's responsibilities and opportunities.

8. This arrangement was changed after long negotiations between OPEC and the companies. See Chapter Seventeen.
9. Personal communication to the author.

Iran Loses Oil Markets

Nationalization resulted in the virtual exclusion of Iranian oil from world markets. The international oil companies with their extensive and far-flung facilities for transporting, refining, and marketing oil and oil products would have nothing to do with it. Lesser companies, anxious to extend the scope of their operations, whether deterred by the magnitude of the readjustments that the purchase and marketing of Iranian oil would necessitate or by the fear that by purchasing Iranian oil they would be buying a lawsuit, likewise kept hands off. The majors quickly filled the gap in world supplies by expanding output in other areas under their control. The Kuwait Oil Company, jointly owned by AIOC and Gulf Oil Corporation, took up most of the slack, increasing Kuwait's daily output from 340,000 barrels in 1950 to 960,000 barrels in 1954.

Increasing world demand made it possible for the consortium members to direct the flow of Iranian oil once more into world channels without serious disturbance to oil markets. But nationalization permanently slowed down the Iranian oil industry's relative rate of growth. The major companies, which had expanded facilities elsewhere or utilized them more effectively, were reluctant to curtail their operations for the benefit of Iran. And the other Middle East countries would have resented their doing so. Before nationalization, Iran was the Middle East's most important oil producer. It has never regained its premier position.

Nationalization Brings Some Compensations

However, nationalization has not been without its long-run compensations. Whether they will make up for the material losses, the political turmoil, and the individual suffering that accompanied nationalization may never be established. As judged by short-run considerations, Iran was the loser in the struggle and paid dearly for it. A lesson it learned was how completely its economic welfare depended on a single foreign oil company. This did not lessen its determination to free itself from such control. The consortium arrangement with its multiple ownership shifted decision-making from a single managerial group to several. To what extent this improved Iran's lot is a moot and complex question.

But with the lesson of nationalization behind it, Iran set about improving its own lot. Nationalization launched Iran on a program of self-help, the full consequences of which both to Iran and to other Middle East oil countries may not be discernible for many years.

This program gave birth to a form of corporate enterprise—the so-called joint venture—new in concession-granting by Middle East oil countries. The 50–50 profit-sharing agreement, which the oil companies like to refer to as a partnership arrangement, is about as much a true partnership as is the relation between the United States government and the thousands of corporations that it subjects to corporate income taxes. While a former General Motors executive found satisfaction in identifying General Motors' welfare with his country's welfare, it was not in "profit sharing" through taxation that he conceived the identity. Taxes are still regarded by corporations as having about the same certainty as death and are looked upon with equal enthusiasm. It is true that since Aramco and other American corporations receive credit toward their domestic income tax obligations for the taxes they pay to foreign governments, the shift from a royalty basis to a tax basis in meeting their obligations to oil-producing countries was virtually painless to them. And it brought tremendous satisfaction to the Middle East governments, engendering temporarily at any rate enough mutual good feeling to justify a change in the conception of the relationship between the host countries and their corporate guests from one of exploiter and exploited to one of partnership. While the oil companies still like the term, for the oil countries its glamor has worn off.

Enrico Mattei and the Joint Venture

Iran's Petroleum Act of July 1957 laid the basis for the joint venture, which in reality constitutes a partnership, a concept in the utilization of which Iran pioneered but to which it cannot lay exclusive claim. To Enrico Mattei, who more than any other man of his generation left an imprint on the international oil industry—and that at a time when corporate enterprise was replacing individual enterprise—belongs the major credit. Mattei, a complex character, was among other things an innovator.[10] After World

10. P. H. Frankel in his brilliant and perceptive biographical profile of Enrico Mattei, *Mattei: Oil and Power Politics,* characterizes him as a "man of many parts; he had in him the makings of a *condottieri* and of a pirate, of an unscrupulous go-getter and of a revivalist preacher" (p. 25); nevertheless a dedicated man whose "mission was to secure for Italy a place in the front rank of industrial nations and also to fight the battle of the little man against the big bosses." (p. 47.)

War II he got into the oil business through his position in Milan as commissioner attached to one of the Fascist national enterprises, Azienda Generale Italiana Petroli, known as AGIP. Organized in 1926 for the purpose of conducting all activities connected with the petroleum industry in Italy, AGIP had established itself before World War II as one of the leading marketers of petroleum products and had engaged unsuccessfully in exploring for oil at home and abroad. As a part of Mussolini's drive for economic self-sufficiency, AGIP had engaged in a costly and at the war's end an unsuccessful search for oil and gas in the Po Valley. When Minister of Industry and Commerce Gronchi (later president of the republic) decided to curtail AGIP's exploration activities, Mattei intervened with a plea to save AGIP. Under Mattei's energetic management and with good fortune AGIP's exploration activities culminated in a dramatic success in finding gas. Profits from gas supplied Mattei with funds with which he greatly expanded the scope of AGIP's activities. Among these were a network of gas pipelines serving the industrial north and a grandiose national network of filling stations displaying the symbol of a six-legged dog bearing the tradename and legend Supercortemaggiore La Potente Benzina Italiana.

But this constituted only one sector of the industrial enterprises that AGIP and affiliated companies were building. When in 1953 Ente Nazionale Idrocarburi, known as ENI, consolidated the control of the several state enterprises engaged in oil and gas activities, Mattei became president of ENI; president of SNAM (Societa Nazionale Metanodotti), engaged primarily in oil and gas transport; president of Azienda Nazionale Idrogenazione Combustible (ANIC), engaged in refining and petrochemicals; president of STANIC, a refining enterprise owned jointly by ANIC and the Standard Oil Company of New Jersey; director of ROMSA, engaged in refining, particularly in the production of lubricating oils and bitumens; and managing director of AGIP.

Mattei Goes Abroad in Search of Oil

As the head of an integrated oil enterprise of this magnitude short in its essential raw material, crude petroleum, of which Italy was barren, it was inevitable that Mattei should look abroad for a source of cheap oil. In his struggle for a larger and more secure place in the empire of oil, he regarded the established oil companies, which he contemptuously referred to as "the cartel of the seven sisters," as his natural enemies. Blocked in his effort to join the select circle, he fought it.[11] In so doing, he developed an individual philosophy of a partnership relationship between an oil-producing and an oil-consuming country. He regarded the international oil companies as unnecessary middlemen. A true partnership would eliminate them and in so doing would make possible larger gains to the producing countries and cheaper oil to the consuming countries. The joint venture was the instrument to achieve these salutary results.

A basic feature of Mattei's conception of the joint venture is that it permits a country, without risking its capital in the hazardous undertaking of hunting for oil, to share in the enterprise on equal terms if the search is successful. Only if and when oil is found in commercial quantities must the producing country provide its share of the capital and meet its share of the costs. It is an attractive arrangement for a poor country. A man of action, Mattei began negotiations to put his ideas to work. In February 1957, Mattei completed a joint-venture agreement with Egypt establishing the Compagnic Orientale des Pétroles d'Egypte (COPE) jointly owned by the International Egyptian Oil Company (IEOC), a subsidiary of ENI, and two Egyptian concerns, one private and one public.[12]

Iran's Petroleum Act of 1957 and the NIOC-AGIP Joint Venture

Article 1 of Iran's Petroleum Act of July 1957 proclaims the act's two purposes: the exploration for and the exploitation of oil as rapidly as possible in all areas outside of those assigned to the consortium; and the development of refining, transportation, and sale of oil throughout Iran and abroad. It designates the National Iranian Oil Company as the instrument for accomplishing these objectives. NIOC, as originally created, was an autonomous commercial and industrial organization, authorized to take

11. Before Iran nationalized her oil resources, AGIP obtained its crude oil from AIOC. According to Frankel, when the "Italian government acquired 'stolen oil' from Dr. Mossadek through a somewhat obscure firm by the name of SUPOR," Mattei remained loyal to AIOC and "would have nothing to do with them." Apparently Mattei had hoped that such loyalty together with ENI's industrial status would entitle ENI to participate in the consortium. His efforts to do so were fruitless. Frankel, p. 95.

12. Muhamad A. Mughraby, *Permanent Sovereignty over Oil Resources* (Beirut: The Middle East Research and Publishing Center, 1966), p. 71. Mughraby's book is a pioneer and penetrating study of Middle East oil concessions and legal change.

over the properties previously held by AIOC and to engage in all phases of the oil industry at home and abroad either directly or through subsidiaries or in association or co-operation with other companies. Its basic character under the Petroleum Act remains unchanged, but its functions are broadened somewhat and at the same time made more specific. For purposes of research, exploration, and extraction, the act authorizes NIOC to divide the territory of the country, including the continental shelf but excluding the area covered by the consortium agreement into districts each containing not more than 80,000 square kilometers (Article 5). At its discretion NIOC may declare any district open for exploration and invite other companies to enter into joint agreements for its exploitation.

It is the nature of the exploitation agreements that constitutes the act's novel feature and which prompted one of the act's administrators to characterize the act as "the most progressive legislation in any Middle East country," one that "has set a precedent for other oil laws in many parts of the world."[13]

The Joint Structures

The agreements contemplated are partnerships characterized in the act as "mixed organizations" or "joint structures." NIOC is to hold not less than 30 percent ownership in any joint structure; it may hold 50 percent or more.[14] In practice NIOC has made no joint-structure agreement in which it holds less than 50 percent ownership. The discussion that follows is limited to 50 percent partnership arrangements.

To prevent monopolization, insure prompt development, and prevent holdings from remaining idle for long periods, the act limits the holdings of any joint structure in any one district to a maximum of 16,000 square kilometers, provides that the joint structure begin drilling for oil within four years from the date of the agreement, limits the agreement to twenty-five years with provision for renewals of three terms of five years each, and requires the return to NIOC within ten years of half of the total area included within the original agreement (Article 7).

13. Modir, head of petroleum districts and oil affairs co-ordination of the Oil Operations Organization, in a speech on "The Petroleum Act and the Petroleum Districts" at the Oil Operations Supervisors' Training Conference, Kermanshah, June 1964. Reproduced in *Iran News Letter,* No. 73, August 1964, published monthly by the National Iranian Oil Company, Teheran. (Hereafter *News Letter.*)

14. The law also authorizes NIOC to make nonpartnership concession agreements covering small areas, an authority which it has never exercised.

To make it unnecessary for NIOC to risk its limited capital in a search for oil, the act provides that the NIOC partners pay an annual rental, a portion of which is to be prepaid in a lump sum (to be recovered as a part of operating costs once oil has been found in commercial quantities) or in lieu of the rental the act provides that the partner meet all of the exploration costs, which he may recover if and when oil is discovered in commercial quantities. If the joint structure finds no oil, the foreign partner alone bears the cost of exploration and development.

If and when the joint structure finds oil in commercial quantities, NIOC supplies its part of the developmental costs and shares in the profits of the operation. The government derives its revenue from the 50 percent income tax paid by the partnership. Because NIOC, a government-owned enterprise, as partner in the venture gets 50 percent of the profits after taxation, the government's total share in profits (taxation and return to capital) constitutes 75 percent.

In addition to creating the general framework and establishing specific provisions within which joint structures may be consummated, the act bestows on NIOC the authority to conclude "any agreement which it deems appropriate . . . not inconsistent with the laws of the country" (Article 2). Armed with this broad discretion, NIOC has proceeded to put the flesh of reality on to the legal skeleton of the joint structure and to inject into it the breath of corporate life.

The NIOC-AGIP Joint Venture

NIOC's first joint venture was with Mattei's AGIP. Although the petroleum act uses the term "joint structure," the partnership agreements subsequently entered into commonly are referred to as "joint ventures," a term which this text will use hereafter. As previously indicated, the law gave play to both Iran's national aspirations and Mattei's commercial ambitions. Whether AGIP and NIOC were fashioning their joint venture to insure its coming within the scope of the act or whether the act was being shaped to legalize a joint venture of the sort that NIOC and AGIP were formulating is perhaps irrelevant to the future of the joint venture. At any rate, the AGIP-NIOC joint agreement was ready almost as soon as the act.

The Petroleum Act bears the date of July 31, 1957. It provides that agreements under it shall be signed by the National Iranian Oil Company, submitted to the council of ministers and, if approved by the council, placed before the legislature for approval. "Agreements thus signed shall be enforceable after approval by the Legislature and with effect from the date of such approval" (Article 2). The legislature approved the NIOC-AGIP agreement on August 24, 1957. The agreement provides that within sixty days NIOC and AGIP shall establish a corporation "to undertake for their mutual interest, operations for the exploitation and production of crude petroleum" (Article 1). The corporation, the Iran Italian Petroleum Company, known as SIRIP, is an Iranian corporation in which AGIP and NIOC each has an equal interest. NIOC and AGIP appoint an equal number of members of its board of directors. NIOC appoints the board's chairman, AGIP its vice-chairman. Although AGIP supplies SIRIP's technical personnel, SIRIP is required to limit foreign personnel to such positions as it cannot fill with qualified Iranian personnel and is obliged to co-operate with NIOC and AGIP to establish a program for the training of Iranian personnel to replace within "reasonable limits" the foreign personnel as soon as possible (Article 33).

SIRIP has exclusive exploration and production rights over 5,600 square kilometers in the northern part of the Persian Gulf lying over the continental shelf, over 11,300 square kilometers located on the eastern slopes of the Zagros Mountains, and over 6,000 square kilometers extending along the coast of the Gulf of Oman.

Initially the whole expense of exploration falls on AGIP, which agrees to spend at least $22 million hunting for oil, $6 million within the first four years, and $16 million within the next eight years (Article 19). If and when AGIP completes a commercial oil well (which the agreement carefully defines in Article 8), SIRIP is obliged to repay AGIP for its exploratory expenditures "within the shortest possible time" but at an annual rate of not less than 10 cents for every barrel of oil produced (Article 11). After it has opened one commercial oil field, AGIP is obliged to explore for others until it has spent the entire $22 million. At its discretion, AGIP can suspend exploration operations at the end of four years or any year thereafter, but in doing so, it is obliged to pay to NIOC one half of any unexpended balance of the obliged $22 million (Article 19).

Once a commercial well has been completed, each of the partners provides half of the capital for exploiting it. SIRIP, as an independent business corporation, must first offer to sell its crude equally to NIOC and AGIP, and if their offers are satisfactory each partner takes title to the crude to dispose of as it sees fit. If the offer of either partner is not satisfactory, SIRIP may sell the crude to other buyers, but not for less than its owners have offered to pay for it (Article 12).

SIRIP is required to post prices at which it will sell its oil generally to all buyers. The posted prices "shall conform with current prices prevailing in the region of the Persian Gulf" for oil of similar quality (Article 13). If necessary to sell its oil, SIRIP may offer it at discounts from its posted prices if approved by a committee of two of its directors, one each selected by AGIP and NIOC.

SIRIP pays no royalty to the Iranian government nor to NIOC. But it pays half of its net profits to the government as a tax (Article 17). Because NIOC gets half of the remaining profits, the total take of the government is 75 percent of net profits, 50 percent as a tax and 25 percent as a return on capital.

The above discussion indicates the basic character of the first joint venture under the law of 1957. It is a partnership arrangement that places the financial burden of exploring for oil on the foreign partner, which also furnishes the technical knowledge and the managerial skills. As the venture becomes a going enterprise, having passed from the stage of oil hunting to oil producing, its operations are taken over by a jointly-owned company (SIRIP) operating as any business concern for the benefit of its partners. Should disputes arise, the agreement provides for a method of settling them designed to protect the interests of both parties (Articles 43, 44, and 45).

Article 39 of the agreement constitutes a declaration of good faith. "The contracting parties agree to carry out the terms and the procedures . . . in good faith and to observe both the spirit and the letter of the said terms and procedures." It is designed to place the agreement beyond the reach of political action. "No general or special measure, legislative or administrative, or any other act of this kind emanating from the Iranian government, central or local, can annul this agreement, amend or change its provisions, prevent or hold up the necessary and effective execution of these stipulations."

NIOC Opens New Districts

The trade press viewed with skepticism the AGIP-NIOC joint venture, suggesting that the liberal terms to which AGIP had subscribed reflected its inexperience, terms to which more sophisticated oil companies would not have agreed. The trade press was wrong. NIOC opened District One to exploration in November 1957. Between the preliminary announcement and the official opening of the district fifty-seven companies from nine countries applied for questionaires, a preliminary requirement of NIOC of prospective bidders. Twenty-two companies completed the questionaires. NIOC pronounced eleven companies 100 percent competent to engage in exploration and ten others as having competence ranging from 10 percent to 75 percent. After the formal opening of the district sixteen of the companies paid NIOC $2,700 each for "documentation" on the district. Fourteen eventually submitted bids.[15]

Pan Am–NIOC Agreement

The successful bidder was Pan American International, a subsidiary of Standard Oil Company of Indiana. The NIOC–Pan Am joint-venture agreement became effective in June 1958. Although it follows the pattern set by AGIP, it differs in some important respects. AGIP got its contract without paying a bonus; Pan Am paid NIOC a cash bonus of $25 million recoverable in ten annual installments after commercial production begins (Article 31.5). Whereas AGIP agreed to spend $22 million over a twelve-year period searching for oil, Pan Am agreed to spend $82 million (Article 30). Whereas SIRIP, the operating company under the AGIP-NIOC partnership, is a business corporation sharing its costs and profits with each of its partners, the Iran–Pan American Oil Company (IPAC) is a nonprofit operating company acting as an agent for Pan Am and NIOC, jointly controlled by them and producing oil for their account.

In both agreements the foreign company initially meets all exploration expenditures, NIOC's obligation in regard to exploration and operating costs beginning only if and when commercial production begins.

IPAC formulates its production program on a basis of the requirements of its joint owners. The production program, as does that of NIOC-AGIP, includes oil for Iran's internal consumption [to be supplied within indicated limits (Article 26.1e) at cost plus 14 cents per cubic meter (Article 26.3)], oil for IPAC's operations, and oil for export to meet NIOC's and Pan Am's sales obligations. Each party is entitled to take one half of the oil available for export. If either party does not take all it is entitled to, the other party may take the excess. If the two partners do not take from IPAC all the oil

15. Modir, p. 6.

available to them on mutually acceptable terms, IPAC as agent of the joint venture has the right to sell the excess to other buyers on terms no "less favorable than those available to the joint structure" (Article 22.1). The value of the oil taken by each party is calculated monthly by multiplying the number of barrels each takes by the relevant posted price less any discounts that the partners may have authorized. If the value of oil taken by one party exceeds the value taken by the other, the party taking the larger amount shall pay half of the excess to the other party (Article 22.2). If IPAC as agent of the joint venture deems it advisable, it may sell oil at such discounts as its owners may authorize (Article 25.2).

The Government Recovers 75 Percent of Profits

NIOC and Pan Am are each obliged to pay the Iranian government taxes at the rate of 50 percent on net income derived from IPAC's operations.[16] As with the AGIP agreement so with Pan Am, the government recovers 75 percent of the profits derived from the agreement, 50 percent as taxes and 25 percent as an investor.

The Pan Am agreement (as does the AGIP) provides that the employment policy of the operating company (IPAC) give preference to Iranians and that it conduct a training program designed to accomplish this end. Specifically, within ten years IPAC's total labor force must be 98 percent Iranian and its executive personnel at least 51 percent Iranian.

The agreement is designed as a partnership in law and in fact with each party obligating itself to carry out the terms of the agreement "in accordance with the principles of good will and good faith and to respect . . . [the agreement's] spirit" as well as its letter (Article 38.1).

Outcome of Early Joint Ventures

Within a year after the passage of its Petroleum Act of 1957 Iran had entered into three joint ventures: the AGIP, the Pan Am, and the Sapphire.

16. Article 31. According to the agreement, NIOC's net taxable income is equal to one half the gross receipts derived from the sale of oil produced by IPAC less one half the costs incurred by IPAC in its production operations. Pan Am calculates its net taxable income in the same way except that for ten years it may annually subtract one tenth of the bonus it paid NIOC and the annual rentals in accordance with Article 32.1 of the agreement and Article 9A of the Petroleum Act.

Sapphire Petroleum, Ltd., a Canadian corporation, signed an agreement with NIOC dated June 1958. Only the Sapphire agreement has proved a commercial failure. AGIP has proved a modest commercial success. Pan Am has done better.

By the middle of 1962 an AGIP joint committee declared SIRIP's Bahregansar offshore discovery suitable for commercial production. By the end of 1967 it had ten producing wells, and crude output averaged 22,000 barrels daily. By the spring of 1968 it had discovered an additional field, Norooz, which was expected to produce 40,000 barrels daily.[17]

The Pan Am–NIOC joint-venture agreement gives IPAC exclusive exploration and exploitation rights to a 16,000 square kilometer area in the northern part of the Persian Gulf. After drilling three dry holes in this offshore concession, IPAC completed in November 1961 its Darius No. 1 with an initial output of 10,000 barrels a day. By the spring of 1962 IPAC had opened a second field with the completion of its Cyrus 2A, which tested at rates up to 12,000 barrels daily. During 1967 IPAC's production from these two fields averaged 98,000 barrels daily. Two other discovered fields had not yet been placed in production.[18]

NIOC Opens Other Districts for Exploration

With the IPAC venture such a success NIOC announced in April, 1963, its intention to declare additional offshore areas open for bidding. Because the Iranian Petroleum Act does not provide for exploration permits before granting concessions, as do the laws of many other countries, prospective concessionaires may have too limited technical knowledge about an area thrown open for bidding to make a really informed bid. That may work to the detriment of the potential concessionaire and of the country calling for bids. In areas about which little is known the element of risk is great. The result may be niggardly bids or none at all. When NIOC opened Iran's District One for bidding in 1957, on much of the area it could offer only scanty geological and stratigraphic data.[19] As a result the fourteen companies that either individually or in groups submitted bids

17. M. Farmanfarmaian, "Iran, NIOC and National Oil Companies," paper delivered at the Second Seminar on the Economics of the International Petroleum Industry, American University of Beirut, May 20, 1968.

18. *Ibid.*

19. Article 9.13 of Iran's Petroleum Act provides that NIOC prepare a booklet containing the necessary specifications such as geographical situation, geological conditions, and other relevant information.

applied for virtually identical acreage located near the mainland adjacent to areas where exploration operations had revealed favorable geological features and where production had been successfully conducted.[20] For 16,000 square kilometers lying directly opposite the Neutral Zone of Saudi Arabia and Kuwait, Pan American's bid was the best of a good lot. But the bidding resulted only in the Pan Am agreement; the greater part of District One remained uncommitted. When it again opened the uncommitted areas of District One, NIOC hit upon a novel plan for obtaining technical information necessary for intelligent bidding without cost to itself. It invited all those interested in submitting bids to participate at their joint expense in financing a comprehensive marine seismic survey. Only those companies that participated would have access to the results of the survey and would be permitted to submit bids. None was guaranteed that its bid would be accepted. Despite the fact that they were only buying the opportunity to submit bids, oil companies eagerly responded to the invitation. On August 10, 1963, at a press conference in Teheran, NIOC's director of oil operations announced the names of companies which had asked to participate and whose requests NIOC had approved. These included eight individual companies, three groups of American companies together representing thirteen different individual companies, one group of French companies, and one group of German companies, each representing five individual companies.

Meanwhile NIOC had contracted with the Western Geophysical Company to make the seismic survey over an area of approximately 48,000 square kilometers (an area four times as large as nearby Kuwait, which rested on the world's largest known oil reserves). The area covered underlay much of that portion of the Persian Gulf over the mineral rights of which Iran claimed sovereignty. It was one of the most extensive surveys of its kind ever undertaken. Its total cost was approximately $3.5 million.[21]

Western Geophysical completed this survey in the late winter of 1964. Its results proved exciting. The *Petroleum Intelligence Weekly,* which has an exceedingly sensitive nose for news and a fine capacity for dramatic

20. *News Letter,* "The Petroleum Act and the Petroleum Districts," No. 73, August 1964, p. 6; Modir, p. 6.

21. The above discussion of the marine seismic survey is based largely on P. Mina and F. Najmabadi, "A New Approach to Full Utilization of Iranian Offshore Prospects in the Persian Gulf," *News Letter,* No. 85, August 1965, pp. 1–9; A. M. Mirfakhrai, "A Case Study: NIOC's Handling of a Marine Seismic Survey," *Ibid.,* No. 83, June 1965, pp. 1–5; and Y. Paran and J. G. Chricton, "Highlights of Exploration in Iran 1956–1965," *Ibid.,* No. 96, July 1966, pp. 13–14.

reporting, stated in its March 16, 1964, issue that "the world's biggest oilfield may lie waiting to be tapped beneath 175 feet of Persian Gulf waters." This impression, although thus far proven false, was not without some substance. The Western Geophysical survey had revealed twenty-seven structural features favorable to the accumulation of oil in commercial quantities. As one "top-level executive" warned, "Every structure, of course, isn't necessarily oil bearing. And talking of its size before actual test drilling is obviously premature, but there's no doubt that the structures and domes uncovered by the seismic work are tremendously exciting."[22]

On September 1, 1964, NIOC officially invited those companies which had shared in the expense of the survey and had been certified by NIOC as competent to participate in the bidding, if they wished to do so, to prepare their proposals in accordance with the procedures that NIOC had laid down.

In November 1964, NIOC announced the names of individual companies and company groups that had submitted bids. Of those that had participated in the seismic survey, only Standard Oil of New Jersey and Continental did not participate in the bidding. Maximum acreage to be granted a successful bidder depended on the type of development contract: 8,000 square kilometers if NIOC owned 50 percent of a joint-venture enterprise; 4,500 square kilometers if NIOC owned less than 50 percent; and 2,500 square kilometers if NIOC did not participate as a partner.

By January 1965, NIOC had signed joint-venture agreements with the following five groups: Royal Dutch-Shell; ENI, Phillips Petroleum, and the Indian Oil and Gas Commission; Atlantic group, including Atlantic Refining Company, Sun Oil Company, Union Oil of California, and Murphy Corporation; French group, including the government Régie Autonome des Pétroles (RAP), Bureau de Recherches de Pétrole (BRP), Société de Recherches et d'Exploitation de Pétrole, (EURAFREP), Cie. de Participations de Recherches et d'Exploitation Pétrolières (COPAREX), and Cie. Franco-Africaine de Recherches Pétrolières (FRANCAREP); and the Tidewater group, including Tidewater Oil Company, Skelly Oil Company, Superior Oil Company, Sunray DX Oil Company, Kerr-McGee Oil Industries, Inc., Cities Service Company, and Richfield Oil Corporation.

Socony Mobil, Compagnie Française des Pétroles (CFP), the Pauley group (including Pauley Petroleum, Allied Chemical Corporation, and Pure Oil), American International and the Pakistan government, and the

22. *Petroleum Intelligence Weekly,* March 16, 1964.

German group (including Deutsche Erdöl, Elwerath, Preussag, Winter-shall, Gelsenberg, Schachtbau, and Scholven-Chemie) were unsuccessful bidders.

By June of 1965 the German group had joined the winners and had signed a joint-venture agreement with NIOC.

Under the law, NIOC can use its discretion in designating successful bidders. In making awards it weighs the various terms of bids and desig-nates as winners those companies whose offers it considers to be more in conformity with the interests of the country with due regard to all relevant factors. The successful bidders were a heterogeneous lot. They included one of the largest international oil companies (Royal Dutch-Shell), two government groups (French and Indian) anxious to obtain a dependable and adequate supply of crude in the exploitation of which they would have a voice, and a number of aggressive American independents the operations of which had hitherto been confined largely to the United States, but which were interested in obtaining low-cost crude supplies that might enable them to penetrate international markets.

The successful bidders agreed to pay cash bonuses aggregating $190 million and to spend more than $164 million over a twelve-year period in exploration and development. Each concessionaire has since established with NIOC a joint-venture relationship as provided in the Petroleum Act of 1957 and the two parties to each of the agreements have formed nonprofit joint stock companies to explore and exploit their joint venture.[23]

In all of these agreements the foreign partner meets the initial expenses of exploration. All agreements provide for an equal sharing of costs and profits. All provide for taxes to Iran at the rate of 50 percent of net earnings and hence provide Iran, as investor and tax collector, a total income equal to 75 percent of pre-tax earnings. None provides for the pay-ment of royalty as such, but all provide that after commercial production the foreign partner's taxes shall equal at least 12½ percent of the value of its production calculated at posted prices. None provides for the expensing of royalties. All provide that 25 percent of the concession be relinquished at the end of five years and a second 25 percent be relinquished at the end of ten years. All provide that exploration begin within six months of the signing of the agreement and that it proceed with reasonable speed. All provide for a total exploration period of twelve years. All limit the conces-sion to a twenty-five year period subject to renewal under specified condi-tions to a total of forty years.

23. *News Letter,* May 1965, pp. 8–12.

By the spring of 1968 two of the more recent joint ventures had proven commercially productive. LAPCO (the operating company of the NIOC-Atlantic group) had drilled sixteen wells, the first three of which had an average daily production of 10,790 barrels. Exports at the rate of 40,000-50,000 barrels daily were scheduled to begin in the summer of 1968 rising to 200,000 barrels daily in 1969. Meanwhile IAPCO had struck oil in another situation with a well testing 5,400 barrels a day.

Meanwhile IMINOCO, jointly owned by NIOC and Italy's AGIP, the Phillips Petroleum, and the Indian Oil and National Gas Commission, had discovered a promising field in its offshore area where production was scheduled to begin in 1969 at the rate of 25,000 barrels daily, rising to 100,000 barrels daily by the end of 1970.[24]

It remains to be seen how profitable these joint ventures will prove, but they signify a new pattern of concession in the Middle East, apparently one that the countries and the companies can live by.

Act of 1957 Expresses a Sense of Urgency

Mossadegh mobilized and dramatized Iranian national sentiment and directed it toward a recapture of the nation's sovereign rights in its natural resources which former rulers had surrendered. Underlying the national aspiration was the conviction that the Anglo-Iranian Oil Company was exploiting Iran's most important source of national wealth and welfare for the benefit of itself and a foreign government. Despite the consortium settlement, Iranian nationalists have never abandoned their goal of bringing the nation's oil resources more completely under national control and using them to promote national ends. The consortium agreement compromised and temporarily submerged their national aspirations. It did not eradicate them. The Petroleum Act of 1957 epitomized a determination to bring the nation's remaining oil resources into exploitation without relinquishing the nation's control over them. And to do it now. The act expresses a sense of urgency on the part of its framers. It is enacted "with a view to extending as *rapidly* as possible . . . research, exploration, and extraction of petroleum throughout the country . . . and of extending promptly the operations of refining, transportation, and marketing," throughout the nation "as well as abroad" (Article 1) (Italics supplied).

24. Farmanfarmaian, *op. cit.*

SIRIP, the joint venture under the NIOC–AGIP agreement, is obliged "to use *all its possible efforts* in order to raise to a *maximum*" the sale of petroleum (Article 12). Similarly, the NIOC–Pan Am agreement obliges Pan Am to "exert its *utmost* efforts to explore" the area assigned to it (Article 12). And it obliges both NIOC and Pan Am to exercise their *"utmost"* effort to sell the *"maximum* possible quantity of petroleum economically justified" (Article 21) (Italics supplied). Each of the more recent joint-venture agreements places similar obligations on the parties to the contract.

The ERAP Agreement

Not content with the rate of progress made under its joint ventures and desirous of increasing its control over production operations and its revenues from oil, Iran has recently entered into a new type of agreement, which NIOC's Chairman Eghbal at a press conference on August 27, 1966, characterized as "revolutionary," and predicted that it would open a new chapter in the history of the Iranian oil industry. Like the NIOC-AGIP joint-venture agreement, the new agreement is between two state enterprises, Iran's National Oil Company, NIOC, and the French-owned Entreprise de Recherches et d'Activités Pétrolières (ERAP). It is designed both to increase "benefits accruing to Iran" and to develop "the economic ties already existing between the two countries" (Preamble to the Agreement).

Under the agreement ERAP is to provide Iran with "capital, technical competence and management skills in an effort to expand the production and export of Iranian crude oil" (Preamble). More specifically, ERAP obligates itself to lend NIOC the money to cover the cost of conducting a geologic and seismic survey over a vast onshore area of approximately 200,000 square kilometers and a smaller offshore area of 20,000 square kilometers. Société Française des Pétroles d'Iran (SOFIRAN), a subsidiary of ERAP created for the purpose and registered under Iranian laws, will conduct the survey. At the end of a year on a basis of the knowledge which the survey yields, SOFIRAN will join with NIOC in selecting 20,000 square kilometers from the land area and 10,000 square kilometers from the offshore area for more intensive exploration. As exploration progresses both the onshore and the offshore areas will be further reduced until at the end of six years the land area remaining for exploitation will include only 5,000 square kilometers (Article 16) and at the end of seven years the offshore area will include only 3,333 square kilometers (Article 14).

ERAP obligates itself not only to lend NIOC without interest funds to cover the exploration costs, but if oil is discovered, to extend interest-bearing loans to cover the costs of operations until "the cash flow accruing to NIOC as a result of the operations . . . shall be sufficient to enable NIOC to provide the financing of appraisals and development operations" (Article 5). If SOFIRAN discovers no oil, ERAP never recovers its exploration loan. If SOFIRAN discovers oil in commercial quantities sufficient to insure repayment of the exploration expenses, NIOC is entitled to set aside as a national reserve 50 percent of the "discovered recoverable reserves." These shall be determined by "mutual agreement" separately for discovered fields in the offshore and onshore areas (Article 21). The other 50 percent of recoverable reserves will be brought into immediate exploitation. SOFIRAN will conduct production operations under a program worked out in agreement with NIOC. Although SOFIRAN, as the "General Contractor," manages the production enterprise, it does so only in close co-operation with NIOC and with Iranian personnel, if competent Iranians are available. All of the production equipment used by SOFIRAN and all of the oil it produces belong to NIOC. If SOFIRAN fails to comply with the exploration obligations within the time period specified in the contract, NIOC can terminate the agreement (Article 13, Sections 9 and 10) without any obligation to SOFIRAN. On the other hand, if SOFIRAN has complied with its obligations and concludes that conditions in the area selected preclude a reasonable chance of discovering oil in commercial quantities, at the end of three years in offshore explorations and at the end of four years in its onshore explorations, it can discontinue operations. If it does so, it pays NIOC one half of any unexpended balance that has been set aside for exploration (Article 13, Section 8), and the contract terminates.

If and when commercial production is established, NIOC is obliged throughout the twenty-five-year life of the contract to sell ERAP at cost plus 2 percent from 35 to 45 percent of the output (depending on the distance of the fields from seaboard) (Article 29, Section 1). From the monies thus realized NIOC will repay ERAP for the exploration expenses at the rate of one fifteenth of the total expenses annually or 10 U.S. cents for each barrel of crude bought, whichever is greater (Article 27, Section 1). On profits from oil so purchased ERAP is obliged to pay the Iranian government the 50 percent income tax calculated as the difference between what ERAP pays for the oil and the "realized price" of Persian Gulf oil as determined by a committee of four experts, two appointed by NIOC and two by ERAP in accordance with a procedure outlined in the agreement (Article 30, Section 1).

After commercial production has been established, NIOC will begin to repay with interest the development loans that ERAP has supplied. It will repay them in five years either in cash or from the proceeds from the sale of oil.[25] If NIOC elects to repay in oil, ERAP, acting as NIOC's broker, will sell a maximum of one million tons of crude oil each year for five years and will retain the proceeds calculated on a basis of "realized price" minus a brokerage commission of two percent. If this does not yield adequate sums, NIOC will make good the deficiency in cash.

In addition to selling NIOC's oil to recover its developmental loans, ERAP agrees to export to world markets for NIOC three million tons of crude each year during a five-year period and four million tons yearly during a second five-year period. For this oil it will pay NIOC on the basis of the "realized price" less a two percent brokerage commission. These sales are contingent on the Iranian government's using the proceeds to buy French equipment, products, or services as mutually agreed on by the two governments (Article 31).

Such are the major provisions of the new contract. In characterizing it as "revolutionary," Chairman Eghbal emphasized that it was essentially a "work contract" under which NIOC hired the services of a "general contractor" with funds initially supplied by ERAP. In comparing it with arrangements hitherto made for developing Iranian oil resources, he emphasized the superiority of the new contract both in its decision-making aspect and in the sharing of revenues. Under the Consortium Agreement the trading companies decide how much oil will be produced, make independently all important managerial decisions, and pay the government only 50 percent of profits as an income tax. Under the joint ventures, NIOC shares decision-making with its partners and the government recovers 75 percent of the profits, 50 percent as a tax and 25 percent as an investor. Under the new contract, NIOC makes the important decisions and the government according to Chairman Eghbal's calculations will receive from 89 to 91.5 percent of the benefits. In arriving at these figures, he takes account of the fact that NIOC will set aside as a national reserve 50 percent of all oil discovered, and that on the 35 to 45 percent of the balance to which ERAP is entitled (which represents from 17.5 to 22.5

25. "Each annuity shall comprise the repayment of loans and interest . . . computed at the rate of discount of Banque de France plus 2.5 percent." (Article 28, Section 5.)

percent of the total discoveries) ERAP will pay a 50 percent income tax, thus leaving ERAP with the profits derived from 8.75 to 11.25 percent of total oil discovered. Such a calculation does not have much meaning as a measure of profitability. Its chief defect is that it fails to take account of the fact that ERAP tax payments are based on realized prices rather than posted prices.

Thomas Stauffer of Harvard University, at the time engaged in a study of Middle East oil, in a provocative paper presented at the Sixth Arab Petroleum Congress meeting in Baghdad in March 1967, challenged Chairman Eghbal's calculations. Using a combination of assumptions and definitions and the technique of the discounted cash flow, he concluded the ERAP contract would yield Iran at best less than the conventional type of OPEC concession, resulting in a split of 45–55 in the profits, and that it represented a "new departure" in that it was equivalent to tax discounting. Stauffer's paper engendered much controversy. Khalid Shair, a graduate of the London School of Economics, with a Ph.D. from Cambridge, offered the most convincing counter-argument in a paper published as a supplement to the *Middle East Economic Survey* (May 5, 1967). Using the same technique but different assumptions, he concluded that the contract promises Iran about 65 percent of the profits.

How profitable the agreement will prove in fact depends, of course, on how much oil is found, at what cost, and what prices it will bring. But it represents a contract more nearly in accordance with Iran's desire to run its own business.

The Consortium Agreement Modified

THE NIOC-ERAP contract may eventually lead to an increase in oil shipments into world markets at NIOC's discretion. Negotiations recently concluded between Iran and the consortium will probably contribute to the same end. Although in 1965 consortium members produced almost three times as much oil in Iran as Anglo-Iranian produced in 1950 and paid the government more than eleven times the amount Anglo-Iranian had paid at the earlier date, government officials became disappointed with the consortium performance. Their dissatisfaction was compounded of numerous elements, three of which in essence are plainly discernible: The government believed that consortium members who had other Middle East and North African interests were pushing developments elsewhere at the expense of Iran. The government was disturbed over the consortium's failure to develop more rapidly the vast area which it held in idleness; it resented the gap between the consortium's potential production and actual output; it was growing restless about the power that the consortium exercised over Iran's economic welfare. The government needed additional revenues and foreign exchange to finance its economic development and defense programs and it regarded the consortium as an appropriate source of these revenues.

In the course of negotiations extending over many months, government officials developed these points publicly and privately and with increasing urgency.[1] Their demands took on more specific shape as their urgency grew. They wanted a commitment from the consortium that it would expand its 1966 crude production by 20 percent and double its production within five years. They wanted the consortium to supply NIOC with oil at favorable prices with which it could meet its rapidly growing committments in world markets. And they wanted the consortium to release substantial portions of its concession area for such use as NIOC wished to put them to.

In putting their claims before the consortium, government officials alleged unfair treatment of Iran for the benefit of younger producing companies. They called attention to the oil companies' failure to increase Iran's output as recommended by the Organization of Petroleum Exporting Countries while they had exceeded OPEC's recommended output for Saudi Arabia and Libya, and to the consortium forecast of an Iranian increase in output of only 9 to 10 percent in the face of OPEC's recommended increase of 17.5 percent. Dr. Manuchehr Eghbal, managing director of the government's National Iranian Oil Company, wrote the consortium on November 2, 1965, indicating that

OPEC Member Countries themselves have come to recognize Iran's special circumstances, and her rightful need for a higher oil income required for the execution of projects planned to develop the vast area and provide her large population with better living conditions . . . [and] they have agreed to allocate some 24% of the estimated increment of oil production over the previous year of the total OPEC area to this country. Such recognition by countries that are in effect vying with Iran for the sale of their oil in world markets is evidence of the firm basis upon which Iran's long-standing claims are founded, namely, that her special circumstances demand that the production of oil from the Agreement Area should be rapidly increased.

In pursuance of the discussions held at the meetings in London in September last we shall be glad therefore if timely arrangements are made for increasing the rate of export of oil so that in accordance with H.I.M. The Shah's wishes Iranian crude oil production in the year 1970 will reach the 200-million ton level.[2]

Iran Resented Consortium's Failure to Develop Its Vast Potential Oil-Bearing Lands

Apparently referring to the consortium's failure to develop more rapidly its large holdings, Prime Minister Hoveyda, commenting on the

1. Official documents detailing the negotiations in full are not available. This discussion is based partly on public documents, partly on reliable press accounts, and partly on the logic of relevant facts.
2. *Petroleum Intelligence Weekly,* November 29, 1965.

NIOC-ERAP agreement in a speech before the Iranian Majlis on October 22, 1966, stated,

> Oil is a national wealth belonging to Iran. Utilization of this wealth must take place in accordance with the needs of the country, capable of responding to its requirements. Oil must not be buried in the heart of the earth.

And he avowed:

> The growth and development of Iranian society must not be subjected to the fluctuations caused in production and export of oil as a result of unilateral decisions taken by foreign boards of directors behind closed doors and outside Iran, where part of oil utilization takes place.[3]

As early as the spring of 1965 Hoveyda had publicly expressed the same sentiments. Iran, he averred, cannot afford to allow its oil to lie idle in the ground. "NIOC's international commitments and the imperative need for funds to meet the huge development costs warrant the stepping-up of oil output."[4]

Iran's Need for Additional Revenues

The press comments and the public statements of government officials on the negotiations express a sense of urgency with regard to Iran's need for additional oil revenues. Iran's need for additional revenues has two aspects: its defense program and its economic development program. Its defense program for which it allotted $400 million in 1966 springs from a feeling of disillusionment regarding its membership in CENTRO and SEATO. Government officials have shown no reluctance to express publicly their growing belief in CENTRO's unreliability. This dissatisfaction apparently stems immediately from the West's failure to support Pakistan, a fellow CENTRO member, in its border conflicts with neutralist India. About this an Iranian government official is quoted by *Kayhan International* for November 22, 1966, as having said, "The trouble between Pakistan and India taught us a tremendous lesson." CENTRO's failure to back Pakistan showed that "these agreements are just pieces of paper."

In a less direct reference the shah in a speech to Majlis deputies on March 1, 1966, stated, "There have been developments in the world recently which have been an exemplary lesson for us. The lesson is that Iran cannot surrender its destiny to the whims of foreigners even if they be

3. *Kayhan International,* an English-language Teheran daily, reported in full Hoveyda's speech in its October 22, 1966, issue.
4. *Ibid.,* May 20, 1965.

very close friends."[5] And in November 1966, Hoveyda stated in an interview with the Associated Press, "We have seen that alliances do not work—your friends just send you telegrams of sympathy."[6]

Soviet-Iranian Accord

It is not the Soviet Union that Iran now fears. In recent months amity and good feeling have replaced hostility and fear in Iran-Soviet relations. Mutual confidence has expressed itself in a general agreement for economic and technical co-operation between the two countries—documents covering which were exchanged between the foreign ministers of Iran and the Soviet Union during the shah's ten-day visit to Russia in the summer of 1965 and were ratified by the Iranian Senate on February 23, 1966[7]—and in a number of special arrangements. The most important was an agreement signed on October 6, 1965, which provided for the extension of Russian credit in building a "steel mill complex," machine-tool plants, and a gas pipeline from Iran to Russia.[8]

Iran Fears Trouble in the Persian Gulf

Developments in the Middle East rather than Soviet-Iran relations are the cause of Iran's uneasiness. As Prime Minister Hoveyda expressed it in an Associated Press interview, "There is potential danger in the Middle East."[9] And more particularly it was in the oil-rich Persian Gulf area that he feared trouble. The shah in his spring speech to Majlis deputies had voiced a similar concern. In a thinly veiled reference to Nasser he lamented that Moslems were destroying other Moslems with tanks, guns, and mortars, "that those who had been defeated in their own lands were trying to find a vacuum somewhere," and he was determined that Iran would defend itself against such a search.

If not a single fish could be found in the Persian Gulf, nor if there was not a single oil well or no commercial ship passed through and if Khuzistan and south

5. *Ibid.*, March 1, 1966.
6. *Ibid.*, November 22, 1966.
7. *Ibid.*, July 4, 1965, and February 24, 1966.
8. *Ibid.*, October 6, 1965.
9. *Ibid.*, November 22, 1966.

Iran and our oil resources were nothing but wasteland, it would nevertheless be our duty to defend every inch of that territory.[10]

He expressed similar sentiments at a joint meeting of the Senate and Majlis in October 1966, when he said,

in spite of the friendships and pacts that exist, we must before anything else, rely on ourselves.

It is in the pursuance of this principle that we have unfortunately no alternative but to acquire the necessary equipment from any source which is in our national interest in order to safeguard our independence and territorial integrity. . . .

We are not prepared to be caught unawares, God forbid, some day.[11]

To prepare against such a contingency is costing money.

Iran Needs Funds for Its Development Program

It is Iran's economic development program that looms largest in its search for additional revenue. A brief review of this program will aid in understanding why Iran demanded more favorable terms from the consortium and eventually threatened unilateral action to obtain them.

The virtual disappearance of its oil revenues occasioned by its nationalization of the industry in 1951 brought Iran's first seven-year plan of economic development, adopted in 1949, to an abrupt halt. Settlement of the oil dispute injected new life into the plan organization and in 1955 it promulgated a second seven-year plan to end in 1962. The second plan, which involved total expenditures of more than a billion dollars, although falling short of its goal, made significant progress. It aided many small municipal development projects, launched a Tennessee Valley Authority–type project for the province of Khuzistan to be conducted by the Development and Resources Corporation under David Lilienthal's management, constructed extensive irrigation systems, and began several smaller regional development projects. Projects not completed by 1962 were continued under a third plan extending from September 1962 to March 1968. Oil revenues provided most of the money for these development projects. For the first six months of the third plan the government allocated 55 percent of its income from oil to meet the cost of its development program and

10. *Ibid.*, March 2, 1966.
11. *Ibid.*, October 8, 1966.

stepped up the allocation annually by 5 percent until by 1967–68 oil revenues were scheduled to provide 80 percent of the cost.

The Shah's Reform Program

Meanwhile under the shah's leadership Iran has been conducting its own "Great Society" program. The shah began the distribution of crown lands to tenants in 1951 and by 1962 when the program was completed he had distributed a total of 200,000 hectares. The shah's voluntary program having met with little response from the other great landowners, parliament in 1960 enacted an agrarian reform statute limiting land holdings to 400 hectares of irrigable land and 800 hectares for dry farming. A 1962 statute confines a landlord's property to a single village. The government has financed this program with 15-year bonds. Speaking in October 1966, the shah, while expressing satisfaction with the pace of the land reform program and characterizing the land reform law as having "earned the praise and respect of the world," indicated his determination to modernize completely Iranian agriculture and maximize the per hectare yield of the land.[12]

He also declared war on illiteracy and disease, particularly in the rural areas, and has established a literacy corps, a health corps, and a development corps.[13]

Iran Speeds Up Its Development Program

After the election of 1964, Iran's development program assumed a broader aspect, took on a new tempo, and acquired a new spirit. In the national election of that year the so-called "Progressive Center," the forerunner of a new political party, the Iran Novin party, obtained a majority of parliamentary seats. To carry on his "White Revolution," a six-point reform program which he had submitted to a national referendum in 1963,[14] the shah appointed as prime minister Hassan Ali Mansur, one of the founders and leaders of the new party. Mansur was representative of

12. Address before a joint meeting of the Senate and Majlis at the inauguration of the 1966–67 legislative session, *Idem.*

13. *Ibid.*, October 5, 1965.

14. For the shah's description of this program, see the Beirut *Daily Star*, March 19, 1964.

Iran's younger intellectuals who had thrown their influence behind the shah's program. He reflected the spirit of the new administration in a stirring address before the Majlis on November 24, 1964: "Iran," he said, "is now on the threshold of an all-embracing economic and social revolution." Its goal is to build a new Iran and, befitting this goal, the revolution needs a banner under which the country can unite. He appealed to all Iranians to join the government in a "National Crusade for Progress" on which they could embark with a religious zeal appropriate to its lofty moral and ethical aims.[15]

When a young religious fanatic, Muhammad Bokharai, assassinated Mansur in January 1965, the shah, undeterred by "the reactionary opposition of the Nationalists and the . . . obscurantism of the Shia clergy"[16] manifested his determination to carry out his program of modernization by appointing as Mansur's successor his intimate colleague and like-minded cabinet member, Amir Abbas Hoveyda.

Hoveyda Carries On

Hoveyda brought to his task "a keen, analytical intellect, an openmindedness, a methodical approach to problems and wide administrative experience."[17] Hoveyda in presenting his cabinet to the Senate on February 1, 1965, promised to "crush all short-sighted elements" and to continue the reform program initiated by his predecessor. "The progressive party program of the Mansur government is the program of this government." Characterizing Mansur as the "outstanding symbol" of the shah's bloodless revolution, he promised to strive unceasingly to attain its objectives in the "Spirit of Mansur."[18] The reform government, which had attracted a group of dedicated and able administrators with a pragmatic rather than doctrinaire approach to Iran's problems, inaugurated a new foreign policy and in the language of the conservative *Economist Intelligence Unit* sparked a "major leap forward in the over-all standard of living at home."[19]

The shah characterized the foreign policy as an "independent national policy" based on "respect of [for the] Human Rights Charter, peaceful

15. *Kayhan International*, November 25, 1964.
16. New York *Times*, International Edition, February 3, 1965.
17. Kazen Zarnejar in *Kayhan International*, January 28, 1965.
18. *Kayhan International*, February 2, 1965.
19. Annual Supplement, *Economist Intelligence Unit, Quarterly Economic Review; Iran.*

co-existence and co-operation among nations with different political regimes." It has expressed itself in a series of economic and commercial treaties with Communist countries. At the same time it has led to closer economic ties with Iran's neighbors, Turkey and Pakistan, which joined Iran in organizing the "Regional Corporation for Development" with Fuad Rouhani as its executive secretary.

The New Development Program

It is beyond the scope of this study to inquire into the economic validity of Iran's "leap forward," which in the industrial sector is directed primarily toward the development of capital intensive industries. But it is relevant to point out that it involves what a contemporary observer characterized as a "massive investment program."[20] The program includes 700–800 kilometers of railroads; the steel complex in the planning and financing of which Soviet Russia is playing a significant role; a $100 million petrochemical project in which Iran's recently organized National Petrochemical Company (NPC) joined as equal partner with the Allied Chemical Corporation of the United States to produce fertilizers, plastics, and other chemicals having a petroleum base; a second petrochemical partnership with B. F. Goodrich in which NPC supplies 74 percent of the capital, a portion of which it will get from the U.S. Export-Import Bank; a 50–50 partnership between NPC and Amoco International S.A., a subsidiary of the Standard Oil Company of Indiana, to establish plants for producing sulphur and liquid petroleum gas; a $450 million gas pipeline extending from the southern field to Iran's northern border where it will deliver gas to Soviet Russia in repayment of the Russian loan for financing the steel complex, supplying en route gas for Iran's major consuming centers; and a plant for making pipe formed with the cooperation of Czechoslovakia.[21]

That the petrochemical projects are well planned is suggested by the experience of the private partners which are participating, the fact that they will utilize two of the world's lowest-cost raw materials—gas and oil—and will supply products which the Eastern Hemisphere is apt to use in progressively larger amounts.

20. Shapour Nemizee in *Kayhan International,* October 10, 1965.
21. *Kayhan International,* October 26, 1965; October 28, 1965; November 8, 1965; February 15, 1966; June 2, 1966.

Iran Looks to the Consortium for Increased Oil Revenues

Although many of its projects are in the planning stage and some will be financed in large part by foreign credit, Iran's development program has occasioned a shortage of government revenues and of foreign exchange. With economic and political conditions as they were, it is not surprizing that government officials looked to the consortium to make good the deficiency. In a "moving speech" Senator Ashraf Ahmadi recently characterized increased oil revenues as a "matter of life and death," essential to the realization of Iran's "unprecedented opportunities."[22]

Iran's dissatisfaction with the consortium's performance was of long standing. The shah as early as 1960 had indicated his concern about the pace of Iran's growth in output and had declared that Iran should receive "at least one half—at the very least" of each year's growth in Middle East output until it has recaptured its former premier place among Middle East producers.[23]

Such complaints were not made the subject of negotiations until some years later. At the outset negotiations for a consortium new deal took the form of informal talks between NIOC representatives and consortium officials. Government officials meanwhile were giving public support to what NIOC was striving to attain through private negotiations. Prime Minister Hoveyda in a statement to the Iranian parliament early in March 1966 urged the necessity of the consortium's increasing its exports by 20 percent. At about the same time the shah himself had pronounced the consortium's 1965 production increase as unsatisfactory and, according to the London *Financial Times,* had hinted that unless Iran's 1966 level of output met its needs, Iran might be forced to seek new sources for her imports.[24]

In his opening remarks at a meeting between representatives of NIOC and the consortium in London in October 1966, Dr. Eghbal spoke more urgently of the need to increase Iran's production to enable her to carry out

22. *Ibid.,* November 1, 1966.
23. Interview with Wanda M. Jablonski, *Petroleum Week,* December 9, 1960, p. 14.
24. The *Financial Times* correspondent interpreted the shah's remarks as a veiled threat to turn from the United States to the Soviet Union for its arms supplies. *Financial Times,* March 11, 1966.

"her economic and social development plans" and to achieve the "nation's aspirations."

Time and again we have stressed the need for action on your part to implement promises given to bring about the realization of Iran's cherished hopes for regaining her rightful position among the producing countries of the Middle East.

He concluded on a more ominous note:

The restriction that is being imposed upon the normal growth of production in Iran in order to enable the fulfillment of other plans elsewhere is most deplorable and I feel it my duty to remind you . . . that overlooking Iran's vital interests in this regard is causing grave concern to my country.[25]

About the same time the shah in an interview with Frank Giles, foreign editor of the London *Sunday Times,* published October 23, 1966, in urging prompt action by the consortium in meeting Iranian demands said,

It comes down to the point when we have to decide whether the oil companies, which are, after all, only commercial undertakings, have the right to refuse to extract from our soil our own resources. This cannot be accepted. And what if no settlement of this difference is possible? We may in that case have to act unilaterally—that is, to depart from our agreement with the Consortium.

On October 21, 1966, Premier Hoveyda sounded a similar warning in presenting a bill to the Majlis covering the NIOC-ERAP agreement. In characterizing oil as "national wealth belonging to Iran" and emphasizing the importance of using it to meet national needs, he said,

Obviously, the Government cannot remain idle even for a moment towards a question which is linked to our national interests. The necessary measures will be adopted and will be brought to the attention of the respected National Assembly.[26]

On October 26 the shah in remarks delivered in a ceremony celebrating his 47th birthday made even more pointed reference to the possibility of Iran's acting unilaterally in modifying the Consortium Agreement. In doing so he contrasted the present situation to that of 1951 when Iran nationalized its oil industry. The earlier movement he characterized as

disingenuous and foreign-directed, leading to the shutdown in the oil industry. . . . It is not the same situation now. The days of getting directions from foreigners and of saying one thing and of doing another are over. We have an open hand. Our case is perfectly legitimate and logical. No one can deny us the right of utilizing our resources.[27]

25. *MEES,* October 14, 1966, Supplement. This gives the full text of Dr. Eghbal's address.

26. *Kayhan International,* October 22, 1966.

27. *Ibid.,* October 27, 1966.

While the shah and the prime minister were suggesting the likelihood of unilateral action, Dr. Eghbal was reported to be in London locked in what the press characterized as "a final confrontation with Consortium leaders."[28]

Press Comments on the Controversy

On October 29 the press reported that Iran had given the consortium one month in which to meet its demands. During the interim the Iranian press, the press of Western Europe, and the radio Voice of Moscow had taken note of the controversy. They spoke with some feeling. Moscow showed no restraint in condemning the "cartels" which, Moscow alleged, had built up vast empires by exploiting the Middle East countries and by robbing the consumers. The cartel members, Moscow charged, sold oil which cost them only $1.50 a ton to produce for twenty times as much.[29]

The Iranian press vigorously supported the shah's threat to take unilateral action, but on the whole it lacked the acerbity with which Middle East commentators usually attack the oil companies. It echoed the arguments that government officials had mobilized in supporting their demands and reflected the same determination that Iran's oil should be utilized for Iran's benefit and the same confidence in the "ultimate triumph of justice."[30]

The British press manifested understanding of Iranian demands but recommended caution. In its October 29 editorial, the London *Times* warned that Iran

seems perilously close to producing a threat like Law 80 introduced by Kassim's government in Iraq. This would produce a situation that the Consortium could not possibly accept and which might therefore precipitate another breach as serious as the 1951 Abadan crisis.

No one would deny that the Iranian Government should pursue the most progressive developmental plan consistent with the resources available to the country. But Iran has already broken two agreements. If it were now unilaterally to take action which, for example, included the confiscation of a proven oil field, it would do immense damage to its credit worthiness with oil companies, bankers, and with other governments on whose good will its future must partially depend. One of the great lessons of Middle East oil is that if in a dispute either side threatens extreme measures, then everyone suffers.

28. *Ibid.*, October 23, 1966.
29. *Ibid.*, October 30, 1966.
30. *MEES,* November 18, 1966, gives a "Comment Round-up on Iran-Consortium Dispute."

The Economist of the same date reflected a similar note of caution and compromise:

Both sides stand to lose so much that no one can be sure how far Iran is actually prepared to go at this stage . . . ultimately each side knows that oil sold in world markets is their only interest.

The Consortium's Position

Meanwhile the consortium had remained rather quiet concerning its position in the controversy. Its members produce oil in many places. Its leading members have other important interests in the Middle East. Their business is to produce and sell oil. Except as they are forced to bow to political pressure or are limited by the terms of their concession contracts, when demand justifies, their interest is in increasing the output of low-cost oil. But they are aware of the relative inelasticity of the demand for petroleum and its products and realize that competition in producing and selling oil may lower its price without appreciably increasing its consumption. They also realize that to sell oil or its products they must meet the prices of competing offers. All of this makes them cautious in their output program. Consortium General Manager J. A. Warder reflected this caution in a statement to *Kayhan International* reproduced in its issue of March 8, 1965. He indicated that any large increase in consortium output was practically impossible without tapping new markets. He emphasized that in the face of tumbling crude prices Iran could not hope "to wedge a stronger stake in overseas markets."

Consortium members recognize that except as demand increases or new markets develop, an increase in any one Middle East country's output must be at the expense of another's and they also know that a concession made to one Middle East country quickly becomes a demand of all Middle East countries. They are slow therefore to grant any country's demands and frequently wait until they assume crisis proportions before effecting a compromise. That politics influences their decision can only make them unhappy; but occasionally they must bow to it.

Talks between the consortium and NIOC had been conducted intermittently for more than a year when the parties scheduled another round of talks in London in the fall of 1966. They later transferred them to Teheran where a team of four men, including J. Addison, general manager of Iranian Oil Participants, Howard Page, seasoned negotiator of the Standard

Oil Company of New Jersey, George Parkhurst of the Standard Oil Company of California, and David Steel of British Petroleum, Ltd., assembled on November 22, 1966. What Addison told pressmen was a "purely social visit" began with an audience with the shah and extended through six days of what the press described as "secret negotiations."[31] Although Page alleged that he had come with "nothing" in his briefcase, he evidently had something under his hat. At any rate a government press release issued at the end of the conference indicated that remarkable progress had been made and an agreement in principle had been reached. On December 15 the parties made an announcement regarding the terms of an agreement.

Terms of the Settlement

The agreement gave Iran much of what it wanted without immediately threatening any serious disturbance in world markets. Although under the 1954 concession agreement the consortium was not obligated to release any of the concession area before 1979 and then only 20 percent of it, the consortium agreed to release within the next three months 25 percent outside of the area of its presently producing fields, the area to be determined by mutual discussion. It agreed to meet a minimum schedule of output during the next two years said to represent an annual increase in output of 13 or 14 percent, and, perhaps more important, it agreed to sell NIOC 20 million tons of crude oil over a five-year period. NIOC agreed to dispose of this oil in barter transactions with specified Eastern European countries (where consortium members do not sell) for refining within the countries to which it is delivered. The parties did not disclose the price which NIOC would pay for the oil it buys but have characterized it as "interesting."[32]

As suggested at the outset of this discussion, the new agreement promises to increase greatly the amount of disposable crude oil in the hands of NIOC. NIOC will be free to enter into new agreements for the exploration and, if oil is found, for the exploitation of the released concession areas. How much oil this will eventually bring on to world markets, and when, only the future will divulge. But NIOC immediately got significant quantities of low-cost oil. It is true that the consortium has endeavored

31. *Kayhan International,* November 22, 1968.
32. See *MEES,* Supplement, December 9, 1966, "The Iran-Consortium Settlement," and *P.I.W.,* December 19, 1966.

to protect its markets by limiting those of the NIOC. But before the agreement NIOC had entered into long-term barter contracts for the sale of oil in Eastern European countries. The new agreement freed a corresponding amount of oil that NIOC obtains from its joint ventures for sale in other markets.

This is the first time that any major Middle Eastern concessionaire agreed to sell oil for export to any national oil company. NIOC is the oldest and most experienced national oil company in the Middle East and in recent years has been aggressively expanding its markets. Virtually all of the major producing countries have organized national companies, perhaps as a first step towards the eventual nationalization of their oil industries, but immediately as instruments for supplying their domestic market with petroleum products and gaining experience in the integrated operations by which petroleum products are brought to the market. The NIOC-Consortium arrangement has set a precedent which other state companies may wish to emulate. The competition which it may eventually engender augurs ill for the co-operation they have introduced through OPEC.

NIOC Enters World Markets

NIOC did not await the new consortium agreement to become an international marketer of petroleum. With SIRIP and IPAC producing oil in commercial quantities and other joint ventures under way, it looked for foreign outlets for the oil and oil products it had and expected to get. This objective involved not only the efforts of NIOC's top management but also of the highest government officials. The shah, who has dedicated himself to the betterment of his people and who is determined to recapture something of Persia's greatness, regards Iran's oil resources as an instrument in achieving those ends. To utilize the instrument effectively called for a reorientation of Iran's foreign policies. As previously indicated, the shah, having lost faith in reliance on the West for his country's security, is determined to transform co-existence from a symbol into a program. He regards peace with the world as essential to progress at home. To promote it he has conducted a series of goodwill tours primarily to those Eastern countries against which the West is aligned in an ideological war. As his program is broad, so have been his travels. In the three years from 1964 through 1966 the shah or other high-ranking government officials headed teams that visited Russia, Czechoslovakia, Yugoslavia, Bulgaria, Hungary, Austria, Poland, Afghanistan, Turkey, Pakistan, India, Argentina, Morocco, Formosa, and Japan. These trips had as their dual purpose the strengthening of cultural and trade ties. They have resulted in long-term trade pacts with the Soviet Union, Roumania, Czechoslovakia, Hungary, and Poland. While the precise role oil will play in this program is not yet clear, it seems likely to be an important one. The shah indicated as long ago as 1957 his intent to make of NIOC a major international oil company participating in oil markets throughout the world. Because Western European markets were fully occupied by the well-established international companies and the American markets were virtually closed by United States restrictions on imports, NIOC has turned to Asiatic, African, South American, and Eastern European markets.

NICO Wins Afghanistan and India Contracts

In invading foreign markets NIOC began with a neighbor. In the spring of 1964 Dr. Eghbal, NIOC's managing director, went with a six-man team to Kabul, Afghanistan, to confer with the Afghan minister of industries and mines about renewing a contract under which NIOC was supplying Afghanistan with about one-third of its annual requirement of oil products. This visit eventually resulted in the renewal of the contract and a new agreement under which NIOC assumed responsibility for the refueling of all aircraft using the Kabul and Kandahar international airports. This contract was a modest one but, because in obtaining it NIOC met the competition of both private companies and Soviet Russia, it was an important one.

More important was a contract won by NIOC and its joint-venture partner, Pan American (through its subsidiary American International) in competition with fourteen international companies to construct for the Indian government a 50,000 barrel–daily refinery at Madras, India. The refinery when completed will cost about $50 million. Under this contract NIOC and American International will lend the Indian government $35 million to aid in the refinery construction. An operating company will be jointly owned by NIOC and American International (each with 14 percent of the equity capital) and the Indian government (72 percent). Over a ten-year period NIOC and American International will supply the refinery with between 150 million and 185 million barrels of crude (at a total cost of approximately $200 million) from their joint venture in the Persian Gulf.

NIOC Looks for Markets in Eastern Europe

It is in the Eastern European market, however, until recently exclusively pre-empted by Russia, that NIOC has found its most attractive opportunity. Several factors have made it so. Russia customarily has sold oil abroad under a two-price system. In selling to buyers in non-Communist countries Russia had to meet the competition of a large number of sellers. In selling to its Eastern European satellites Russia has been the sole supplier. It has adjusted its prices accordingly. Under its two-price system Russia has been selling to its Eastern European satellites at from 50 to 80 percent more than to Western countries. Soviet trade statistics for 1967 as reported by the Radio Liberty Committee indicate that Soviet export prices at the frontier or port for oil shipped to Czechoslovakia averaged 15.4 rubles per metric ton; to Poland 14.6 rubles; to Hungary 15.3 rubles. During the same period export prices for oil to France averaged 8.5 rubles, to Italy 9.6 rubles, to Japan 9.3 rubles, and to West Germany 9.2 rubles.

This practice has been growing less satisfactory to both Russia and her satellites.

At least one Soviet economist, I. Dudinskii, writing in the Russian economic journal *Voprosy Ekonomiki,* pointed out that Russia's sales of crude in satellite markets are by no means all to the good economically. Thus, Russian exports of raw materials to Eastern Europe continue to exceed in value the exports of industrial products, while the Eastern European countries have been increasing their exports of finished industrial products. Yet in the international socialist market raw material prices tend to be unduly low. Thus, per unit of foreign exchange earned in socialist markets, the Soviet Union invests from five to eight times as much in raw materials as in heavy machinery exports. For oil the ratio is even higher. Moreover, because the heavy machinery imported from Eastern Europe is frequently of inferior quality, Russia, he avers, has to supplement it with better machinery bought from capitalist countries.[33]

With similar implications Poland's leading Communist party weekly, *Polityka,* has questioned the reliance of Eastern European countries so

33. *Voprosy Economiki,* April 1966, pp. 84–94.

exclusively on coal as an energy resource. Coal currently supplies about 90 percent of Poland's energy needs but the demand for gasoline and diesel oil is expanding rapidly as motor transportation develops. By mutual accord the Eastern European countries are committing themselves to less reliance on Russia for oil as an energy resource. As their demand for gasoline and diesel oil is expanding as motor transportation develops, they are confronted by a growing reluctance of Russia, whose oil consumption is increasing and whose needs for foreign exchange are growing, to supply their demands. Both Russia and the Eastern countries apparently welcome their looking for new sources of supply.[34]

Iran Quickly Seizes the Opportunity Eastern Markets Offer

Iran is the first Middle East oil-producing country to recognize the opportunity this situation has created. The shah's recent visits to Eastern European countries, as indicated above, have had a double objective. While emphasizing the importance of peaceful coexistence among countries of conflicting ideologies in an atomic age and Iran's interest in closer cultural ties with Communist-bloc countries, he has laid the basis for trade treaties which may open new and promising markets for NIOC's oil.

One of the first concrete arrangements to grow out of Iran's closer ties with Soviet satellites was the economic co-operation agreement between Iran and Roumania signed on October 25, 1965, by the Iranian minister of economy, and the Roumanian undersecretary of trade in the presence of the prime ministers of both countries. Under this agreement Iran will export to Roumania over a ten-year period about $100 million worth of oil and import machinery and goods required in Iran's development program. The agreement contemplates continuing close economic ties between the two countries, to insure which they created a joint ministerial committee in August 1966.

The commitments of other satellite countries to buy Iranian oil are less precise. The Czechoslovakian and Polish trade pacts, however, have provided Iran with long-term credits at low interest rates, and representatives of both countries have expressed their intent to buy Iranian oil. So also have representatives of Bulgaria and Yugoslavia, and in the fall of 1966 representatives of Hungary and Iran were conducting talks looking towards

34. Robert Campbell, *The Economics of Soviet Oil and Gas,* (Baltimore: Johns Hopkins Press, 1968), pp. 243–249.

Hungary's granting long-term credits to Iran and closer trade ties between the two countries.

These developments seem destined to open up markets for NIOC that are likely to grow in importance. The Eastern European countries want more and cheaper oil, and they want to become less dependent upon the Soviet Union in getting it. Apparently they would prefer to buy from state enterprises rather than from the major international oil companies, which they have been taught to regard as profiteering monopolists. And they would prefer to barter than to pay cash. Iran wants both to sell its oil and to buy capital goods that will hasten its industrialization. Their interests are complementary.

NIOC's Purchase of Consortium Oil Will Intensify Competition

It was primarily to supply the Eastern European market without using its joint-venture oil that led Iran to demand cheap oil of the consortium. Although the consortium has tried to protect its members' markets by restricting NIOC's freedom in selling oil, its sale to NIOC increases NIOC's capacity to bring additional oil on to already crowded world markets. NIOC remains free to sell its joint-venture oil where it can. It seems determined to sell it. In selling its oil in competition with established international marketers, it must make attractive offers. With the world's producing capacity in excess of world consumption, NIOC, like other companies, must sell oil below its posted prices. Although in private transactions sellers try to keep their prices private, in important transactions, particularly where competitive bidding is involved, prices have a way of becoming public. When commercial production was established in their Darius field, Pan Am and NIOC posted a price of $1.63 a barrel. This was approximately the Kuwait Oil Company's Persian Gulf posted price for crude of similar quality. According to trade reports, NIOC and American International will supply the Madras refinery Darius crude at about $1.35 a barrel. Some rumors say $1.32. This indicates a discount of about 20 percent.

Meanwhile, NIOC has been trying aggressively to break into the Japanese market, where buyers over the past several years have been getting their oil for substantially less than Persian Gulf posted prices. After a month's visit to Japan and Nationalist China early in 1966, NIOC's managing director was sufficiently encouraged to report that "NIOC has now emerged as a company of international status." The Madras refinery and the Afghan contracts, he averred, "are only the beginning of NIOC's expansion in foreign countries, an expansion that will include both refining and marketing."[35]

What effect this and similar developments in other Middle East oil countries will have ultimately on the structure of the international oil industry, its pricing behavior, and the effectiveness of OPEC cannot be precisely determined. The significance of the trend can be seen more clearly.

The major Middle East oil-producing countries have as their ultimate objective the control of their oil resources. They wish to recapture the sovereignty over oil which they surrendered in their original concessions. Given political stability within their borders, none is apt to repeat the mistake Mossadegh made in 1951. All realize that the most likely result of nationalization now would be the loss of foreign markets for their oil. Saudi Arabia, Iraq, and Kuwait have all followed the example that Iran set and have organized national oil companies.

35. *News Letter,* June 1966.

9

Iraq Seeks a New Deal

ALTHOUGH the incorporation of the 50-50 profit split into the Middle East oil concessions increased greatly government revenues, it did not bring peace and harmony between the companies and the countries. It brought dissension and controversy. Before the adoption of this principle Iran, Iraq, Saudi Arabia, and Kuwait all received their major payments from their concessionaires as royalties. Each country was in the position of a landlord who had entered into a long-term rental contract for the use of his land, under which his rent varied not with the price of the product but with the yield of the land. Receiving a fixed royalty in shillings per ton,[1] the

1. The royalty provisions varied slightly in the different concessions and from time to time. See Article 10 of the Iraq Petroleum Company agreement; Article 11 of the Basrah Petroleum Company agreement and of the British Oil Development Company agreement; Article 3 of the Kuwait Oil Company agreement; and Article 14 of the Arabian American Oil Company agreement. The IPC, BPC, and BOD agreements provide for four shillings per ton royalty for twenty years after the completion of a pipeline to a port for exports. For the next ten years the royalty was to be increased or reduced in any year depending on profits of the company but were to be no less than two shillings per ton nor more than six shillings. The Kuwait agreement provided for a royalty of three rupees. In 1950 this was changed to 12½ percent of the sterling equivalent of the posted price. The Anglo-Persian agreement provided for a royalty of four shillings per ton with a guaranteed minimum and nine shillings per ton as commuted taxes. It also provided for payment to the government of 20 percent of any distribution which the company made to ordinary stockholders in excess of £670,000 in any year. Gordon H. Barrows (P.O.B. 1591, Grand Central Station, New York) has reproduced the basic oil laws and concession contracts of the Middle East countries in several volumes entitled *Petroleum Legislation*.

government was concerned not with the cost of producing oil nor with the prices at which it was sold but with the amount of oil produced. This determined the annual rental. The government's position was as that of a rentier, not an entrepreneur.

Royalties Combined with Taxes

The 50-50 principle changed the relationship between the governments and their concessionaires. Ostensibly the concessions as modified still recognized the rentier relationship. They did not abandon the royalty principle but they compromised it by combining the royalty with what on its face was an income tax, calculated to guarantee the government an income which together with the royalty payments amounted to 50 percent of the concessionaire's net profits. The result is a hybrid product—part royalty, which in customary accounting practice is regarded by the entrepreneur as a cost, and nominally part income tax, which according to traditional economic theory is regarded as a distribution of earnings.

With the concessions so modified, the governments of the producing countries necessarily acquired a vital interest in what had hitherto been an exclusive entrepreneural function. Their revenues now depended on both the price at which oil was sold and the cost of producing it. As custodian of its country's welfare, no government could remain indifferent to the concessionaire's cost-accounting methods nor its decisions on oil prices. A corporate income tax has no meaning if deductions from gross income are left entirely to the party taxed.

Aramco Tries to Protect Its Independence in Costing

Despite the new relationship that the 50-50 principle inaugurated between the governments and their concessionaires, Aramco endeavored to protect itself against unwarranted governmental interference in its cost accounting by writing protective clauses into the agreement. The Aramco agreement confirms the company's "policy of conducting its operations in accordance with first-class oil field practise and its accounting in accordance with generally recognized standards" and the Saudi government confirms its "confidence in the Management of Aramco in conducting Aramco's operations."[2] Such mutual expression of confidence was a shaky reed on which to tie a policy of nongovernmental interference.

Cost Accounting Provisions of the IPC Agreement

The 1952 agreement between the Iraqi government and the Iraq Petroleum Company group (Iraq Petroleum Company, Ltd., Mosul Petroleum Company, Ltd., and the Basrah Petroleum Company, Ltd.; hereafter IPC, MPC and BPC when referred to separately, and the "companies" or IPC when referred to collectively) apparently was aimed at protecting the companies' entrepreneural functions and at the same time eliminating unnecessary controversy over cost-accounting matters by defining costs and fixing the rate at which capital assets should be depreciated.[3] Article I of the agreement states,

"Actual costs" means the aggregate costs determined by sound and consistent accounting methods fairly and properly attributable to the operation of the companies in Iraq in respect to
(i) operating expenses and overheads and
(ii) depreciation of all physical assets in Iraq
at the rate of ten per centum per annum and amortization of all other capital expenditure in Iraq at the rate of five per centum per annum until such assets and expenditures are fully written off.

These specific definitions give content to the general agreement but leave plenty of room for misunderstanding and controversy.

The agreement goes a step further in defining cost. Article 9, paragraph (b), subparagraph (iv), specifies the companies' "actual costs" for 1951 as twenty-three shillings a ton and for 1952 as seventeen shillings and sixpence. Thereafter "actual costs" are tentatively set at thirteen shillings, which is also designated as "the fixed costs." When actual costs in any year differ from fixed costs by more then 10 percent, actual costs shall apply. Actual costs are thus recognized as flexible, subject to determination in the first instance by the companies under the principle laid down in Article 1, but subject to agreement with the government. If agreement fails to be reached, costs become subject to arbitration.

The arbitration provisions of the 1952 agreement are all-inclusive.

2. Paragraph 8 of the agreement concluded December 30, 1950.
3. The Iraqi government concluded a single agreement on February 3, 1952, with IPC, MPC, and BPC, all of which have a common ownership; 50-50 profit sharing was made applicable to IPC's operations for 1951 and to MPC's and BPC's as soon as exports began.

Article 13, paragraph (c) provides that "If any doubt, difference, or dispute shall arise between the Government and the Companies concerning the interpretation or execution" of the contract that cannot be settled by agreement, they shall be settled by arbitration in the manner prescribed in the contract.

The Saudi-Aramco agreement with its vague and general provisions on cost and the Iraq-IPC agreement with its more specific provisions have both been followed by years of controversy. The Iraq controversy is of particular importance to anyone concerned with the political and economic aspects of Middle East oil. It deepened, widened, and increased in bitterness over the years and in April 1961 led to the government's ordering the companies to discontinue all exploration work. Eight months later the government unilaterally reclaimed 99.5 percent of the concession's acreage on which the companies' rights were not to have expired for another half century.

IPC's Corporate Status Raises Accounting Problems

The corporate relationship between IPC and its sister companies, MPC and BPC, and their owners gives a special significance to the government's interest in cost. IPC is not a profit organization. It sells its oil to its parents at cost plus a shilling a ton. To enable IPC to carry out its production program, the parent companies contribute capital in proportion to their ownership. The parents' profits are dependent on the difference between IPC's per barrel cost (which is their buying price) and the price at which they sell the oil less the cost of selling it. As long as the government's revenue was based on the tonnage sold, whether a cost fell on IPC as a producer or on the parent companies as marketers was of no concern to the government and of relatively little concern to the parent companies or IPC. The impact of shifting an accounting cost from the parent company to the operating company fell on the amount of capital which the parents would have to provide IPC in carrying out the production program which they had devised for it rather than on the profits of the company. A cost shifted backward would be covered in the price the parents paid for their oil. A cost shifted forward would lessen IPC's need for working capital. Where costs are placed makes no difference in the ultimate profits of a company whose operations are fully integrated.

Nevertheless, both economic and accounting theory afford a basis for properly allocating costs and determining their amount. Costs are the accounting equivalent of economic functions. If the functions are performed, their costs must be met. Good accounting procedure requires that they be met by the agent performing the function and that they reflect the market value of the function performed. These principles may be ignored without serious consequences where a succession of operations is performed by an integrated ownership, although ignoring them tends to conceal the profitability of the specific stages of the integrated operations.

As indicated, these principles were of no concern to the Iraq government so long as its oil revenues were calculated on a tonnage basis. After the 50-50 profit-sharing agreement came into effect, whether a cost was assessed against IPC or its parent companies became a proper and indeed a vital concern of the government. To assess a cost encountered by its parent companies against IPC decreased the annual revenues that the parents were obliged to pay to the government.

Discounts Reflected in IPC's Selling Price

For the purpose of calculating profits that the companies were to share with the government, the 1952 agreement specified the border value of 36° API crude oil of Kirkuk quality and provided a method of determining the border value of all Iraq crude (Article 9, paragraph b). The border value of oil shipped by IPC pipelines and exported from its Mediterranean terminals made allowance for a marketing discount of seventeen shillings and sixpence from the price paid by the parent companies for each ton bought. The border value for oil shipped by BPC to the Persian Gulf terminals made an allowance of thirteen shillings for each ton of oil bought by the parent companies and exported from the Persian Gulf terminals. IPC had initially justified these discounts, which approximated 20 percent of the oil's selling price, as costs encountered by IPC in selling its oil.

Not long after the signing of the 1952 agreement the government questioned the propriety of discounts, contending that because IPC sold its oil to its parent companies it encountered no selling costs and that if the parent companies encountered selling costs in disposing of the oil, they, not IPC, should bear them. As a result of negotiations IPC and the government signed an agreement on March 24, 1955, whereby IPC lowered the discount to 2 percent of the selling price, or to approximately two shillings per ton.[4]

4. Dr. Nadim al-Pachachi, *Iraqi Oil Policy August 1954–December 1957,* (Baghdad, May 1958; translation from the Arabic by the Research and Translation Office,

Later, on learning through its exchange of information that the Saudi Arabian government had negotiated a more favorable settlement of a similar complaint with Aramco, the Iraq government negotiated a further reduction of these discounts to one percent of the selling price. IPC made the one percent applicable at the beginning of 1954. The government contended that the lower rate should apply to the 1953 operations and, when IPC disagreed, requested that the issue be referred to arbitration. In lieu of arbitration both parties agreed to settle it by a suit against IPC in the Queen's Bench Division of the High Court of London. Before the case came to trial IPC offered to settle the matter by paying the government £7 million for the year 1953. The government agreed and withdrew its claim before the court.[5]

Government Challenges IPC's Cost Accounting

The government manifested its concern with the companies' general cost-accounting methods in 1956 when Dr. Pachachi, minister of national economy, proposed to the Iraqi director on IPC's board that the government at its own expense subject the company's accounts to the scrutiny of a reliable, independent firm of chartered accountants. Consistent with this proposal, the minister wrote the companies on July 1, 1956, requesting detailed information on their physical assets and capital expenditures as of the end of 1955 and the manner in which they had been depreciated, and on receipt of this information turned it over for analysis to an accounting firm of the government's selection. The government scrutinized similarly the companies' accounts for 1956 and 1957. As a result of these analyses the government challenged the company's treatment of four items: fixed costs, dead rents, exploration costs, and drilling costs.

Fixed Costs

It will be recalled that the 1952 agreement specified that actual costs for 1952 and later years be assumed to be thirteen shillings per ton, an

---

Beirut, Lebanon, June 1958), pp. 3, 10–11. This is a record of correspondence between various governmental officials concerned with oil matters and between governmental officials and IPC, MPC, and BPC. It was offered by Dr. Pachachi as a record of his stewardship as Iraq's minister of national economy from August 1954 to December 1957. The translation extends through 76 mimeographed pages. Hereafter it will be referred to as Pachachi, with appropriate page references.

5. *Ibid.*, pp. 23–24.

amount which was thereafter to be designated as "fixed costs." The agreement provided, however, that if actual costs should differ in any one year from fixed costs by as much as 10 percent, the actual cost should apply, subject, however, to agreement by both parties and, failing an agreement, to arbitration. Pending arbitration, fixed costs should apply.

The accountants' scrutiny revealed that in their 1956 and 1957 cost accounting the companies had made substantial upward revisions in their fixed costs to arrive at more realistic figures. In these years IPC's exports of crude had been substantially reduced and fixed costs per ton correspondingly increased as a result of the dynamiting of pipelines and pumping stations, a reverberation of the Anglo-French-Egyptian dispute which closed the Suez Canal. The government rejected the companies' cost figures and gave notice of arbitration. The case was never arbitrated, and the companies, recalculating profits to be shared with the government on a basis of the thirteen shillings specified in the agreement, paid the government an additional £4,341,000.[6]

Dead Rents

Dead rents are payments that the companies had agreed to make to the government during the period between the signing of the concession and the discovery and exploitation of oil. Accountants are not agreed as to their proper treatment. If oil is not discovered and a concession is abandoned, such "rents" are "dead" in the sense that they will never be recovered. If oil is discovered, how the concessionaire treats them is of no concern to a government so long as it receives its payments as specified amounts per barrel of oil produced. Under the 50-50 profit-sharing agreement of 1952, how the companies treated them affected directly the revenue of the government. The accountants' analysis revealed that the companies had

6. A spokesman for the companies has stated that by failing to carry through arbitration, the government "has effectively secured an extra £4,341,000 from the companies while retaining the ostensible position of a dissatisfied and aggrieved partner." This quotation is from page 4 of an 11-page printed pamphlet by "an experienced outside observer" writing at the request of the IPC group of companies. The pamphlet, issued in February 1962, presents the companies' position on each of the issues in dispute in the unsuccessful negotiations between the companies and the Iraq government, which ended in the government's reclaiming 99.5 percent of the companies' concessions. It will be referred to hereafter as *Statement of the Companies.* The pamphlet was reprinted as a supplement to the *Middle East Economic Survey's* (MEES) issue of February 16, 1962.

capitalized these payments and were recovering them as production costs at the rate of 5 percent a year, a procedure they had followed before the 1952 agreement. Most of the dead rents under the IPC concession had been amortized before 1952, but most of the BPC and MPC rents were still to be recovered. The companies' accounting procedure reduced the government's annual revenues by 2½ percent, or one half of the dead rents. In effect, the government was gradually returning to the companies one half of the dead rents it initially received. This the government contended was improper. Dead rents, it argued, were intended to be bonuses paid the government as a partial consideration for its having granted the concession.[7] The government contended, moreover, that when negotiating the 50-50 agreement the companies had proposed that dead rents be treated as capital expenditures to be recovered out of subsequent earnings and had included a provision to this effect in the second draft of the agreement. On the insistence of the government they deleted it.[8]

Exploration and Drilling Costs

Exploration costs are costs incurred in hunting for oil in areas where it has not been found previously. They may include the cost of reconnaissance and geophysical surveys, core-drilling, and any of the numerous procedures that modern petroleum engineering has devised for locating oil reservoirs. Drilling costs as distinct from exploration costs represent expenditures for drilling wells for the recovery of oil once a reservoir has been located. Until the precise limits of an underground oil reservoir and the thickness and depths of the production horizons have been determined, it may be difficult and meaningless to distinguish between exploratory drilling and operational drilling. Just as the distinction between the two types of operations is not always clear, neither is accounting practice regarding their treatment always uniform. An established company constantly exploring for new reserves while exploiting those already found may choose to treat all such expenses as production costs, chargeable against the year in which they are encountered. Or it may choose to capitalize one or

7. The government sets forth its position on the several issues in dispute between it and the companies in a statement issued by the Baghdad press following the breakdown of negotiations in April 1961. *MEES* translated the statement from the original Arabic and summarized it in a supplement to its issue of May 19, 1961.

8. Pachachi, p. 39.

the other type of expenditure and amortize it over some time-period appropriate to the anticipated life of a field.

The chartered accountants retained by the Iraq government agreed with the companies that all exploration expense encountered before the discovery of oil should be treated as capital expenditures recoverable at a yearly rate of 5 percent. To this the government could not very well object. The accountants also agreed with the companies that exploratory and drilling costs encountered after the discovery of oil should appropriately be treated as production costs chargeable against the year's operation in which they occurred. With this practice the government disagreed, insisting that all drilling and exploration expenses, whether before or after the discovery of oil, should be treated as capital expenditures recoverable at the rate of 5 percent yearly. Whether treated one way or the other would affect somewhat the size of the government's revenue over the short run, reducing it in the early years, but would have little effect over the long run. When Qasim after the July 14, 1958, revolution began negotiations for sweeping changes in the agreement, the cost-of-drilling issue was still unsettled. The companies eventually agreed to treat these costs as the government proposed, but inasmuch as no general agreement was reached, they remain unsettled.[9]

Other Accounting-Cost Differences

While fixed costs, dead rents, and exploratory and drilling costs have given rise to the major cost-accounting disputes between the companies and the Iraqi government, they do not exhaust the list of differences that have inevitably arisen as the government has sought larger revenues through cost control. The Middle East governments generally and Iraq in particular challenged the allocation of company overhead expenses to Middle East operations, expenditures on public relations, corporate grants, and the like. The Iraqi government protested specifically the allocation of one half of the London office's expenses to operations in Iraq, contending that this is excessive inasmuch as the London office's responsibilities cover operations outside of Iraq. The government also objected to the companies' charging expenses incurred in publicity in both Iraq and London as operating cost, contending that such expenses are unrelated to production and should rest exclusively upon the parent companies. Although these are minor griev-

9. *Statement of the Companies*, p. 4.

ances, they have gone unsettled for years and have contributed to the persistent suspicion and hostility that have characterized the government-company relationship over the years.

Concessions and the Price of Oil

The Middle East governments' concern with prices is even more vital than their concern with costs. The chief function that posted prices have served in recent years is that of a benchmark for calculation of governmental oil revenues. They have never been true market prices in an economic sense, although they have been by no means entirely insulated from market forces. But whether they have moved in response to the decision of oligopolists or as more recently in response to the forces of a freer market, their impact on government revenues is identical. If they move up with no significant variation in costs, the government's take per barrel of oil moves up. If they move down, the government's oil revenue per barrel declines. It has been of little solace to the Middle East governments, whose obligations have generally mounted more rapidly than their incomes, that recent price cuts have been more than compensated by an increase in oil production.

Inevitably the 50-50 profit-sharing agreement meant that thereafter the Middle East governments would regard posted prices for oil as their proper concern.

IPC Agreement on Posted Prices

The IPC-Iraqi Agreement of 1952, clearly recognizing the government's legitimate concern with posted prices, provides for its participation in their determination. Specifically Article 1 of the agreement defines "posted prices" as

the prices (expressed in shillings per ton) f.o.b. seaboard terminal for Iraq crude oil of the gravity and quality concerned arrived at by reference to free market prices for individual commercial sales of full cargoes and in accordance with the procedure to be agreed between the Government and the Companies or if there is no free market for commercial sales of full cargoes of Iraq crude oil, then posted prices shall mean fair prices fixed by agreement between the Government and the Companies or in default of agreement by arbitration having regard to the posted prices of crude oil of similar quality and gravity in other free markets with necessary adjustment for freight and insurance.

When the government and the companies signed the 50-50 profit-sharing agreement, the BPC had just begun to export crude oil. Thereafter the parent companies posted prices for Basrah crude at the Fao terminal at the upper end of the Persian Gulf. Apparently the parent companies regarded the posting as tentative, final determination awaiting a more mature judgment based on experience with freight rates, refinery yields and other factors of significance to Basrah crude prices in relation to those of other crudes. It was not until March 1955, after the companies and the government had agreed on a working memorandum covering posted prices,[10] that the government finally accepted the posted prices then prevailing.

When in 1956 the parent companies announced a reduction of five cents a barrel for crude exported from the Fao terminal effective February 9, 1956, the Iraqi minister of national economy protested. His concern was apparently not so much with the amount of the reduction as with the fact that it had been made without consulting the government. On February 21, 1957, he wrote the presidency of the Diwan (administrative staff of the council of ministers) stating that "acquiescence in the principle that the Companies have the right to give separate treatment to Basrah oil and to cut its price in an arbitrary manner may be considered a dangerous precedent which might result in great harm to the interests of Iraq."[11]

In the ensuing correspondence that extended over the greater part of a year the companies justified the price reduction as necessary to make Basrah crude competitive with other crude of similar character, and at the same time they denied responsibility for making the cut. In their words,

Whilst the Iraq, Mosul and Basrah Petroleum Companies are, of course, responsible for conducting the operations by which the crude oil is made available for export, we should like to make it clear that they do not, and cannot, intervene in the establishment of Posted Prices. In accordance with the provisions of the Agreement and the Working Memorandum the term Posted Price means the Sellers' Posted Price. Each selling company posts a price which in its commercial judgment reflects the highest competitive price at which the output can be disposed of to buyers in a free market.[12]

The minister of national economy accepted neither the logic of the reduction nor the propriety of a unilateral decision by the parent companies and recommended to the presidency of the Diwan that the issue be referred to arbitration. There the matter rested.

10. Pachachi, p. 48.
11. *Ibid.*, p. 55.
12. *Ibid.*, p. 50.

Demand for a New Deal

While dead rents, the treatment of fixed costs and of exploratory and drilling expenses, and the rights and responsibilities of the government on prices remained unsettled, more fundamental issues had arisen. What had begun as a dispute over the terms of the Iraq-IPC agreement and their application broadened eventually into a demand for a new deal. The immediate cause for this lay in large part outside of Iraq.

In July 1957, Iran passed its national petroleum act insuring that the National Iranian Oil Company (NIOC) would play a major role in granting oil concessions. In August 1957, NIOC made its first joint-venture agreement with Azienda Generale Italiana Petroli (AGIP), subsidiary of Ente Nationale Idrocarburi (ENI), the Italian government's petroleum agency. A year later it engaged in an even more attractive partnership with the Standard Oil Company of Indiana's foreign subsidiary, Pan American. In December 1957, Saudi Arabia signed an agreement with Japan's Arabian Oil Company for the development of its rights in the offshore area of the Kuwait-Saudi Arabia neutral zone, providing for royalty payments of 20 percent and a total share in company profits of 56 percent.[13]

The impact of these developments was prompt. In resuming negotiations on the unsettled issues, Iraq representatives now demanded more. Inasmuch as the companies' concessions covered virtually the whole of the country, Iraq was left no opportunity to improve its lot by inviting other companies to participate in the exploitation of its oil resources. The government accordingly demanded that the companies release substantial areas of their concessions. It also demanded a better deal on those portions of the concessions that they would retain. It based this demand on a rather vague commitment IPC had made in a letter addressed to the government (and never released for publication) in connection with the negotiations eventuating in the 50-50 profit-sharing agreement. According to IPC, it had committed itself to a re-examination of the profit-sharing provision of the agreement if and when "any neighboring country receives a larger revenue per ton" of oil produced than does Iraq.[14] Iraq representatives

13. For a detailed discussion of this agreement see George W. Stocking, "Arabian Oil Company: Progress and Prospect," *MEES Supplement,* September 25, 1964.
14. On October 17, 1961, the Baghdad Radio broadcast the minutes of the negotiating meetings held under Qasim's leadership on September 28, October 8, and October 11, 1961. The British Broadcasting Corporation translated these minutes and

interpreted this as a commitment to re-examine should a neighboring country obtain an agreement with a concessionaire which granted more favorable terms than did the IPC-Iraq agreement.

Negotiators Confront New Issues

Accordingly, when the government and IPC resumed negotiations in June 1958 to settle the issues long in dispute, IPC was confronted with new and more fundamental issues. The government was not only dissatisfied with the company's interpretation and application of the terms of the concession but was now questioning the fairness of these terms. Moreover, it had become dissatisfied with the rate at which IPC was developing its vast concession and producing oil from it.

Evidences of mounting dissatisfaction throughout the last weeks of 1957 and the first six months of 1958 are numerous. In the fall of 1957 the Iraqi minister of national economy instructed a group of lawyers to examine the company agreements for loopholes that might enable Iraq to end their monopoly of Iraq's oil resources.[15]

The decline in oil revenues in 1957, occasioned by the destruction of company pipelines in Syria during the Anglo-French-Israeli attack on Egypt, directed public attention to the financial insecurity of a country dependent largely on a single industry for its major source of revenue. But it also contributed to government dissatisfaction with the amount of its oil revenues. In January 1958, both the Arab and the Western press reported that the government was urging the oil companies to increase Iraqi output substantially.

A demand for nationalization, dormant since the 50-50 profit-sharing agreement, was also stirring again. Members of the chamber of deputies were manifesting concern about the unsettled issues, a concern aggravated

---

Platt's Oilgram News Service, November 24, 1961, published the translation in a special supplement. On October 17, 1961, the Iraq Ministry of Oil published in pamphlet form its English translation of these minutes. The translations, although not identical, differ in only minor respects. I have used one or the other of these translations depending on which in my judgment more accurately expresses the views of the negotiator quoted. The above quotation comes from the remarks of Fisher as published in *Platt's Oilgram News Service,* p. 3. The Ministry of Oil's translation quotes Fisher as having said, "In case any neighboring country receives more revenues than you do, then you will have the right to demand an increase."

15. *MEES,* December 10, 1957; *Le Monde* (Paris daily) November 28, 1957.

by reports that production costs had risen and that this heralded a further reduction in revenues. More important was the growing conviction that the ENI and Pan-Am agreements with NIOC justified a change in the 50-50 profit-sharing agreement. As early as September 1957, Muhammad Jawad al-Khatibi, a member of the chamber of deputies, had publicly demanded that the Iraqi government ask for a larger share of profits. Muhammad Hadid, chief Iraqi delegate to the Afro-Asian Solidarity Conference in January 1958 and later Qasim's chief oil adviser, reported in an interview with the Cairo weekly *Rose el-Youssef* that public opinion in Iraq as a result of the Italian-Iranian agreement had forced the Iraqi minister of national economy to concern himself with this issue.

Formal Negotiations Begin

Whether inspired by top governmental officials or prompted by independent thinking of those lower in the governmental hierarchy, the demand for a change was so strong that something had to be done about it. Early in 1958 the Iraqi prime minister communicated the prevailing Iraqi view to IPC's executive director and early in April high-level official talks began with IPC representatives. In the initial discussions the minister of national economy and two other high officials from the directorate general of oil affairs represented the government, and IPC's joint managing director and its chief representative in Iraq represented the companies. Later M. J. Rathbone, president of Standard Oil (New Jersey), and Prime Minister Nuri al-Said joined the discussions. These talks apparently were preliminary in nature and were concerned primarily with the issues of cost accounting that had remained unsettled for long. Meanwhile the publication in May 1958 by Dr. Pachachi of his account of his stewardship on oil matters during the period August 1953 to December 1957 (the first time in Arab producing countries that the details of government-company negotiations were released to the public) raised doubts in the public mind regarding the basic equity of the concession agreements. As parliamentary concern was heightened, a three-man ministerial committee began an official study of the Saudi Arabia–Japanese agreement and of the NIOC agreements with ENI and Pan Am as a basis for further governmental negotiations with IPC.[16] It soon became apparent in informal circles both inside and outside Iraq that the government would present demands for a revision

16. *MEES*, June 27, 1958.

of the profit-sharing agreement. Prime Minister Nuri al-Said, while in London during the latter part of June, although denying that he was there for the purpose of discussing the oil agreements, "left little doubt . . . that an Iraqi demand for better terms would not be long delayed."[17]

On July 4, 1958, IPC's managing director, G. H. Herridge, accompanied by legal advisers, arrived in Baghdad to renew formal talks; they began the following day and continued with occasional interruptions until July 12. Although the government made no official announcement of the scope of the talks, press reports, perhaps government inspired, indicated that they would include a re-examination of the 50-50 profit-sharing principle, relinquishment of a part of the concession, increase in production, final settlement of the dispute over the company's cost accounting, and improvement in the opportunity of the country's citizens for employment in the oil industry, particularly at the higher levels, an issue of growing concern to all the Middle East oil-producing countries. On July 12 the negotiations were adjourned, apparently to enable company representatives to consult with their principals in London. Meanwhile press reports indicated that the talks had been conducted in a friendly atmosphere and that considerable progress had been made. Official comment was limited, but on July 13 the minister of national economy stated that the companies had agreed to double oil production by 1960 and to surrender parts of the areas under concession. He indicated that the government would invite foreign companies to explore and exploit the surrendered areas.[18]

The Revolution Takes Over

These achievements were lost in the wreckage of the Revolution of July 14, 1958. As General Qasim and his cohorts seized the reins of government, frenzied mobs let loose their pent-up fury and army detachments surrounded the palace. Before order was restored, they had murdered King Faisal II, his cousin, Amir 'Abd al-Ilah, the regent, and Iraq's strong man, Nuri al-Said, who more than any other individual for more than a quarter of a century had dominated political life and molded policy in Iraq, and who to Arab nationalists had become a symbol of "Western imperialism" and domestic feudalism.

The revolution, with no philosophical basis, but dedicated to unification

17. *Financial Times* (London), June 27, 1958.
18. *Iraq Times* (Baghdad English daily), July 11, 1958.

of the Arab world, improvement in living conditions at home, and elimination of all traces of "Western imperialism," but relying largely on shibboleths rather than a program to capture popular support, had a special significance to the oil industry, long the symbol of imperialism throughout the Arab world, and to the consuming countries dependent on it. Qasim, however, recogizing the importance of oil revenues to the revolution's political security and the country's economic stability, and perhaps influenced by the landing of United States Marines in Lebanon and the alerting of British troops in Aden, endeavored to put to rest Western Europe's fears over the continuity of its oil supplies and IPC's over the security of its properties. On July 18 in a Baghdad radio broadcast Qasim stated,

> In view of the importance of oil for the Iraqi national wealth and the world economy, the Iraqi Government announces its wish for continued production and export of oil to world markets. It also upholds its obligations to all parties concerned. The Government has taken the necessary measures to preserve the oil fields and oil installations. It hopes the parties concerned will respond to the attitude taken by it toward the development of this vital source of wealth.[19]

Other reassurances were soon forthcoming. On July 21, spokesmen for the Iraqi government in London and New York asserted that their government did not intend to nationalize its oil industry. The London spokesman, Colonel Abdul Faik, senior assistant military attaché at the Iraqi embassy, was particularly comforting when he said,

> It is not the intention of the Republic to think about this, because we believe that if the oil flow continues to the usual markets, it will be for the benefit of both parties—you get your oil and we get our pounds.[20]

This was language that businessmen understood. Following this announcement Western officials breathed more easily, and shares of British Petroleum, Burmah Oil, and Shell, which on news of the revolution had declined on the London market some £90 million, rose briskly.

Although continuing to deny any intention of nationalizing the oil industry, various government officials, including Qasim himself, made it clear from the outset that to carry out the aims of the revolution—one of which was to improve the standard of living of its citizens—Iraq would need and expected to receive larger oil revenues.

Convinced that nationalization was of no immediate threat, IPC reaf-

19. *MEES,* July 25, 1958.
20. *Financial Times,* July 22, 1958.

firmed its intention to carry through an expansion program designed to increase substantially its Iraq production and exports over the long run. Specifically, in January 1959, it announced investment plans that would increase Iraq's annual export capacity within two years from 34 million tons to 57 million tons.[21] Occurring in a period of world oil surplus, this announcement may be reasonably interpreted not merely as an expression of confidence in the revolutionary government but as a strategic move in negotiations on demands that had not yet been fully formulated nor formally presented, but about which rather awesome rumors continued to circulate.

In the spring of 1959, amid rumors that Iraq intended to nationalize its oil industry, informal talks were conducted between representatives of the company and the government, in which Lord Moncton, chairman of IPC, and Prime Minister Qasim participated. These talks put the rumors to rest and confirmed the companies' conviction that the issues in dispute could be resolved within the framework of existing agreements and that nationalization of the industry was not Qasim's aim. Following these talks, Dr. Ibrahim Kubbah, minister of national economy, at a press conference on April 8 stated,

The question of oil nationalization was never discussed between the oil companies and the Iraq Government. In fact, the Government still adheres to the oil policy declared on the first day of the Iraqi Revolution. This policy is based on upholding the oil agreements with the oil companies operating in Iraq and the possibility of amending them only with the full consent of these companies.[22]

Lord Moncton on his return to London, in declaring that he saw no immediate danger of nationalization, stated,

I do not remember any discussions conducted in a more amicable spirit, and they were continually saying that they wanted to work with us to our mutual advantage.[23]

Despite these expressions of mutual confidence, some evidence indicates a difference of understanding as to just what the conferees had agreed to. Dr. Kubbah in a second press conference stated that the companies had

21. *Herald Tribune* (Paris edition), February 1959, Monthly Economic Review; *Times* (London), February 6, 1959; *Financial Times,* February 7, 1959.
22. *Iraq Times,* April 9, 1959.
23. *Times,* April 24, 1959.

indicated that their expansion program already underway would double Iraq's production and export of crude oil by 1962.[24] He also stated that the companies had conceded the necessity of establishing oil refineries in Iraq, whose profits should be divided equally between the companies and the government. On the following day the companies issued a statement confirming their intention of doubling Iraq's capacity to produce and export in the next three years but making it clear that the actual increase in production would depend on market conditions. They also made it clear that they had no intention of establishing a refinery in Iraq.[25]

Qasim Takes Charge of Negotiations

Although Dr. Kubbah in his capacity as minister of the national economy until July 13, when Qasim abolished the ministry, and as minister of agrarian reform and acting minister of oil thereafter, continued to make public statements from time to time about his hopes for Iraqi oil and what was being done to realize them, he apparently participated in no further negotiations on oil matters. When negotiations were resumed in the fall of 1959, Qasim had taken personal charge. But his activities were cut short on October 7 when a would-be assassin's bullet pierced his shoulder and sent him to the hospital for almost two months. On December 2, the day before he left the hospital, Qasim conducted a six-hour press conference during which he discussed the achievements of the oil ministry, revealing that the negotiations had come to grips with more basic issues than those the government had raised immediately before the revolution that had brought him to power. He stated,

As a result of . . . negotiations, we managed to obtain a decision on the part of the company to relinquish 90,000 sq. km. of its present concession area in Iraq. However, we did not accept this. We want the company to relinquish 60 percent . . . and we also wish to have a voice in selecting these areas.

While acknowledging the importance of building up "a friendly relationship to replace the doubts and distrusts which result from unsolved issues," he also indicated that he had taken a firm stand on them.

24. *Times* (International edition), April 28, 1959; *Herald Tribune* (Paris edition), April 29, 1959.
25. *Financial Times*, April 29, 1959.

I had set the company a limit of 30 days from the end of the negotiations which finished only six days before those treacherous bullets were fired at me . . . [and] had informed the company that unless we received its reply before the appointed deadline, we could no longer be committed to adhere to the principles agreed upon in the course of the negotiation. . . . I now have the company's reply, the contents of which I have just disclosed.[26]

The company's reply did not satisfy Qasim, but he took no immediate action to achieve his ends. Meanwhile, press statements indicated that formal negotiations would resume in April 1960. But the spring negotiations, if indeed they had been planned, never took place. Government officials continued to confer informally with representatives of the companies in Baghdad or London and, as evidenced by subsequent events, to prepare demands to present in formal negotiations originally set for June 30, 1960, but later postponed until August 15.

The August meetings convened in a quite different atmosphere from that described by Lord Moncton as having characterized the spring conferences of 1959. Two unfortunate developments had transformed a formal cordiality into bitter hostility: a dispute over port cargo dues and a reduction in the posted prices of Middle East crude oil.

The Dispute over Cargo Dues

As early as November 1959, the government had set up a committee consisting of representatives of the director general of ports and the ministries of finance, oil, and communications to study the question of increasing Basrah port cargo dues to obtain revenue to carry out its harbor development plans. In December the government dispatched a three-man committee to make a comparative study of port dues levied by Kuwait, Saudi Arabia, Bahrain, and Iran.[27] As a result of this investigation the port authority announced an increase in cargo dues from 23.4 fils to 280 fils a ton applicable alike to oil and all other cargoes. BPC's response to this was sharp and prompt. On July 20 it notified the Iraqi government that it had stopped production from the Rumaila field and for the time being would confine production to the Zubair field at an annual rate of eight million tons. This brought an equally prompt and sharp reaction from government officials and the Baghdad press.

On July 21 Qasim in an extended statement to the Iraqi News Agency

26. *Ittihad al-Sha'b* (Baghdad daily), December 5, 1959, as reported in *MEES,* December 18, 1959.
27. *Al-Dunia* (Beirut daily) as reported in *MEES,* December 18, 1959.

expressed the fear that the company's action represented its intention to ignore its obligations under the 1952 agreement to export at least eight million tons of oil a year from the BPC concession and declared that the

. . . arbitrary economic pressure which the oil companies are exerting on Iraq in order to halt the rise in its people's standard of living and the country's development, will not under any circumstances serve those companies' interests.

He advised the company to reverse the decision and stated:

Moreover, we reserve for Iraq those rights to which it is entitled owing to the company's failure to fulfill its obligations and because of its arbitrary action against Iraq.[28]

Two days later the Iraqi director general of ports, Major General Mazhar al-Shawi, echoed Qasim's protest, charging that BPC's action was "an open political manoeuvre to exert pressure on the Iraqi Republic." He characterized a 1955 agreement fixing oil cargo dues at 23.4 fils per ton as "a farce played by the Basrah Petroleum Company and the British personnel who were then in charge of the Iraqi Ports Administration."[29]

As a result of these protests and following a conference with Qasim, IPC's chief representative in Iraq, F. C. Ryland, notified the government that IPC had resumed production at a small rate in the Rumaila field and expressed the hope that operations would soon be normal. When later asked whether this meant that BPC had accepted the increased port dues, Ryland answered, "Very definitely not."[30]

At a second press conference Qasim announced that the question of port dues would be discussed at the forthcoming negotiations but that meanwhile he had obtained a written promise that the minimum rate of production from the Basrah oil field would be raised from eight million to twelve million tons a year.[31]

The Company Explains Curtailment of Production

The company in notifying the port authority that it was stopping production in the Rumaila field explained its action on purely economic

28. *Al-Thaurah* (Baghdad daily), July 22, 1960, as reported in *MEES*, July 29, 1960.
29. *MEES*, July 29, 1960.
30. *Ibid.*, August 5, 1960.
31. *Idem.*

grounds. The increased offtake price of BPC oil at Basrah had resulted in "reduced tanker liftings."[32] In short, buyers confronted with higher costs were turning to cheaper sources of supply. This explanation is, of course, economically sound, but its soundness does not rest on the logic of competitive pricing. The evidence does not indicate that posted prices in the Persian Gulf conformed to those that would prevail in a truly competitive market. On the contrary, they reflected a considerable element of monopolistic profits. The soundness of the company's explanation rests on the logic of oligopolistic pricing. The offtakers were primarily the international companies that owned IPC. As participants in concessions elsewhere, they could readily turn to other sources of supply from which their profits would be greater. To do so was good business but, coming as it did on the eve of the August negotiations, this move did not make agreement easier on the issues to be negotiated. It deepened and broadened the area of disagreement.

Companies Reduce Middle East Prices

The August 1960 reduction in posted prices for Middle East crude announced by Esso Export Corporation, a New Jersey Standard affiliate, on August 9 and quickly followed by the other international majors had a similar though more marked effect. It aroused the indignation and heightened the hostility of all Middle East oil-producing countries and of the whole Arab world. Of more significance, it gave birth to the Organization of Petroleum Exporting Countries, which heralded the substitution of united action among the leading oil-producing countries in their effort to resolve differences with and obtain better terms from the international oil companies in whose custody they had placed the exploitation of their oil resources, and it eventually shifted negotiations from the political plane on which they had thus far developed to an economic level of technical competence.

It served to aggravate immediately the differences between IPC and the Iraq government as they resumed negotiations on August 15.

Formal Negotiations Begin Again

For these negotiations Qasim delegated his top governmental officials including the ministers of planning, oil, and commerce, the former minister

32. *Idem.*

of finance, the director of planning at the ministry of defense, the head of the general petroleum authority, the governor of the Central Bank, and the director general of the political department in the ministry of foreign affairs. Supporting these were the legal counsel to the Iraqi government, a representative of the government's chartered accountants, and a representative of the government's solicitors.[33]

As its team IPC delegated its managing director, G. H. Herridge; its executive director, N. M. Eskerdjian; and its chief representative in Iraq, F. C. Ryland.

On the day that the conference convened, Qasim, in commenting on the negotiations in a speech before the Army Reserve College, said,

> We have prepared ourselves to discuss each problem in a peaceful and fair manner. We do not wish to encroach on the rights of others, but we shall seek to secure our rights through peaceful negotiation. . . . The objectives we have in view will serve the best interests of the country. . . . We shall be guided in all our actions by a spirit of fairness and patience. Patience is the means to success.[34]

However disarming this statement may seem, subsequent events indicate that the government had formulated a set of far-reaching specific demands on which they were to insist with increasing tenacity.

The negotiations that began in August, with their extended recesses, lasted more than a year. Periods of intensive negotiation were three: August 15–31, 1960; April 2–7, 1961; and August 24–30, 1961. The August 1960 negotiations, which recessed on September 1, 1960, resumed on September 20, and with intermittent meetings continued until December 19, 1960. The August 1961 negotiations recessed on August 28 until September 28. At a final meeting on October 11 negotiations were ended for good.

The government made public the minutes of the critical sessions of each of these periods of negotiation, and both the government and the companies issued statements after the severing of negotiations. These several documents make clear the issues in the dispute and reveal the final position of both parties on each of them.[35]

33. *Ibid.*, August 19, 1960.
34. *Al-Thaurah,* as reported in *MEES,* August 19, 1960.
35. *Iraq Times,* beginning with its issue of October 2, 1960, and continuing through October 12, 1960, published in full the minutes of the seven-hour proceedings of August 31, 1960. *Financial Times* published a resumé of the minutes of the meetings of April 2 and April 16, 1962. As previously indicated (footnote 14), *Platt's Oilgram News Service* and *Iraq Ministry of Oil* published the minutes of the meetings of September 28, October 8, and October 11.

The Government's Position

The issues were largely the same as those raised but left unsettled by the government before the July 14 revolution. Nor had the government changed its position significantly on most of the issues. It insisted that dead rents were properly paid to the government in consideration of the concession; that the agreement made no provision for the company to recapture them; that in doing so the companies had violated the principles of equity and sound accounting.

It argued that equity and sound accounting demand that all exploration and drilling costs be considered as capital expenditures recoverable in accordance with agreement at the rate of 5 percent annually. It contended that posted prices were artificial prices serving solely as a basis for calculating the government's share of company earnings and, therefore, a proper concern of the government. It protested unilateral action by the company on prices and advocated pricing under a definite formula arrived at by agreement.

It reiterated the old government's claim that any discount to cover selling costs was fictitious, since sales were made at no expense to the parent companies. What happened thereafter was no concern of the government. It contended, as had the prerevolutionary government, that the company had been dilatory in complying with Article 29 of the 1925 convention, which provided that the company employees should as far as possible be Iraqi subjects. It went further, however, in asserting that the company by its tactics had violated the spirit, if not the letter, of the provision. About this it said,

It became apparent from further dealings with the companies in this matter that they were trying to prevent the government from participating in company administration on the ground that this is the sole prerogative of the company. But the companies thereby chose to overlook the fact that Iraq has two indisputable rights: as a partner in company profits to have a say in company policy, and as a sovereign state to issue legislation governing employment in the country.[36]

It called attention to the exchange of letters in connection with the profit-sharing agreement of 1952 in which the company agreed to reconsider the division of profit should the governments of Iran, Saudi Arabia, or

36. *Statement of Government,* pp. 8–9.

Kuwait obtain agreements that yielded their governments a greater revenue per ton of oil produced than Iraq received. And it contended that the more favorable agreements entered into by Iran with AGIP and Pan Am, and by Kuwait and Saudi Arabia with the Arabian Oil Company of Japan, justified a revision of the 50-50 profit-sharing agreement.

It expressed resentment that the companies' concessions covered virtually the whole of Iraq, that while monopolizing "all Iraqi territory for more than a quarter of a century" the companies "failed to develop more than 3 percent of the total area involved,"[37] and it demanded that the companies return to the government most of the area covered by the concessions. The government also demanded that IPC agree to pay its oil revenues in convertible sterling or in other convertible currency instead of exclusively in pounds as contemplated in the agreement, contending that the companies' refusal to do so conflicted with its Foreign Exchange Control Law of 1950, which obliged the companies to value the exported oil in foreign currencies agreed on by the government.

Government Makes New Demands

In addition to these claims, all of which the previous government had advanced, the government made four new demands of IPC:

1. Calling attention to Articles 35, 42, and 44 of the convention of March 14, 1925, which entitled the government to appoint two directors of the company's board, who were to enjoy the same rights and privileges as other directors, it argued that the intent of these provisions had been circumvented by the bylaws of the company, which assign only limited functions to the board while granting wide power to the executive directors on whom rests the responsibility for the company's management. One other fact, it argued, had contributed to robbing government directors of any influence in the management of the company: basic decisions—approval of the budget, drawing up of plans and projects, fixing of prices—were made by the directors of the parent companies, not by IPC. To give to the government a genuine voice in managerial functions, the government demanded that one of the two Iraqi directors be made executive director of IPC. It also demanded that a joint government-company committee be formed to approve company expenditures and determine their validity.

2. Desiring to broaden the base of its revenues and to share in the

37. *Ibid.*, p. 10.

profits of the integrated operation of the oil business, the government demanded that the company agree to give priority in the shipment of crude oil to a tanker fleet that the government was planning to establish.

3. Alleging that the burning of surplus gas constituted a waste of Iraq's natural wealth, the government demanded that the company either pay it for gas being flared or give it to the government.

4. And finally, the government demanded that it be allowed a 20 percent participation in the ownership of IPC.

The Company's Position on the Minor Issues

The issues of dead rents, drilling and exploration expenses, and sales discounts, although of long standing, were relatively unimportant, and in the course of negotiations IPC, without conceding the justice of the government's claim, acceded to two of its demands. It agreed to discontinue treating dead rents as capital expenditures pending arbitration on terms suggested by the government. It also eventually agreed to treat all exploration and drilling expenses as capital costs to be amortized over a period of years rather than as operating expenses to be written off in the year encountered.

Available records do not reveal that the parties reached an agreement on sales discounts, but there is no reason to believe that the company would have permitted this issue to have blocked a settlement of the more important issues before the negotiators.

Although the record does not reveal IPC's willingness to bestow on one of the Iraq directors executive functions, the companies stated that

> When the Government . . . expressed a wish for one of its directors to be more closely and actively associated with the Companies' affairs, the Companies agreed. But the Government has not yet taken advantage of the arrangements made at its request; indeed, the ordinary Government seats have remained unoccupied for many months past.[38]

On the demand that the companies accelerate the replacement of foreign personnel by Iraqis, the companies agreed with the government's objectives but said the problem was "one of formulating a practical way between the Minister of Oil" and the companies.[39]

38. *Statement of the Companies,* p. 9.
39. Minutes of August 1960, meetings, *Iraq Times,* October 3, 1960.

In response to the government's contention that the agreement does not provide for independent company action on prices, the companies pointed out that the parent companies established the prices and that the operating companies could not interfere. Moreover, Fisher asserted that in the "Working Memorandum" signed in 1955 it was agreed that posted prices should be the prices posted by the seller of oil.[40]

In response to the government's demand that an Iraqi tanker fleet receive priority in the shipment of IPC crude, IPC conceded that where the government chose to receive its royalty in kind it had the right to ship the oil in its own tankers. IPC took the position, however, that to obtain the right to ship any other oil, it would be necessary for the government to negotiate with the transport subsidiaries of the parent companies. IPC expressed a willingness to intercede in such negotiations and the opinion that "the Iraqi tankers once built . . . will not be without work."[41]

On the government's demand that payment due it be made in convertible pounds or other convertible currency, IPC advised the government to take this matter up with the British government.

On the Government's Demand for a Larger Share in Earnings

In response to the government's argument that under recent agreements Kuwait, Iran, and Saudi Arabia had obtained more favorable profit-sharing terms than those provided in the Iraqi agreement and that in accordance with the exchange of letters following the 1952 agreement this required a reconsideration of the Iraqi agreement, the companies rested their position on the language of the letters exchanged between the two parties. These provided for a reconsideration of the profit-sharing provisions whenever Iran, Kuwait, or Saudi Arabia should receive higher revenues per ton of oil produced than does Iraq. This condition had not in fact arisen. The companies correctly pointed out that revenue per ton depended as much on the size of the profit margin as on the proportions in which it was divided; that the profit margin depended both on costs and selling price; that costs varied with numerous factors—character of the reserves, rate of flow, the

40. Minutes of 1961 meetings, *Iraq Ministry of Oil*, p. 49.
41. Minutes of August 31, 1960, meeting. *Iraq Times*, October 5, 1960. The government presents a slightly different version of IPC's position on the use of Iraqi tankers, stating that IPC agreed to consider the use of government tankers if needed and if the parent companies approved. *Statement of Government*, p. 11.

distance of the well from the terminal, the quality of the oil, and the like—and that crude prices varied from terminal to terminal. No government was as yet receiving under any contract a higher revenue per barrel of oil produced than was Iraq. Not until this condition prevailed was the government justified in demanding a reconsideration of the profit-sharing provisions.

On Natural Gas

The first evidence of the government's concern with the wastage of natural gas appeared in February 1957. At that time Dr. Pachachi directed the attention of the council of ministers to the problem, stating that an estimated 80 million cubic feet of gas rich in sulphur and valuable both as a fuel and as a source of petrochemicals was being flared daily in the Kirkuk field and another 20 million cubic feet in the Basrah field. "Both this Ministry and the Ministry of Development," he wrote, "have studied the possibility of stopping the squandering of this natural gas, and have consulted both foreign and Iraqi experts to this end."[42] On a basis of this study the minister recommended that the development board establish a sulphur recovery plant in Kirkuk and a fertilizer plant in Basrah and transport the surplus gas by pipeline to Baghdad for use both as a fuel and as a basis for a petrochemical industry. Before making the recommendation, he had asked the companies for the precise composition of the gas and how much was available above their operational needs. In supplying this information, the companies informed him that they were willing to sell the surplus gas (then being burned) for ID 0.006 (approximately 2½ cents) per thousand cubic feet. The minister, insisting on the government's right to use the surplus gas without payment, recommended that if the companies insisted on charging the government for the gas the government pass a law "forbidding the burning of natural gas or its waste in any other way and requiring the Companies either to utilize it commercially or reinject it into the producing reservoir as now happens in Venezuela."[43]

The Pachachi correspondence on this problem ends in June 1957 with no evidence of a change in the companies' position. Between this date and the breaking off of negotiations in October 1961, however, the government changed its demands and the companies their offer. For some time IPC had

42. Pachachi, p. 58.
43. *Ibid.*, p. 59.

been supplying free of charge any quantity of surplus gas for which the government could find use in the country. It had resisted the government demand that either IPC use all surplus gas or pay the government for it. In the negotiations it had patiently explained the complexity of the problem of utilizing all the gas being produced in its oil operations. Since the gas was dissolved in the oil and released as it drove the oil to the surface, it was not possible to produce the oil without the gas. Whether reinjecting gas into an underground reservoir impedes or assists in the recovery of oil depends on the nature of the oil-producing horizon. In general, "technical conditions of the known Iraq fields do not favor gas reinjection, with the possible exception of the Zubair field where studies are still going on."[44] It contended that the remoteness of its operations from the major markets for gas either as a fuel or a basis for petrochemicals made its commercial disposal at that time impossible. It agreed to put at the government's disposal for either domestic use or foreign sale the surplus gas then being burned—150 million cubic feet daily at Basrah and 96 million cubic feet at Kirkuk. Either the government or the companies should have the right to dispose of any future surplus of gas, depending on which first found a market for it.[45] This offer was substantially the same as that which the Shell Petroleum Company and the Kuwait government had agreed to for Shell's development of its offshore concession and was liberal as judged by contemporary standards. Nevertheless, the government refused it, insisting that the companies pay it for all gas that they continued to flare.

The Demand that the Government Be Allowed a 20 Percent Share in IPC

In defending its demand that the government be allowed to subscribe to 20 percent of the companies' capital stock, the government pointed out that Article 34 of the 1925 convention provides that

Whenever an issue of shares is offered by the Company to the general public, subscription lists shall be opened in Iraq simultaneously with lists opened elsewhere, and Iraqis in Iraq shall be given a preference to the extent of at least twenty percent of such issue.[46]

44. *Statement of the Companies*, pp. 7–8.
45. Minutes of meeting of October 11, 1961, *Iraq Ministry of Oil*, p. 44.
46. *Statement of the Government*, p. 12.

When IPC pointed out that this situation had not arisen, the government countered with the argument that it could not arise without a change in the company's charter, which established it as a private, not a public, corporation. The inclusion of Article 34, the government argued, was evidence of bad faith, the article having been designed to deceive the government in the belief that it might one day become a stockholder while in fact the corporation charter would never permit it to do so.

In denying that such was a fact, the company as evidence of its goodwill proposed that it set aside a part of its concession to be exploited by a new corporation jointly owned by the government and the company. The government rejected this proposal.

Relinquishment of a Portion of the Concession

The government and the companies had reached agreement on the principle of relinquishment before the Revolution of July 1958 but had worked out no details. This was one of the first issues that Qasim had discussed with the companies. It will be recalled that he announced at his December 3, 1960, news conference that the companies had agreed to relinquish 90,000 square kilometers, but that he had demanded that they release 60 percent of their concession areas and that the government be given a voice in selecting the relinquishments. As the companies increased their offer, Qasim raised his demands. Before negotiations broke down the companies had agreed to relinquish 75 percent of the concessions immediately. They also agreed to release an additional 15 percent after seven years and to organize a new company to exploit it in which the government would be entitled to a 20 percent share. In response to this offer, Qasim renewed his demand that the government be allowed a 20 percent participation in IPC and a voice in selecting the areas to be released. Later Qasim, making what he termed a "final offer," indicated he would drop his demand for participation in present producing areas if the companies would agree to increase the government's share in them, relinquish 90 percent of the concession areas immediately, and allow the government to participate in the remaining 10 percent on new principles of profit-sharing. Fisher dismissed this offer with the remark that it was a "sudden proposal."

After prolonged negotiations extending over a period of three years with no agreement having been reached on the major issues, both parties recognized that further negotiation would be futile. Qasim brought the meeting to a close with the announcement that the companies could continue to exploit existing wells as they wished but with the ominous statement, "I am sorry to tell you that we will take the other areas according to legislation we have prepared, so that our action will not be a surprise to you. Thank you for your presence here."[47]

47. Minutes of meeting of October 11, 1961, *Iraq Ministry of Oil,* p. 61.

10

The Revolution's Failure and the Negotiations' Collapse

DURING the more than three years between the July revolution and the final collapse of negotiations in October 1961, the government gave almost continuous attention to the oil controversy. Company and government officials held no less than twenty-eight meetings including two periods of intense negotiations in an effort to reach an accord. Why did they not succeed? Why, as the companies made substantial concessions, did Qasim increase both the scope of his demands and the intensity with which he pushed them? An examination of the revolution's aims and achievements may throw some light on these questions.

The Revolution's Aims

The Revolution of July 14, 1958, was a revolution without a philosophy and initially without a program. It was essentially a revolt against a Western-oriented political regime, born of Western decisions and nurtured under Western guidance; a regime in which power had become concentrated in the hands of a governmental clique that frequently changed leadership without losing its character or modifying significantly its makeup; a regime supported by a privileged landowning class whose support was conditioned on a continuation of their privileges; a regime remote from the people and intolerant of their liberties; a regime that had become increasingly isolated from the Arab world and its political aspirations; a regime that, despite the relative abundance of its oil revenues and however sound its long-run development plans, had done little to improve the lot of the great majority of its citizens, who lived close to a subsistence level, impoverished, illiterate, and diseased.

The revolution's aims were not precisely articulated nor clearly defined. They ran largely in terms of symbols and slogans. Vague and intangible as they were, they may be loosely grouped into three interrelated categories: freeing the government from the influence of the West, uprooting "Western Imperialism and colonialism" in all of its aspects; unifying the Iraqi people, whose loyalty would be tied to the newborn republic and its leader, and achieving a closer political union with all Arab states; insuring social justice to all of the people and improving their economic lot. These aims temporarily captured the loyalty of the masses, but they afforded a poor amalgam with which to bind the revolution's leaders, who lacked a community of ideals and were united only by a common hatred of the old regime.

Iraq's Break with the West

Of its three broad aims the revolution clearly realized only the first. At the outset even the break with the West was not clear cut. In a Baghdad broadcast on the day of the revolution, the revolutionary government made known its intention to honor all foreign commitments consistent with the interests of the country and to conduct the affairs of state in accordance with the principles of the Bandung Conference. Despite its declaration of neutrality, by a series of moves Iraq retreated from the West and drew closer to the Soviet bloc. A week after signing an economic and technical aid agreement with the Soviet Union on March 17, 1959, the Iraqi government announced its withdrawal from the Baghdad Pact. In making the announcement over the Baghdad radio, a government spokesman stated, "As of today our country becomes an independent sovereign republic after shedding the last vestige of imperialism." Thereafter in quick succession Iraq withdrew from the sterling area, ordered British air force units out of Habbaniya base, and canceled the Point Four Agreement with the United States. On the anniversary of the revolution, Qasim ordered a cabinet reorganization, dismissing some of the more conservative ministers and replacing them with known Communist sympathizers. Not content with these measures, a leftwing "palace guard," organized within the government, conducted its own purge of those opposing Qasim's increasingly leftist regime.

These developments not only reflected but aggravated hostility to the Iraq Petroleum Company, symbol of "Western imperialism." As the government became more definitely within the Soviet sphere of influence, a satisfactory outcome of the oil negotiations became more remote.

Unifying the Iraqi People and the Arab World

Qasim, as head of a revolutionary government free from Western influence and friendly to Russia, sought to develop an independent Iraqi "brotherhood" bound by an "unselfish devotion" to the Iraqi republic, "working together in co-operation and solidarity in the interests of the people," a brotherhood that would eventually transcend geographic boundaries to encompass the whole Arab world. Qasim often publicly expressed these sentiments. In a talk in November 1960 delivered to the Second Congress of Oil Workers and Employees Trade Unions, but addressed to his "Brothers, sons of the victorious Iraqi people . . . all over the country," he declared,

Our country is a powerful, unbreakable and enduring entity. The south, the north, the east, the west are integral parts of the Immortal Iraqi Republic for which we are ready to die. We will bring together the ranks of the triumphant Iraqi people . . . and the ranks of the whole Arab world.

But dissension and disunity had already broken out among the leaders of the revolution. The dismissal of Staff Colonel 'Abd al-Salam 'Arif from his post as deputy commander-in-chief of the armed forces, his subsequent arrest as a disturber of the peace and a plotter against the homeland, his sentence to death by the Special Supreme Military Court (the so-called "Peoples Court") and subsequent pardon by Qasim; the recognition as a national hero of Rashid Ali al-Gailani, leader of the shortlived revolt against the old government in May 1941, his subsequent sentence to death by the Peoples Court, his pardon by Qasim; the Mosul revolt in March 1959 led by Arab Nationalist Army officers who were sympathetic to Nasser and alleged that Qasim had been unfaithful to the revolution; the Communist-inspired Kirkuk revolt in July 1959 that turned Arab against his Turkmenian neighbors and took the lives of more than one hundred, including many women and children; the attempt on Qasim's life in October 1959, believed to have been inspired by Nasser sympathizers; Qasim's cabinet reshuffle in February 1960 and dismissal of Dr. Ibrahim, pro-Communist intellectual and architect of Russian-Iraqi co-operation. All these events reflect Qasim's inability to harmonize the aims of the revolution's leaders and to solidify the support of the people.

Arab Unity

Nor was Qasim more successful in knitting the Arab world into a common brotherhood. Egypt's hostility was total. Following the trials of Arif and Gailani and the Mosul rebellion, Nasser unloosed a verbal battle against Qasim, marked by an increasing flow of bitter propoaganda that ended only with Qasim's eventual overthrow.

Relations between Iraq and other Middle East Arab countries were less turbulent but far from cordial. King Saud of Saudi Arabia, although never friendly with the Iraqi branch of the Hashimite family, could scarcely view with equanimity the revolutionary overthrow of a neighboring kingdom.

The July 14 revolution, which had seized the reins of government over the dead body of King Faisal II, cousin of King Husain of Jordan, and of their common uncle, Amir Abd al-Ilah, killed in its infancy the Arab Federation, organized in February 1958, to give to Iraq and Jordan common foreign policies, a common educational system, a unified army, and ultimately to remove all economic barriers between them. It was not until October 1, 1960, that King Husain officially recognized the Iraq Republic.

The increasingly hostile conflict between Nasser and Qasim shaped Iraq's attitude toward Syria. Qasim could scarcely acknowledge the permanence of the United Arab Republic, dominated by a leader dedicated to his overthrow. Accordingly he revived the Fertile Crescent scheme embracing Jordan, Syria, Lebanon, and Iraq. Because Nuri al-Said, the "tool" of "Western imperialism and colonialism" had first proposed the idea, Qasim felt it necessary to explain that

This project was an imperialist project when Iraq was a strong imperialist base and support, but now that Iraq has become a free, liberated fully sovereign and independent country which struggles for its independence and sovereignty, this project does not constitute a danger.[1]

1. Benjamin Shwadran, *The Power Struggle in Iraq* (New York: Council for Middle Eastern Affairs Press, 1960), p. 50.

This scheme served only to widen and deepen the breach between Egypt and Iraq and to intensify Nasser's campaign of vilification against Qasim.

The Kuwait Affair

A series of fast-moving events in the summer of 1961 made almost total Qasim's alienation from the rest of the Arab world. On June 19 Great Britain canceled its 1899 protectorate agreement with the sheik of Kuwait. On the same date Britain recognized Kuwait as a sovereign state and agreed to provide, when requested, military assistance to protect its independence. On June 25 at a press conference in Baghdad Qasim startled the Arab and Western worlds by solemnly announcing that Kuwait was and always had been a part of Iraq. He proclaimed that "The Iraqi Republic will never cede a single inch of this land." And he threatened, "When we say this, it means we can execute it."[2]

In rapid succession, Kuwait requested British protection, British troops occupied Kuwait, Britain introduced a resolution in the Security Council of the United Nations demanding that all states respect Kuwait's sovereignty and promised to withdraw British troops at Kuwait's request. Iraq introduced a counter resolution affirming its peaceful intentions toward Kuwait and demanding that Great Britain immediately withdraw its troops.

When the Security Council failed to approve either resolution, the Arab League took up the issue. It eventually adopted, over Iraq's protest, a resolution providing that the British troops be withdrawn and that the Arab countries preserve Kuwait's independence by dispatching an Arab military force to its assistance.

Thereupon Iraq withdrew from the Arab League, making its isolation from the Arab world complete.

Establishing Social Justice and Improving the People's Economic Lot

Social and economic justice are vague concepts. They provide no precise standard by which to measure a revolution's success. Nevertheless, although some progress was made on both fronts, Qasim's promises exceeded his achievements. Despite the generality of these concepts, the new

2. Benjamin Shwadran, "The Kuwait Incident," *Middle Eastern Affairs*, January 2, 1962, p. 2.

government's program to achieve them contained three concrete elements: improved educational opportunities, agricultural reform, and industrial development. It was in the area of education that the revolution's success was most marked. Between 1958 and 1961, when oil negotiations collapsed, the new government increased substantially educational expenditures, the number of schools, the number of teachers, and the number of pupils, with no apparent lowering of standards.

Agricultural reform, the most important single element in the revolution's economic and social justice program, did not progress so well. In truth, its future became more remote as its past receded. Before the revolution Iraq's land and agricultural system was probably as near the feudal prototype as any extant. A relatively few absentee landlords held most of the country's agricultural lands. Whether recently settled nomadic tribal leaders or long-time town dwellers, the owners of the land lived apart both sociologically and geographically from the people who farmed it. Absentee ownership with its wealth and idleness had become a symbol of prestige. Following the enactment of an agrarian reform law, Qasim "on the eve of September 29, 1958, announced to the Iraqis and to all the peoples of the world the approaching end of feudalism in Iraq."[3]

What the slow processes of history had achieved over several centuries in Western Europe, Qasim hoped to achieve overnight. It was too much to expect. Land distribution on a large scale requires time. The work of implementation is difficult and detailed. A shortage of engineers and surveyors, a lack of maps and topographical surveys, and the necessity of settling legal claims inevitably retarded the program. The influence of Communists, both within and without the government, in whose scheme of things small-scale peasant land ownership had no place, is said to have exerted a similar influence.

By December 31, 1961—three and one half years after the revolution —the government had distributed 601,949 donums to 10,672 peasant families.[4] The ministry of agrarian reform characterizes these accomplishments "as of historic importance in the line of the progress of the beloved Republic." In view of the complexity of the ministry's task, it may be that such characterization is justified. The achievement did little or nothing, however, to improve the lot of the peasant. Most of the agricultural land subject to confiscation still remained in the hands of the absentee landlords;

3. Higher Committee for the Celebrations of 14th July Revolution, *The Iraq Revolution in its Third Year* (Baghdad, 1961), pp. 404–407.
4. *Ibid.*, p. 414.

of that confiscated, much remained in the hands of a "Temporary Management," itself an absentee owner, although perhaps a more considerate one.

Unfortunately, and perhaps inevitably, the new peasant owners frequently found themselves in worse condition than they had been in as sharecroppers, and the country as a whole was definitely worse off. The irrigation system had been designed for large estates, not small holdings. The government was in no position to perform promptly and efficiently the functions that the landlords had previously performed, however great the cost in living standards had been to the peasants. Many peasants without adequate irrigation for their newly acquired lands and bereft of the organizing influence of the estate managers became discouraged and migrated to towns. And the landlords, anticipating loss of their holdings, frequently stopped large-scale production. These factors, together with a succession of poor harvests, converted Iraq's customary surplus of agricultural products into a deficit and in 1960 the government found it necessary to suspend exports of barley and import some $45 million in wheat.

Progress on the Industrial Front

Nor had the revolution made more gratifying progress in its economic development program. To raise the standard of living of the Iraqi people by developing the country's resources, the old regime had created a development board in 1950, to which it allocated 70 percent of its annual oil revenues.

In shaping its program, the development board sought the advice of Western experts. These included at one time or another the International Bank for Reconstruction and Development; Sir Arthur Salter, economic consultant to the development board; the industrial consultants, Arthur D. Little, Inc.; and Professor Carl Iversen, who made a report for the National Bank of Iraq on monetary policy in Iraq. These several reports cautioned against forced industrialization aided by tariffs or government subsidiaries. They emphasized the importance of projects designed to increase agricultural output—irrigation, flood control, drainage, use of fertilizers, and the like—and an industrial program designed to complement agricultural expansion—fertilizer plants, cotton textile plants, vegetable oil extraction— and to increase the supply of consumer goods.

These several reports served as a guide to the development board's six-year plan covering the years 1955–56 to 1960–61 and contemplating a total expenditure of ID 500 million. Although realization had fallen short of expectations, the board's development program was moving forward when the July 14 revolution took over.

Qasim Abolishes the Development Board

The Qasim regime first crippled, then abolished the development board. Early in 1959 the government announced that as a "temporary measure" it was decreasing the development board's share of annual oil revenues from 70 percent to 50 percent.[5] Charging that the development board was the handwork of the "imperialists" who had "undermined . . . genuinely constructive projects" and created "an atmosphere that camouflaged their intrigues" and that it had painted "an outwardly bright picture . . . with which to deceive the Iraqi as well as world public opinion,"[6] in July 1959, the government abolished the board. It then transferred its functions in part to an economic planning board composed of the prime minister and the heads of the newly created ministries of planning, works and housing, petroleum, and industry, and certain of the old ministries including finance, agrarian reform, commerce, communications, social affairs, and such other ministries as the prime minister might designate. Each of the ministries thereafter created its own planning division. By these changes the work of the old development board, which had considerable autonomy, was decentralized and at the same time was brought directly under the supervision and authority of Qasim, who as prime minister was chairman of the Economic Planning Board.

The new development program differed in three important respects

---

5. *Financial Times,* April 28, 1959; *Times,* April 28, 1959. Subsequently in an interview with the editor of the Indian magazine *Blitz,* Qasim justified the cut in the development board's budget as follows: "With respect to reallocations, the former Government had allocated 70 percent of oil revenues for spending on projects in the country. The truth is that only 10 percent was spent, while the balance was smuggled outside so that foreign countries could benefit from it. The wealth of the country was robbed. Now we have seemingly allocated 50 percent to the Ministry of Development to carry out the projects in vital industries. The remaining 20 percent we have allocated to the light industries, which are not the responsibility of the Ministry of Development but the responsibility of other ministries. The total allocated is 70 percent as has been the case previously. But we have divided the projects into two parts: vital and secondary. The total is just the same." *Iraq Times,* May 17, 1959.

6. *The Iraq Revolution in its Second Year* (1960), p. 103.

from the old. It placed greater emphasis on industrialization as a means of insuring the economic independence of the country. It looked to Russia and its satellites for financial assistance, advice, and guidance. And it placed final decision-making in the hands of a single individual.

The immediate effect of the revolution was to bring the long-range development projects of the old regime to a sudden halt. The new regime canceled contracts with Western companies and dismissed Western experts. It filled their places with Soviet and Czech technical advisers who, whatever their technical competence, brought to their task an ignorance that only operational experience could overcome. While they learned, the plan foundered.

Qasim, the Decision-Maker

Perhaps the greatest obstacle to developmental progress was Qasim himself. Equipped neither by training, experience, nor personality for the complex task of selecting projects essential to a sound developmental program and timing and co-ordinating them into a developmental plan, he eventually assumed final authority for virtually all decisions. Few projects, important or unimportant, moved forward without his blessing. Whether it was a major irrigation or flood-control project, the widening of a road, or the letting of a contract, Qasim made the final decision. The results were not only delay and confusion, but frequently the designation of projects ill-designed for improving the country's economy or the peoples' living standards.

Despite its defects the development program slowly moved forward. What it might eventually have achieved had Qasim survived cannot be said. What can be said is that when the oil negotiations broke down in the fall of 1961, neither the agricultural reform program nor the industrial development program had fulfilled the aspirations of the revolution. On March 25, 1959, Qasim promised a delegation from Mosul,

> Three or less years from today the standard of the people will be great. From this country goods will be exported to the neighboring states. The country will be excellent from the aspects of building, education and health. We will have heavy factories, and the farmer will be dignified and no one will remain without work, but we will be importing workers from abroad to employ them because our dear workers will not cover the requirements.[7]

Qasim's achievement gap was large.

7. Shwadran, *The Power Struggle in Iraq,* p. 31.

Qasim, a Mystic with a Mission

Faced with disunity and unrest at home, a growing animosity on the part of Iraq's sister Arab states, the failure of his agrarian reform program, and the slow progress of Iraq's economic development, Qasim, to save the revolution, needed a dramatic victory over the oil companies, symbols of the Western imperialism that he had vowed to uproot. In assuming full responsibility for the revolution by acknowledging himself as its "Sole, Faithful and Honest Leader," Qasim apparently believed that he was playing a role to which destiny had called him. A mystic with a mission, he had long believed that he had been chosen by a power greater than himself to shape and fulfill the aspirations of his people. Shwadran relates that long before he came to power, when his cousin, Fadil Abbas al-Mahdawi, later head of the "Peoples Court," interceded at the request of Qasim's parents to prevail upon him to forsake bachelorhood and rear a family, Qasim, in rejecting the appeal, responded simply, "I have an aim." He never forsook it.[8]

An unstable, highly emotional man, he brought to his task a dedication and sincerity that could scarcely be concealed. A special correspondent to the *Times* (London) says of him:

When he spoke, his exalted sense of dedication to Iraq and its people was too intense to be misunderstood. He expressed his thoughts in the simplest of words and metaphors and simplified many problems by looking at them through the lens of his own patriotic fervor and good will.[9]

In dedication to his people he continually recognized the aid of a divinity who guided him. Thus, in a talk before the first educational conference in the Iraq Republic in September 1960 he acknowledged, "We have been created for this age to render a service for you, sisters and brothers, for our coming generation."[10]

Although recognizing the role of providence in guiding and protecting his leadership, he brought to God's aid his own military might. He ate, slept, and worked in the Ministry of Defense, heavily guarded at all times. Close by on permanent exhibit he kept the "bullet-riddled American station wagon" in which he had narrowly escaped assassination and his bloody

8. *Ibid.,* p. 15.
9. *Times,* August 16, 1960.
10. *Iraq Times* (Baghdad), September 19, 1960.

shirt as "a symbol of his belief in his destiny, testimony of God's protection, and the certain failure of all plots against him and his regime."[11]

In fulfilling his mission, he did not hesitate to utilize other worldly devices when he deemed them helpful. In support of his dedicated determination to wrest a dramatic victory from the oil companies, he took his fight to the literate public through the press. Ordinarily in the absence of a free press and a literate citizenship, public opinion is apt to exert little influence on public policy. Yet Qasim was able to create public support for his negotiations, tactics, and aims so solid and insistent that it eventually made politically impossible any substantial retreat from the stand that he had taken.

Qasim Leads Press Attack over Basrah Port Dues

In his public statements in the Basrah port dues dispute in the summer of 1960 after the BPC had shut down production in the Rumaila field, he placed full responsibility on the company, charging their chief representative with having abandoned "the spirit of amicable co-operation" with the government and deliberately trying to "precipitate a dispute." His charges served as the cue for a vigorous and vitriolic attack by the press on the "imperialistic oil companies" who had been "bleeding the country for many years."[12]

As the August 1960 negotiations got under way, the press criticism of the companies became more informed and specific, but no less vigorous. The Baghdad papers displayed such an intimate knowledge of the issues before the negotiations began as to suggest that they had been carefully briefed. Their criticisms reflected their diverse political affiliations, but they were unanimous in their support of the government. What they said and how they said it was designed to soften the companies' negotiators and stiffen the government's. Although Qasim had initially remained discreetly silent about details of the earlier negotiations and cautioned the need for continued patience, he soon let it be known that he expected to get results. The Baghdad daily *al-Thaurah* reported an interview with him on August 28, 1960, in part as follows:

11. David Holden, "Iraq Revolution Loses Its Way," *The Guardian,* August 1, 1961.
12. Al-Hurriyah, *Baghdad Daily,* July 13, 1960, as reported in *MEES,* July 22, 1960.

The Beloved Leader said that he hoped that the negotiations would be crowned with complete success, whereupon he would announce the good news to the people who would remain the steadfast guardians of their rights, of which no part, however small, would be lost.

Al-Bayan, organ of the National Progressive Party headed by Muhammad Hadid, former minister of finance and a member of the government's negotiating team, greeted the temporary suspension of negotiations at the end of August 1960 with an article headlined "The Oil Companies Must Realize that Present National Government Will Make No Concessions as Regards the Peoples' Rights to Their Oil Resources."

Doubtless influenced by French policy in Algeria, a number of Baghdad dailies of different political persuasion demanded the immediate nationalization of the French share in IPC, held by Compagnie Française des Pétroles—a demand that mounted as negotiations remained suspended.

Qasim Becomes More Belligerent

Meanwhile Qasim had become more belligerent in his public remarks. In an address before the Iraqi Education Conference in Baghdad on September 15, 1960, he reiterated his dedication to the people and his determination to make no compromise in their battle with "the monopolistic oil companies."

No one can tempt us with offers of money, rank, or position. The money belongs to the people, and the rank and position are employed in the service of the people. . . . The free and immortal Iraqi Republic has become an independent sovereign state against which no one can prevail.

On November 12, 1960, in addressing the Second Congress of the Iraqi Oil Workers and Employees Trade Union, he indicated that the companies' delay in settling the dispute was both widening the gap between the parties and strengthening his determination to protect the country's interests:

Brothers, we shall make for you a vast wealth which the imperialists and the foreigners used to take for themselves—your wealth. . . . At the present time we are struggling on your behalf and are involved in arduous negotiations with the oil companies. . . . Every day we make new demands on the oil companies because the people are entitled to these demands and rights. In a matter of days we shall announce to the world what we have been able to obtain . . . and what we shall obtain in the future.

But the days dragged on into weeks and the weeks into months. Meanwhile even the faraway London *Financial Times* detected "a new mood of militancy" among the oil-producing nations and advised that the oil companies be flexible on Qasim's demand for the right to participate in selecting the areas to be relinquished, and on other basic issues in dispute.[13]

On the eve of the resumption of negotiations in April 1961 an increase in gasoline prices in Baghdad precipitated a strike of taxi drivers and sparked a series of disturbances so serious that Baghdad's military governor general imposed a curfew on a part of the city and banned gatherings of more than seven people. Qasim manifested his increasing belligerency toward the oil companies by implicating them in the disturbance. He said,

Every time we have tried to enter into negotiations with the foreign companies, imperialism has made use of its stooges and agents to commit irresponsible acts against our Republic. . . . The Revolution took place in order to save the people and raise the standard of living of the poor. We are now sufficiently strong to stand up to the greedy, the imperialists and their stooges, and to defeat their plots against our immortal Republic. We are engaged in a struggle with imperialism and the foreign companies which refuse to recognize our rights until we have fought for them with all our might.[14]

At about the same time Qasim indicated at a press conference his determination to "wrest the full rights of the people" from the oil companies, a position promptly echoed in editorial comment.[15] The April negotiations were short lived. Apparently by this time it was all or nothing with Qasim. When the companies refused to agree in full with his demands on dead rents, a relatively minor issue on which the parties were not far apart, he broke off negotiations (April 6, 1961), informing the companies that they must discontinue all exploration and drilling operations outside the producing areas of the concessions until an agreement was reached that would safeguard the Iraqi peoples' interests "and put an end to the company's unlawful exploitation and monopolistic practices in Iraq."[16] And he warned the companies that if they curtailed production, the government would take other measures to safeguard its interests.

The minutes of the April negotiations as subsequently summarized in the *Financial Times* reveal the temper of Qasim's mounting belligerency:

13. *Financial Times,* November 12, 1960.
14. *MEES,* March 31, 1961.
15. Al-Bayan, April 3, 1961, as reported in *MEES,* April 14, 1961.
16. Statement of Iraq Oil Ministry, April 11, 1961, *MEES,* April 14, 1961.

I hope you know that since the Revolution no one can detract from our rights, even if we have to stop the pumping of oil from wells by force. It is enough that you have swindled the people and the former Government for a long time.

Discontinuance of All Exploration Work

The Baghdad press greeted with prompt approval Qasim's order that the oil companies discontinue all exploration work. Some editors argued that further negotiations would inevitably prove futile and advised that the government without further delay take whatever action was necessary to "protect the peoples' rights." *Al-Thaurah,* echoing Qasim, urged him to reclaim by decree all undeveloped portions of the concession area—an action which he took some six months later.[17] Support for Qasim in stopping all exploration activities was wide and unreserved. Individuals and organizations throughout the nation, and some beyond its borders, expressed strong approval in telegrams or public statements.[18]

Companies Request that Negotiations Be Resumed

With the nation's press mobilized in Qasim's support, it was evident that if negotiations were resumed, it must be on the initiative of the companies. They were not long in acting. On April 28 Lord Moncton, chairman of IPC and associated companies, cabled the Iraqi oil minister expressing regret that a difference of opinion on a mere technical accounting problem (dead rents) should be allowed to deadlock negotiations and suggesting that preparations be made for their renewal.[19]

The government's response was both cautious and demanding. As a condition for resumption Qasim insisted on a clear written statement by the companies on the government's several demands, and the minister of oil suggested that the companies appoint a negotiating team of greater responsibility. The companies complied in part on both suggestions. On June 9 IPC's managing director, G. H. Herridge, advised Qasim by letter that the companies accepted in full his demands on dead rents, on which the April

17. *MEES,* April 21, 1961.
18. Among them, locals of the General Federation of Iraqi Trade Unions from Basrah, Kirkuk, Mosul; Iraqi Marxists; the Central Committee of the Iraq Communist party; the National Progressive party; representatives of the influential Shammar tribe; Hasan ibn Ibrahim, head of the Yemen delegation to a Baghdad meeting of the Arab Economic Council; Sheik Abdullah Tariki, at that time articulate and influential minister of oil in Saudi Arabia, *MEES,* April 28, 1961.
19. *Financial Times,* May 1, 1961.

meetings had floundered, and announced a new negotiating team of responsible representatives of the parent companies.[20]

Negotiations were eventually set for August 24. Meanwhile, the government gave no evidence that it would recede a single step on its far-reaching demands. In truth, Qasim, by abandoning the diplomacy and compromise of the conference room for the public arena with its verbal weapons of patriotism and prejudice, had made retreat politically impossible. The government made clear its mood in an official Baghdad radio broadcast shortly before the negotiations resumed when it declared that "cancellation of the companies' concessions will be the natural outcome of the oil talks if the companies persist in their tyranny and disregard of Iraqi rights."[21]

Qasim had clearly gone beyond the point of no return.

The Companies' Role in the Breakdown of Negotiations

Had the revolution brought prosperity and political unity to the Iraqi people and loyalty and security to their leader, Qasim might have found it unnecessary to mobilize the press in a battle against the "imperialistic monopoly" which, whatever its shortcomings, was in large measure the principal source of such material progress as Iraq had realized.

The truculence and animosity that Qasim eventually manifested at the conference table and in the public forum, however sound its institutional and historical basis, did not create an atmosphere favorable to successful negotiation. His dedicated intransigence affords adequate explanation of the negotiations' failure, but IPC is not blameless.

Logic of the Cost-of-Sales Discounts

The companies, if not guilty of sharp dealings, had not always followed practices calculated to inspire confidence in skeptical minds. The so-called "cost-of-sales" discounts, which the companies had made over the years, illustrate the point. It will be recalled that the border prices fixed in the 1952 agreement for the purpose of calculating the profits to be shared with

20. The full contents of the letter have never been revealed, but apparently it also made company concessions on other demands. *Financial Times,* June 20, 1961; *Times,* June 20, 1961.

21. *MEES,* August 25, 1961.

the government reflected discounts of seventeen shillings and sixpence a ton on oil shipped by pipeline to the Mediterranean and 13 shillings on oil shipped by way of the Persian Gulf. The company initially justified these discounts as covering the cost of selling the oil. Inasmuch as the operating companies disposed of all their oil to their parent companies, cost of sales, if any, was nominal. Good cost-accounting procedures would have required any cost of sales that the parent companies thereafter encountered to be borne by them. The discounts do not in fact seem to have been "cost of sales" discounts at all. Herridge, in explaining them in the 1960 negotiations, said that they represented discounts made by the parent companies in selling their oil on long-term contracts during a period when market weakness necessitated sales below the posted prices. Were this explanation valid, good accounting practice and equity would seem to require that this "cost" be borne by the parent companies who encountered it.

But the historical facts lend little credence to Herridge's explanation. It will be recalled that as a result of negotiations the company reduced the discounts in 1955 to 2 percent and in 1957 to 1 percent. Herridge explained this reduction on the grounds that the state of the market in 1955 no longer justified the larger discounts. The fact is that when the discounts were originally incorporated into the agreement, the market for crude oil was unusually strong, reflecting the inflationary demands growing out of the Korean War. The discounts seem difficult to justify, and they were substantial.

Dead Rents and Natural Gas

Nor were the companies' practices with regard to dead rents better calculated to inspire confidence. In the April 1961 negotiations Qasim asked Herridge the specific question, "Does the 1952 agreement provide for recovering dead rents?" To which Mr. Herridge replied, "That agreement did not provide for their nonrecovery." But he added, "We believe that its general terms permit recovery."[22]

When this statement is set off against Dr. Pachachi's undisputed claim in 1957 that the companies had proposed that provision for recovery of dead rents be included in the agreement but that the government rejected the proposal, it puts the companies' position in a distinctly unfavorable light.

The same may be said of the position initially taken by the company on

22. Minutes as reported in *Times,* April 26, 1961.

the government's request that it be allowed to take without charge as much as it wished to use of the natural gas then being flared. The company agreed to supply the gas at a calculated cost price of ID .006 per 1000 cubic feet, with the proviso, however, that any products made from its use be confined to local markets. Although not itself using the gas, it apparently wanted to protect any foreign markets from its competition.

Company Negotiators Lacked Authority

Although IPC was a corporate entity with a responsibility and authority at law separate from that of its corporate stockholders, it was in fact a servant of their interests. As such, its negotiator-representatives lacked the authority essential to effective bargaining, the essence of which is mutual compromise and adjustment. Lacking authority, company negotiators frequently brought negotiations to a standstill while they sought new instructions. Thus, after two weeks the 1960 negotiations were suspended on August 31 to enable Herridge, IPC's managing director and head of its negotiating team, to confer with his principals in London. Between that date and December 19, when Herridge and his fellow negotiators conferred for six and a half hours with Qasim in Baghdad, they had conducted negotiations in a piecemeal fashion with Herridge sandwiching short conferences with Qasim between longer stays in London.

After the abortive meetings in the spring of 1961 the negotiators did not meet again until August 24, 1961. Realizing that the negotiations had reached a critical stage, IPC's owners then sent a more authoritative, responsible, and impressive group of negotiators to Baghdad. H. W. Fisher, a director of Standard of New Jersey, headed the new team, with F. J. Stephens, chairman of Shell Transport and Trading, and Herridge as fellow members. They brought with them a large group of advisors. But when Qasim made demands beyond the concessions they had been authorized to meet, Fisher and his colleagues adjourned the meeting on August 30 to consult with their respective boards in London, New York, and the Hague. They returned to Baghdad on October 6 and ended the meetings on October 11.

World Oil Surplus

But although the companies' practices and their negotiators' lack of authority—the one by magnifying the obstacles, the other by weakening the instruments by which to overcome them—both contributed to the failure of negotiations, they were not the major stumbling blocks to an acceptable compromise. More substantial factors dissuaded the companies. They were, or seemed to be, negotiating from strength. The world's oil reserves had been mounting more rapidly than demand, and changes in the industry's structure had weakened the control which the majors had previously exercised over prices. Paradoxically, as their market control weakened, their bargaining strength seemed to have increased. The companies in their annual report for 1959 pointed out that "there is . . . estimated to be in the world today an excess of producing capacity over world requirements of some five million barrels a day, the equivalent of 250 million tons of oil a year." All of the parent companies held other sources of supply to which they could readily turn should Iraq oil not be available to them. Although they could not lightly dismiss the danger of unilateral action, they realized that Iraq could scarcely have been unmindful of Iran's unfortunate experience with nationalization, which had shut it off from the world's market and eventually triggered a coup d'etat which overthrew the Mossadegh government. The Iran crisis occurred when surplus capacity was far less and the oil industry had readily met it. The IPC parent companies were in a much stronger position in 1960 to meet a similar crisis. This no doubt made them less willing to adjust to Qasim's no-compromise attitude.

Reluctance to Disturb Existing Concession Patterns

Perhaps more important to the companies was the realization that any major concession they made to Iraq would be promptly demanded by other Middle East governments, constantly guided in their relations with their concessionaires by the "most favored nation" principle. The companies were not ready apparently to upset entirely the pattern of Middle East concessions, already greatly weakened by the advent of newcomers.

Clash in Aims and Attitudes

But more important than all the foregoing factors is the fundamental clash in aims which animated the companies and the government. The companies were private corporate enterprises created to foster the business goals of their parents. Their sense of fair play and equity was consistent with the business codes to which they subscribed. They professed to seek no unfair advantage over the government, but they were uninfluenced by Qasim's plea that "justice is to give peoples their rights." As Fisher put it, "this is a commercial arrangement between the two parties. Much as we feel the hardships of the Iraqi country and people, we should . . . adhere to the commercial basis."[23]

In the recess of negotiations between August 24, 1961, and September 29 the companies' negotiators had prepared an economic and statistical analysis of the contemporary world petroleum situation and its impact on their parent company profits. By outlining the state of the oil industry, they had hoped to convince Qasim that the government was getting not only a fair share but a bigger share of profits than were the companies. "The government's share is the bigger because it is based on posted prices" which are no longer being realized. This being so, Fisher concluded, "[we] consider it our duty to point out that the fifty-fifty profit-sharing formula cannot possibly be altered."[24]

To Qasim, dedicated to freeing the country from the taint of colonial imperialism, seeking to gain prestige for his regime throughout the Arab world, which his isolation had cost him, committed to a revolutionary program which had not gone well, and through his public utterances to a dramatic victory over the oil companies essential to his domestic security and his nation's prestige, such arguments were of no avail.

We must not lose, since we are the custodians of the right in this country. We have nothing now. We feel that the negotiations have broken down because of the stubbornness of the companies . . . and because they have not agreed to concede Iraq's rights.

With a threat and a gesture of politeness, he brought the negotiations to a close. "We cannot keep quiet for long over the loss of Iraq's rights. . . . However, I wish you well." To which Fisher replied that the companies would "have to adopt the necessary steps to safeguard our interests."[25]

Qasim Hints at Far-reaching Unilateral Action

It will be recalled that Qasim had terminated the deadlocked negotiations of April 1961 with an order that the companies discontinue all

23. Minutes as reported by Platt's *Oilgram News Service,* November 24, 1961, p. 3.
24. *Idem.*
25. *Ibid.,* p. 12.

exploration and drilling operations outside of the producing areas of the concessions until such time as the government and the companies should reach an agreement. He had accompanied the order with a warning that the government would take whatever measures were necessary to enforce it. Through the oil ministry he also warned that if the companies persisted in disregarding Iraq's "legitimate demands," the revolutionary government would be "compelled to handle the matter in a manner which will safeguard the rights of the Iraqi people. . . . The Iraqi Republic is determined to defend its rights, safeguard its freedom, and defeat imperialism."[26]

On closing the final round of negotiations on October 11, 1961, Qasim had indicated more definitely the character of the action he intended to take. In a speech before the education ministry on November 12 he stated that the government contemplated a series of measures "if the imperialists did not speedily concede Iraq's legitimate rights." The first measure, he warned, would have "tremendous repercussions." At the same time he uttered a polemic broadside against the British, at whose door he laid the Kurdish revolt, charging that it represented an effort "to safeguard their imperialistic interests in Iraq and elsewhere in the Middle East, including the decadent regimes of the Gulf States." With reckless abandon he prophesied that Britain would not long endure after it had been deprived of the wealth of Iraq and other countries on which it subsisted, and concluded with the ominous threat, "The only way to expel the imperialists is by force."

Law 80

On December 11, 1961, the Iraqi government issued Law 80, defining the areas of exploitation permitted each of the oil companies in accordance with a schedule appended to the law (Article 2). Article 3 of the law authorized the government at its discretion to allocate to each of the companies an additional area no larger than that originally allocated. Article 4 relieved the companies of all rights they previously held to territory not specified in the schedule. Article 5 required the companies within three months to make available free of charge all geological, geophysical, and engineering data they possessed relating to the confiscated areas. Law 80 permitted the companies to retain only 1,937.75 square

26. *MEES*, April 14, 1961.

kilometers or approximately 0.5 of one percent of the original concessions.[27]

Explanatory Statement

The government accompanied Law 80 with an explanatory statement in five parts. Part I gave a "Historical Background to the Concessions"; Part II summarized the "Exploration and Development Operations" conducted by the companies under their concessions. Part III reproduced the statement issued by the oil ministry on April 10, 1961, in which the government set forth the issues in the dispute and reviewed their history. Part IV reproduced a statement issued by the oil ministry on October 17, 1961, bringing the history of negotiations up to date and announcing that the government "finds itself compelled, after prolonged negotiations, to take legislative measures . . . to safeguard Iraq's interests without injuring the reasonable interests of the oil companies."

In its statement of the historical background of the concessions (Part I), the government alleges that (1) after World War I a fierce competition among the "covetous imperialists" was settled when Iraq was not a sovereign state but under the direct control of the British mandatory power, by an agreement that delayed Iraq's development and denied it a fair return essential to improving the living standards of its people; (2) the first concession limited the operations of the company to twenty-four rectangular plots totaling 192 square miles, on which the concessionaires were required to drill a specified footage within a specified period on penalty of losing their concession; (3) the company had failed to meet its obligations, but through the influence of the British government, it had obtained a one-year extension; (4) through the same influence it had obtained a modification of the concession which permitted it to explore and develop oil in an area of 91,000 square kilometers instead of the 497 square kilometers (192 square miles) originally contemplated with no provision for relinquishment and no guarantees regarding rate of exploration and drilling; (5) when the government, desiring a competitor to IPC, granted the British Oil Development Company a concession covering 107,000 square kilometers west of the Tigris and north of the 33rd parallel of

27. *Ibid.*, December 15, 1961, *MEES* published an English translation of the law and the explanatory statement, and its issue of December 27, 1961, reproduces the schedule of the areas the companies are permitted to retain.

latitude, IPC defeated the government's purpose by buying all of BODC's shares; (6) in 1938 IPC, by acquiring a concession through BPC covering 226,000 square kilometers, brought all of Iraq under its monopolistic control except a small tract on the Iranian border under concession to the Khanaqiun Oil Company, owned by Anglo-Persian.

Part II of the explanatory statement, summarizing IPC's drilling and exploration activities, alleges that the companies, concerned with their own, not Iraq's, interests, were slow to develop their concessions until the July 14 Revolution and that they speeded up operations only when confronted with the demand that they relinquish the unexploited area.

Part V of the explanatory statement attempts to justify the cancellation of all but 0.5 of one percent of the concession on the ground that if the concession had included "the customary provision for gradual and periodic relinquishment" of those portions of the concession area not being actively exploited, IPC would hold only those areas which the law permits it to retain as of the date of the law.

Companies Declare Their Rights and Demand Arbitration

In a public statement on December 13, responding to the cancellation proclamation, the oil companies expressed regret at the government's action and declared that their rights embodied in the concessions, which they reserved in full, could not be altered by unilateral action. On January 4 they presented a formal protest to the Iraqi government, declaring that Law 80 and the April order which preceded it, stopping all exploration work under threat of force, were not only grave violations of the companies' rights under their agreements but serious breaches of well-established principles of international law. They demanded that the issues in dispute be submitted to arbitration under procedures provided for in the agreement and requested the government to nominate its arbitrator. At the same time they notified the government that they had appointed as their arbitrator the Rt. Hon. Lord McNair, formerly president of the International Court of Justice and one of the world's leading international lawyers.

Both the United States and the British government dispatched diplomatic notes to Qasim urging him to accept the companies' arbitration proposals. On January 6 at a reception on the forty-first anniversary of the creation of the Iraqi army, Qasim alleged that the British had accompanied their note by the pressure of "troop concentration and movements" and that while the United States had brought no direct pressure, it was relying on Britain as its ally. Qasim was apparently referring to British movement of troops to Aden a month earlier following renewed threats by Iraq against Kuwait. He belligerently characterized such moves as "forces of aggression that will be destroyed by lightning blows from us and the nations that support us." He proclaimed, "The use of pressure against a free Republic which rose to power to restore the rights of the people will not be tolerated by us or the other nations of the world."

The Companies Supply the Government with Data

During the next several months Qasim ignored the companies' request for arbitration. Nevertheless, on March 5, one week before the deadline set under the law, the companies notified the oil ministry of their intention to comply with Article 5 of Law 80, which required them to supply the government without charge with all the geological, geophysical, and engineering data they had collected over the years on the explored but unexploited concession areas. The data, massive in amount, filled 160 boxes ranging between 6 and 12 cubic feet in size. In agreeing to submit the data, an IPC London spokesman denied that this signified the companies' acquiescence in Iraq's seizure of the undeveloped areas, stating, "The company and its associates have always been supplying the Iraqi government with information relating to their operations in the country."

Government Prepares New Law

Meanwhile the government had been quietly preparing a new law designed to bring the confiscated areas under active exploitation. On September 29, 1962, at a three-hour press conference Qasim submitted the draft of a proposed law to the journalists present with extensive commentaries on its provisions. This unique procedure, Qasim explained, would give politicians, journalists and the public generally an opportunity to criticize and suggest improvements in a law of vital significance to the country's welfare and to the oil industry. Reviewing the steps by which the Iraqi people had regained most of the concession areas which the old regime had parceled out for "underhand plundering . . . by the monopolistic oil companies," he urged the collaboration of all loyal Iraqis.[28] Pending their response, he said that the law would remain in draft form.

The records indicate only a limited public response to this unique opportunity in law-making[29] and the draft was still in abeyance when the February 1963 revolution brought the Baathists to power and sent Qasim to his grave.

The Government Considers Arbitration

Strange as it may seem, available evidence suggests that in delaying promulgation of the draft law, Qasim was less interested in obtaining advice from his people than in coming to some agreement with the oil companies. Despite his unceasing fulminations against them and his persistent affirmation of Iraq's ability to get along without them, apparently he had begun to doubt whether his actions had best served his own and his country's interests.

Two developments may have caused him to re-evaluate the situation: first, he was soon brought face to face with the fact that IPC by its control of oil production could exercise an effective control over his country's welfare; second, events forced him to realize that it would be difficult either to replace British technicians in the operation of the already developed properties or to find newcomers willing to exploit the undeveloped lands.

Although IPC had committed itself to an investment and expansion program in the spring of 1959 that promised to double Iraq's production capacity by 1962 and had completed much of this planned expansion, by the beginning of 1962 it was increasing production at a snail's pace. Total Iraqi output for the first three months of 1962 exceeded 1961's first three months output by only 1.8 percent. During the same period Iran's output had increased by 10.4 percent and Saudi Arabia's by 3.4 percent. Iraq's production for the whole of 1962 was up less than one half of one percent, while Kuwait's had increased by 11.5 percent, Iran's by 12 percent, and Saudi Arabia's by 9.2 percent. In 1962 Iraq produced only 48 million tons; the capacity envisioned earlier was 70 million tons. August output in the

28. *MEES*, October 5, 1962, presents an English translation of the proposed law and accompanying statement.

29. According to the *Iraq Times*, October 8, 1962, the government requested several Baghdad University professors to review and comment on the proposed law and the National Democratic party made an extensive criticism of it.

southern field dropped more than 30 percent. This production slowdown greatly disturbed Iraq's officials, as indeed may have been intended. On July 14 in an address before the Baghdad Military College Qasim charged that the companies were endeavoring to pressure Iraq by deliberately curbing output, and he warned,

If they do not increase their production, we will take action to obtain our rights both as regards increased production and the other issues which the companies call subjects of dispute but which we call matters of trespass . . . on the rights of the people.

On August 7 Minister of Oil Muhammad Salman officially requested IPC to step up its production. Soon thereafter the company dispatched a representative to London to consult on the matter with the IPC Board of Directors.

As Qasim was learning by experience the significance of IPC production control as a bargaining instrument, he was also learning the difficulties of obtaining aid from other companies. Although the government had called for bids on Iraq's offshore Persian Gulf rights in July 1960, two years later none had been accepted. Shortly after Iraq had unilaterally reclaimed 99.5 percent of the original concession areas through Law 80, Qasim in a talk before the Fourth Engineers' Conference in Baghdad on January 16, 1962, boasted that the experts had estimated that from the confiscated areas Iraq could obtain an income of ID 350 million a year, more than three times what it was then receiving from IPC, and that it would spend this "on the sons of the people . . . and in liberation of the usurped portions of the Iraqi homeland." Seven years later the area lay idle.

While matters were developing in this manner, officials of the oil ministry had quietly invited a group of British lawyers to consult with the government on legal aspects of the dispute.

On February 2, 1963, Oil Minister Salman in a Baghdad interview with Edouard Saab, special correspondent of the Beirut semiweekly *Le Commerce du Levant,* explaining the status of the proposed oil law, indicated that matters stood just as when the proposed law was submitted to the public in October 1962, and that he had discussed the situation with the oil companies many times since. His statement reflects a faint note of pleading in contrast to the old note of defiance:

The other day, and for the umpteenth time, I was trying to explain to the company representatives just how negative and absurd their attitude was. You are making strenuous efforts, I told them, to help the underdeveloped countries by granting them long-term loans and technical aid. You are doing this, it seems, to raise their living standards and to shelter them from subversive doctrines. We don't ask this much from you for Iraq. . . . Just give us our due. Try at least to behave in good faith. You take a malicious pleasure in misleading us, and in depriving us of our most legitimate rights. In the minds of the people, all this may finally rebound against you and against the principles of what is still called "The Free World."

Turning to a discussion of the Organization of Petroleum Exporting Countries, of which he was one of the founders, Salman left the impression with Saab that he hoped that the dispute might be settled through that agency.

One week after the oil minister's interview the "Immortal Revolution" had come to an ignominious end and its "Faithful Leader" lay in his grave, unwept, unhonoured, and unsung.

11

IRAQ-IPC Relations Since Qasim's Overthrow

NEITHER Law 80 nor Qasim's death settled the controversies between Iraq and IPC. Qasim through Law 80 recaptured by unilateral action the oil rights over most of the nation's vast area that previous governments had granted, but in doing so he left IPC embittered and the confidence of the international oil industry in Middle East contracts seriously undermined. Nor had Qasim developed a program by which the oil potentialities of these reclaimed lands could be determined and realized.

Negotiations Begin Anew

Qasim's overthrow presented IPC with a new opportunity. In trying to come to terms with a new government, new ways might conceivably be found to settle old differences. However, the prospects for the repeal of Law 80 were dim. The Iraqi Baathists had much to do with engineering the revolution, and the army officers who took over control of the government were friendly to their cause. The leaders of the Baathist socialist party were social philosophers whose major doctrines were Pan-Arabism and social-ism. As socialists, one of their major tenets was that the mineral resources of the Arab countries should be nationalized and exploited solely for the countries' benefit. Within twelve days after the overthrow of Qasim, Michel

256

'Aflaq, Secretary General of the Baath party and its main architect, arrived in Baghdad at the head of a delegation of leading Syrian and Lebanese Baathists. They are said to have been accorded a reception appropriate to visiting heads of state. Three ministers and a detachment of the national guard welcomed them at the airport. At a press conference in which 'Aflaq discussed many things, he said of oil that although he stood for the complete nationalization of the petroleum resources of the Arab world, "in practice such measures should only be taken after thorough planning and a long and detailed study of the problems it involves."[1] Meanwhile Iraq's new president had indicated that oil must continue to flow.

The first task of the "National Council of the Revolutionary Command" was to set up a government. It designated Colonel 'Abd al-Salam 'Arif as president of the republic and he chose Brigadier General Ahmal Hasan al-Bakr as prime minister. In the new cabinet of twenty-one members army officers in addition to holding the premiership, headed the ministries of defense, communication, municipalities, planning, industry, and state. 'Abd al-'Aziz al-Wattari was the new minister of oil. Wattari, aged thirty, a petroleum engineer with a degree from the University of Texas, had successively held the posts of acting director general of oil affairs and director of government oil refineries and had represented Iraq at meetings of the Arab oil experts in Jiddah in 1957 and in Kuwait in 1960. Wattari's training and experience had given him a grasp of the technical problems and some familiarity with the commercial problems of running an oil industry.

Negotiations Settle the Dispute over Basrah Port Dues

Oil men hoped that Wattari would be an easier man to deal with than was Qasim. In keeping with President 'Arif's declaration that oil must be kept flowing, the oil minister and representatives of Iraq Petroleum Company soon established communications. Shortly after the new government was established, IPC's executive director, L. F. Murphy, joined W. W. Stewart, IPC's chief representative in Baghdad, in examining the possibility of settling the long-standing controversy over Basrah port charges and cargo dues, a dispute that Wattari had indicated might be handled effectively apart from other more substantial matters at issue. Wattari's judgment was correct.

1. *Middle East Economic Survey,* February 22, 1963.

On June 21 an IPC spokesman announced in Beirut that "an interim arrangement has been reached with the Iraqi government on cargo dues and port charges at Basrah" which will "encourage offtakers to take more oil from Basrah." He expressed the hope "that this limited arrangement may be a first step toward a settlement of larger outstanding problems between the government and the companies."[2]

It will be recalled that during the Qasim negotiations the government raised cargo dues (a form of export tax) on oil loaded at Basrah from 23.4 fils (6.5 cents) a ton, the rate fixed in an agreement reached in 1955, to 280 fils (78.4 cents) a ton. This increased the effective price of Basrah crude by 11 cents a barrel with the result that shipments from Basrah dropped from about 12.0 million tons in 1959 to 8.6 million tons in 1962 with a corresponding drop in government revenues. In addition to so increasing the cargo dues, the Qasim government, five days before its overthrow, issued an amendment (No. 33) to the "1953 Schedule of Basrah Port Dues and Charges" (published in the *Iraqi Official Gazette* of February 3, 1963) fixing *port charges* at the new deep sea terminal at Khor al-Amaya at 280 fils (78.4 cents) per ton of gross registered capacity, graduated downward to 70 fils (19.6 cents) per ton when exports of crude from the terminal in any fiscal year equaled 10 million tons.

IPC regarded these charges as excessive. Negotiations about them between the new government and IPC resulted in an interim agreement apparently satisfactory to IPC.

Negotiations Encounter Obstacles

As negotiations on the Basrah dispute got underway both the London *Financial Times* (May 23, 1963) and the *Times* (May 25, 1963) commented on improvements in the climate of negotiations and suggested that the long climb toward an eventual settlement of all issues between IPC and the government had begun. Their optimism regarding the settlement of the Basrah port charges and cargo dues issue was justified. However, the climb toward an eventual settlement of outstanding IPC-Iraqi issues which began with the Basrah settlement proved tortuous and deceptive. Frequent cabinet reshuffles, the death of President 'Arif, and two abortive coups blocked the upward climb and with the summit almost in sight the Syrian-IPC

2. *MEES,* June 28, 1963.

pipeline dispute and the Israeli-Arab six-day war shattered hopes of a settlement.

Negotiations began soon after Wattari became oil minister. As long as he was in office he succeeded in keeping discussions in the conference room, where they responded to the ebb and flow of rational argument, and out of the public arena, where they were likely to be buffeted by the nationalistic winds of propaganda and chauvinism. From time to time the oil minister announced informally that negotiations were going well and might soon be completed. Meanwhile they had assumed a two-pronged character: one prong pointed toward the issues involving the interpretation and modification of the basic agreement; the other was directed toward the role of the Iraq National Oil Company (INOC) in the exploitation of the expropriated area and its return to the concessionaire. It soon became evident that the new government had no intention of repealing Law 80 and that it expected to transform Qasim's paper plan for an Iraq national oil company into a living document and to utilize INOC as an instrument for conducting integrated oil operations under government control.

Government Establishes National Oil Company

On February 8, 1964, on the first anniversary of the new government, Wattari announced the formation of the Iraq National Oil Company. The law establishing the new company, Law 11 of 1964, followed in the main the provisions of the draft law of Qasim's regime.[3] But it contained some important modifications. Law 11 stated that the objectives of the company shall be to "engage both inside and outside Iraq in the oil industry in any or all of its phases including exploration and prospecting for oil and natural hydrocarbons; production, transportation, refining, storage, and distribution of oil and oil products and their derivations (petrochemicals)" (Article 2.1). The draft law had stated the objectives in similar terms (Article 2.1). Both gave INOC the right to organize subsidiaries or to form companies with other parties or to operate with existing companies in carrying out its functions. Neither the law nor the proposed law gave INOC the right to distribute petroleum products for consumption within Iraq (a

3. The following references to Law 11 are to the official translation of the law as published in *The Weekly Gazette of the Republic of Iraq*, No. 10 of March 4, 1964. The references to the draft law are to the English translation published by the *MEES*, October 5, 1962.

right reserved to the Government Oil Refineries Administration) and Law 11 denied it the right to refine oil for local consumption, both prohibitions to prevail "as long as there exist other government organizations having legal monopolistic control over such operations." (These prohibitions have since been removed.) Law 80 left to the companies only those areas from which they were producing oil when the law was proclaimed. At that time the companies had explored and discovered oil in areas that they were not actively exploiting. Article 3 of Law 80 authorized the government at its discretion to allocate to the oil companies additional territory as reserves in an amount not exceeding the areas that the law specifically reserved to the companies. Law 11 provided that "the Company shall select the areas where it may desire to carry out its operations" (Article 3.2) and that it submit an initial application for defining these areas within six months from the date of the law (Article 3.3) for the approval of the minister of oil and the council of ministers (Article 3.2). Law 11 raised the company's capital, which was to be provided by the government, from ID 20 million, as provided in the draft law to ID 25 million (Article 4.1). Both permitted the council of ministers to increase the capital on recommendation of the company's board of directors, and allowed INOC to borrow funds at home or abroad up to three times its capital, subject to conditions to be determined by the council of ministers. Both laws exempted the profits of INOC from taxation during a limited period, the draft law for ten years (Article 7.1) and the law for five years (Article 8.1), but Law 11 laid down the principle that INOC pay to the government annually 50 percent of its net profits (Article 7.2). The law as enacted provided that INOC's payments to the government should be treated "for income tax purposes . . . as part of the operating expenses." A subsequent amendment deleted this requirement.[4]

The law provides that INOC "shall be administered by a Board of Directors which shall be financially and administratively independent and composed of nine members" (Article 9.1). By providing that only one third of the directors should be senior government officials (instead of the two thirds as in the draft law (Article 8.1) and that the six full-time members may be experts specialized in oil, economics, legal, or technical affairs, the law seemed designed to insure technical competence in INOC's administration. On the other hand, the law provided that the council of ministers appoint the full-time members, select the chairman of the board

4. Law 89 of 1966 published in the *Official Iraqi Gazette*, October 25, 1966.

and the company's managing director (Article 9), approve any partnership or participation that INOC may wish to make with another party, approve any loan that the company may desire to make (Article 11). The law attached INOC to the Ministry of Oil, obligated it to adhere to the "general oil policy of the State" (Article 12), provided that the Council of Ministers decide any differences between the minister of oil and the company in regard to policy (Article 12), and approved its budget (Article 13).

These provisions would seem to indicate that INOC would not be entirely insulated from the political winds of transitory governments.

Wattari Expresses Optimism about INOC's Future

At a press conference following the oil minister's announcement of the formation of the Iraq National Oil Company, Wattari emphasized that organizing INOC was not a first step towards nationalizing the oil industry. He was quite optimistic about INOC's future as an independent company operating "in the public interest" quite separately from "existing concessions." He disclosed that the government had received offers from twelve foreign companies to explore for oil in Iraq—companies that INOC might invite to participate in joint ventures. He predicted that INOC would be producing and marketing oil within four years. Time has proved him wrong.

He also expressed the hope that IPC and its associates would co-operate with INOC, asserted that the government would try to prevent difficulties from arising in their relationship and find a solution to all their problems through peaceful negotiations. He warned, however, that "we are ready to face them . . . the companies must accept the national oil company: this will be in their own interests as well as ours." Negotiations with IPC, he said, would begin shortly.[5]

Attitude of IPC toward INOC

IPC and its associated companies viewed the matter differently. On February 11 they issued a statement from London in which they said,

If the new law does in fact take the form reported in the Baghdad press and over the radio, then it would inevitably constitute a further violation of the companies' agreements with the Iraq government.

5. *MEES*, February 14, 1964.

No official text of the law announced on the Baghdad Radio this weekend is yet available, but reports of its promulgation and the form it is alleged to take have come as a disappointment to the companies in view of recent public indications that discussions were to take place shortly on many outstanding problems, including those created by the promulgation by the Kassim regime of Law 80 in December, 1961, and for which the companies have repeatedly asked for arbitration.

They concluded by stating that the companies were giving careful consideration as to what further action to take to protect their interests.

Before the companies revealed any contemplated action their representatives had begun negotiations with Minister Wattari in an effort to reach a compromise settlement. On June 11, 1965, Wattari announced that they had reached an agreement which he would present to his government for review and approval.

Wattari Outlines the Terms of the Agreement

A few weeks earlier in a statement quoted in the Baghdad daily *al-Jumhuriyah* (May 22, 1965), Wattari discussed the main features of the agreement. Although the agreement was never approved nor published, informed sources have revealed its major provisions. It consisted of two parts. Part One dealt primarily with the issues in controversy before Law 80 and the issue which the decree itself raised. Part Two dealt with the relations of INOC with the IPC participants. Together the two parts constitute in effect a recognition by oil companies of the validity of Law 80 under which Iraq reclaimed most of the concession areas.

Part One restored to the companies an area equal to that which Law 80 had permitted them to retain, 1,937.75 square kilometers. Law 80 permitted the IPC to retain all the company's producing fields but expropriated the now famous North Rumaila field, estimated to contain large oil reserves, which IPC's subsidiary, BOC, had discovered but had not developed. The agreement provided for the return of the North Rumaila field to the companies. The companies agreed to restrict their rights of exploitation and production to an area of about 3,875 square kilometers, representing about one percent of their original concession, but containing all their known fields. Article 3 of Law 80 had provided for this contingency by specifying that the government could at its discretion return a larger area than that initially allowed them. The companies agreed to pay the government an indicated sum (said to be about £20 million) in settlement of all issues outstanding when the Qasim negotiations began—dead rents, exploration and drilling costs, alleged arrears because of the companies' accounting methods less monies owed to IPC for services and supplies—and to substitute actual costs in calculating the government's share of profits for the former fixed-cost system in vogue since 1955. They also agreed to make available to the government natural gas in surplus beyond their own needs. The agreement left for later discussion between the partners the specific issues raised by Resolution 32 of the Organization of Petroleum Exporting Countries, which aims at a restoration of posted prices to the pre-August 1960 level, and its Resolution 33 which calls for expensing of royalties. It accepted the settlement concessionaires had worked out with other OPEC members on the question of marketing discounts—that selling expenses be limited to one half cent per barrel effective January 1, 1962. For a discussion of OPEC and these issues see Chapter Seventeen.

The companies agreed to raise their minimum annual production obligation from 30 million tons to 45 million tons. In addition the companies agreed to produce five million barrels from the new areas restored to them. They also undertook to "use their best endeavors," without any legal obligation, to dispose of 66 million tons of oil in 1966 and 70 million tons in 1967. Had the agreement eventually been signed, these figures would presumably have been updated.

The second part of the agreement represented its most dramatic features. All of the IPC participants with the exception of Standard Oil of New Jersey agreed to establish a joint venture with INOC to explore and develop an area covering 32,000 square kilometers, an area said to include the most promising exploration prospects now known in Iraq. Compagnie Française des Pétroles, Iraq Shell, Ltd., British Petroleum Exploration Company, and Mobil Oil Development Company were each to have a 15⅚ percent interest in the joint venture and the Gulbenkian interests would have 3⅓ percent. INOC was to have 33⅓ percent.

The tax obligations of the participants would have been those now customary in the older Middle East concessions under their 50-50 profit-sharing plan. Iraq Oil Development Co., Ltd., owned by the member companies, would be the guarantor of their tax obligations. The participants in the joint venture agreed to pay the government an annual royalty of 12½ percent, which in principle would be expensed. The terms for implementing this expensing and the rates of discount allowed from posted prices would have been those applicable to IPC, Basrah Petroleum Company, and Mosul Petroleum Company when exports from the joint venture began.

The Baghdad Oil Company (BOCO) was to have been incorporated to operate the joint venture's concessions. It would own nothing and earn nothing and would engage in no activity outside the agreement. It would propose annual budgets and programs for developing the joint venture. Responsibility for the management of the joint venture was to have rested with BOCO's board of directors. INOC would have appointed one director and the participants one each. The board was to appoint the general manager from outside its ranks. The government would have had the right to appoint one director as an observer and one of the INOC members would have served as chairman of the board. The company members were to appoint the deputy chairman.

The joint venture was obliged to spend $30 million in exploration in the first six years and $20 million in the next six. The risks of exploration fell entirely on the companies. They were to advance INOC's share of exploration costs. If oil were discovered in commercial quantities, INOC would repay the advance in six installments starting twelve months after exports began.

Until INOC had established adequate outlets for its one-third share of the joint venture's production, the member companies agreed to buy INOC's excess oil within certain limits at a price midway between costs and posted prices but not exceeding 60 percent of posted prices. The companies were to pay income taxes on the difference between the posted price and the price they paid INOC. The agreement also provided that INOC alone (or in partnership with one or more member companies) might develop at its sole risk a separate field within the concession area if it had a market for its own oil in excess of its share of the joint venture's production. If INOC should develop a separate field at its sole risk, the member companies would no longer have been obliged to buy oil from it.

Significance of the Agreement

Had the government approved this agreement, it would have marked a major change in the relationship between IPC and a Middle East government. The companies by agreeing to exploit a relatively small area of the original concession in partnership with the government were in effect acknowledging the government's right to expropriate most of the concession without compensation. The agreement would have set a concession pattern that would have enabled the Middle East countries to assert with impunity their sovereignty over their oil resources.

The *Middle East Economic Survey* characterized the draft agreement as the work of "hard-headed, realistic experts on both sides" giving no quarter and expecting none. About it the *Survey* said,

The decision to seek a settlement with the companies was presumably a collective one taken by the government as a whole. And it would surely seem to an unbiased observer that the settlement . . . embodied the optimum terms that Iraq could reasonably expect to obtain in the prevailing circumstances.[6]

This seems a reasonable judgment.

Agreement Engenders Controversy

Nevertheless, as the terms of the agreement became generally known they initiated a bitter controversy that spread rapidly throughout much of the Arab world. Early in July six cabinet members, leaders in the so-called Arab Nationalist Movement, resigned because of sharp conflicts over various issues, one of which was oil. While the oil settlement was a side issue, in the words of the *Middle East Economic Survey* "it gave the ANM leaders a convenient cudgel with which to belabor their former cabinet colleagues."[7] Once out of the government they began a campaign of propaganda and criticism against the pact both inside and outside the country. Meanwhile Abdullah Tariki, former Saudi Arabian oil minister, now self-exiled in Beirut and self-appointed spokesman on Arab oil matters, assailed the draft agreement with his customary lack of restraint. Writing in the August 2 issue of the Beirut weekly *al-Hurriyah,* he alleged that the acreage to be returned to IPC contained ten billion barrels of proven reserves that INOC (which had been created for the purpose) might develop without risk. To return these areas to the companies would circumvent Law 80 and represent a "set-back for oil policy in the entire Arab world." He argued that in a joint venture with IPC, the IPC participants would retard INOC's development, preferring to get their oil from areas 100 percent under their control. The daily *al-Thaurah al 'Arabiyah,*

6. *MEES Supplement,* September 10, 1965.
7. *Idem.*

Baghdad organ of the Arab Socialist Union, lamenting that the oil ministry had conducted its negotiations in secret, expressed similar sentiments in an editorial on July 19.[8]

The draft agreement, however, was not without its public defenders. The Baghdad *Review of Arab Petroleum and Economics* in an editorial in its June–July issue (1965) strongly defended the oil ministry's method of negotiating, lamented the political storm that opponents of the draft agreement were creating, and urged the importance of INOC's getting on with the business for which it had been created.

In an editorial in its August issue the *Review* defended nationalization of Iraq's oil industry, for which the clamor was growing louder, but only if it were a part of a collective plan followed by all Arab producing countries. Condemning the "Tarikite tactics of slogan mongering," the *Review* averred that Tariki's brand of piecemeal nationalization would "call for national suicide."

These editorials evoked a sharp response from *al-Thaurah al 'Arabiyah* which in its August 9 and 10 issues accused the *Review* of subservience to foreign powers in the Nuri al-Said tradition. *Al-Thaurah* for some time had been highly critical of government policy. Its criticism grew sharper after the resignation of the six Nasserite ministers in July 1965. To still its criticism and the public clamor which the draft agreement was arousing, President 'Arif, as ex-officio head of the Arab Socialist Union, dismissed *al-Thaurah*'s editorial staff.

Political Instability Contributes to Controversy

Meanwhile Oil Minister Wattari, who as chief sponsor of the draft agreement was an advocate of peaceful coexistence with the companies, in interviews first in Baghdad with *al-Manar*[9] and later in Cairo with *Rose el-Youssef*,[10] stated that every detail of the agreement would be made public before final approval.

On September 6, only ten days after President 'Arif had established a new cabinet, General Tahir Yahya resigned as prime minister. Despite frequent cabinet changes, Yahya had held the premiership continuously since the Baath party was forced out of the government in November 1963.

8. *MEES,* July 23, 1965.
9. *Ibid.,* August 13, 1965.
10. *Ibid.,* August 20, 1965.

In the cabinet reorganization that followed, President 'Arif chose as prime minister Brigadier 'Arif 'Abd al-Razzaq, dropped Wattari from the oil ministry, and appointed Dr. 'Abd al-Rahman al-Bazzaz as deputy prime minister, foreign minister, and acting oil minister. Only ten days after the new cabinet was formed, Prime Minister al-Razzaq led an abortive coup designed, according to press reports, to overthrow President 'Arif and set up a regime dedicated to full union with the United Arab Republic. With the coup's failure al-Razzaq and nine high-ranking army officers fled to Cairo. In the cabinet reorganization that followed Dr. al-Bazzaz became prime minister and Shukri-Salih Zaki became minister of finance and acting oil minister. Dr. al-Bazzaz was the first civilian to head the government since the resolution of 1958. A dedicated Arab Nationalist, he had been imprisoned three times because of his political views—under the British in 1941, under Nuri al-Said in 1956, and under Qasim in 1959. Despite his firm convictions, he is regarded as a moderate in politics. As former Secretary General of OPEC, he clashed with Tariki at the Fifth Arab Petroleum Congress, appealing for a rational not an emotional approach to oil problems.

Prime Minister al-Bazzaz remained in office less than a year. At the outset of a regime marked by moderation he made clear that his policies would be designed to insure the country's social and political stability. At a press conference in Baghdad on September 23, 1965, shortly after becoming prime minister, he said, "The people are fed up with successive coups and regimes that have plagued Iraq." Although dedicated to socialism, he made it clear that he was no Marxist. "We believe in socialism as a means to . . . social justice and economic well-being."[11] Nevertheless, he advocated no general nationalization program but strove for a mixed economy with peaceful coexistence between private and public enterprise.

During his year in office and, more particularly, in the months since, the status of the oil pact became increasingly uncertain. Subsequent months were marked by the death of President 'Arif in an airplane crash, by the selection of his brother, Major General 'Abd al-Rahman 'Arif, as his successor, by an abortive coup following Dr. Bazzaz's peaceful settlement of the Kurdish problem, and by successive cabinet changes. As these changes took place, the government shifted from the center to the left, and a similar change took place in government oil policy. As dissatisfaction with the Wattari settlement intensified and criticism mounted of the alleged

11. *Ibid.*, October 1, 1965.

secrecy with which Wattari had conducted negotiations, Iraq's new president, General 'Abd al-Rahman 'Arif, indicated that before approving the agreement, he would throw it open to public debate. In his first interview following his brother's death, almost a year after Wattari had approved the draft agreement, he stated,

> In fact the new agreement has not come up for discussion yet. As soon as it comes before the Council of Ministers we shall study it most thoroughly, determine its good points and bad points, and put it before the people so that they may pass judgment on it. It is our intention to arrange debates on the agreement on television so that the people may learn all about it.[12]

This proposal to shift debate to the public arena boded ill for the fate of the Wattari agreement. With the change of government following the abortive coup after Premier Bazzaz negotiated the settlement with the Kurds, the new prime minister, Major Naji Talib, became acting minister of oil. Although Dr. Bazzaz had ostensibly submitted his resignation voluntarily, he is said to have been under pressure from President 'Arif to do so.[13] The new prime minister not only placed the government more completely in military hands, but his doing so represented a shift to the left in economic affairs and a move towards closer affiliation with the United Arab Republic.

On August 21, 1966, in his first comprehensive policy statement Premier Talib had this to say about oil:

> As far as oil policy is concerned, the government will study the results arrived at in previous negotiations with the oil companies operating in Iraq, while emphasizing its resolve to safeguard the country's oil resources, its adherence to Law No. 80 of 1961, and its support for the national oil company so as to enable it to achieve the objectives for which it was set up.[14]

Shortly before Premier Talib's statement the Arab Socialist Movement, representing a merger of the Arab Nationalist Movement, which had led the opposition to ratification of the draft agreement, and a group known as the Socialist Unionists, made a comprehensive statement on oil policy. In it the movement pronounced its adherence to Law 80 in all of its aspects and

12. Interview with the Beirut weekly *al-Usbu' al-'Arabi,* April 24–29, as reported in *MEES,* April 29, 1966.

13. A London *Observer* article reproduced in the San Francisco *Chronicle,* January 30, 1969, includes D. Bazzaz as among 65 Iraqis awaiting trial on charges of espionage.

14. *MEES,* August 26, 1966.

its opposition to surrendering any proven acreage to the oil companies. It advocated that INOC be empowered to engage in production and marketing of oil from the proven acreage; that the IPC group step up its production to match the increase in the whole of the Middle East; that the government support OPEC and promote inter-Arab co-operation in all phases of the oil industry, including the establishment of an Arab oil company to explore for, produce, and market oil, and an Arab oil tanker company to transport it, and the construction of an Arab-owned pipeline linking the Arabian Gulf with the Mediterranean.

Shortly after the Arab Socialist Movement's statement the United Arab Republic's state-controlled news agency *MEN* which interviewed Prime Minister Talib on November 9 quoted him as having said,

We do not recognize the existence of an actual agreement. The present government has its own oil policy which stems from the spirit of Law No. 80 and depends for the implementation of this policy on the Iraq National Oil Company which will soon be allocated the acreage it will need to carry out its operations as provided for in Law No. 11.[15]

Whether this was intended as a prelude to the abandonment of the Wattari agreement and a declaration of the government's intent to put INOC in charge of the companies' proven area reclaimed by Law 80 is not clear. Meanwhile Syria's unilateral abrogation of its 1955 pipeline agreement with IPC and its stoppage of the flow of IPC's oil to the Mediterranean greatly complicated IPC-Iraqi relationships. The Israeli-Arab war changed their whole complexion.

15. *MEES*, November 18, 1966.

12

The Syria-IPC Pipeline Controversy

On September 10, 1966, at the request of the Syrian government representatives of the Syrian government and the Iraq Petroleum Company (IPC) began negotiations toward a modification of the agreement of November 29, 1955, establishing annual transit and loading fees to be paid by IPC to the Syrian government for oil transported through its pipelines crossing Syria. After futile negotiations extending over a period of three months, the Syrian government issued a decree establishing rates for the future and providing for retroactive payments. These rates were unacceptable to IPC; this resulted in the stoppage of the flow of oil across Syria. The stoppage, which was complete by December 13, resulted in substantial losses to Syria and IPC. Although not a party to the dispute, Iraq, the source of the oil transported over the pipelines, suffered most severely since that oil had been supplying Iraq with most of its governmental revenues. The flow of oil was not resumed until three months later when the parties announced on March 3, 1967, that they had settled their dispute. Because the dispute illustrates the complexity of the economic factors on which an agreement may be based, the discretionary role of corporate accounting in applying a contract, how tenuous are relations between concessionaires and Middle East governments, how bitter the controversies may become, how political factors may aggravate them, and how widespread their results may be, a detailed discussion of the issues, the course of the controversy, and the eventual settlement is important.

The 1931 Pipeline Agreements

Originally IPC delivered oil from its northern Iraq fields by pipeline by two land routes to the eastern Mediterranean terminals. One route extended from Kirkuk to the Iraq-Jordan boundary and then across Jordan and Palestine to Haifa. The other route paralleled the Iraq-Jordan route until it crossed the Euphrates; there it turned west to the Syrian border and across Syria to the Mediterranean. Before reaching Lebanon the line forked, one segment crossing Lebanon to Tripoli, the other fork extending westward through Syria to Banias.[1] Under "conventions" that these three countries signed with IPC in 1931, each agreed that IPC would be free from any property or income taxes except on profits from selling its products locally, (Article 12), from any import tax or other tax on petroleum in transit (Article 4), and from import duties on materials utilized for transportation purposes (Article 5).[2] Except for a loading fee of two pence on every ton of oil loaded at the terminals, IPC's operations were tax exempt.

After the 50-50 profit-sharing agreement became the pattern in the oil-producing countries, Syria began negotiations with IPC for the application of this principle to IPC pipeline operations within its borders.

The 1955 Agreement

In November 1955, the Syrian government and IPC signed an agreement supplementing the agreement of March 1931. It provided for substantially increased payments to the Syrian government. Although based on the profit-sharing principle, the agreement does not mention profits. It provides for three types of payments: transit fees calculated at the rate of 1 shilling 4 pence per 100 ton-miles; a loading fee of 1 shilling 1 pence per ton; and

1. Since the Palestine-Israel War of 1948 IPC has delivered no oil to the Haifa terminal, so this discussion takes no further account of the IPC pipeline crossing Palestine.

2. See *Conventions Made between the Government of the Syrian Republic and the Iraq Petroleum Company, Ltd., November 11, 1955, Approved by Law No. 128, December 6, 1955, with the 1931 Convention in French, Convention Entre L'Etat de Syrie et L'Iraq Petroleum Co., Ltd.,* Damascus, March 25, 1931.

an annual payment of £250,000 for protection and other services provided by the government.[3]

IPC as a nonprofit service organization operating for the benefit of its owners, to whom it delivers oil at cost, has no profits accredited to its operation. Hence to share profits derived from the operation of the pipeline originating in Iraq and crossing Syria raised two important questions: What are the total pipeline profits? and What part of the total is appropriately attributable to the Syrian segment of the line? In answering the first question, IPC determined what in the jargon of Middle East oil countries has come to be called "notional" profit. To do this it began its calculation by taking the price of crude oil at Basrah at the head of the Persian Gulf, from which IPC regularly exports its crude produced from its southern Iraqi fields. IPC's owners post this price and in doing so presumably set a price competitive with that of similar Persian Gulf oils as posted by other Middle East producers. IPC owners also post a price at the eastern Mediterranean ports, including Banias, with which this analysis is particularly concerned. The Banias price was roughly equivalent to the Basrah price plus the tanker cost of shipping oil from Basrah down the Persian Gulf, through the Indian Ocean, up the Red Sea, and through the Suez Canal. This difference IPC took as its gross receipts properly accredited to its Iraq-Syria pipeline operations.

IPC owners actually sell little of this oil at the price they post. It is either taken by IPC's owners to feed their so-called downstream integrated operations or on long-term contracts at negotiated prices to other integrated companies. However, the Banias posted price is supposed to represent the price at which the IPC owners are willing to sell the oil to any buyer. Oil sold to third parties today is sold at a discount from posted prices, but in 1955 posted prices were virtually identical with realized prices except those negotiated under long-term contract.

The logic of IPC's calculations is simple. Crude oil delivered at Banias, which is closer to the European market, is worth more than oil at Basrah by the cost of getting Basrah oil to the eastern Mediterranean. From these gross receipts (equal to the tanker cost multiplied by the amount of oil transported by pipeline) IPC deducted the cost of its pipeline transportation to arrive at its notional profit.

In answering the second question as to what is the appropriate share of the total pipeline profit to be accredited to the Syrian segment of the

3. "Supplemental Convention between Government of Syrian Republic and IPC, Ltd., November 29, 1955," *Petroleum Legislation*, "Middle East Basic Oil Laws and Concession Contracts, Supplement 11."

pipeline, IPC calculated the overland mileage from Basrah to Banias and got a figure of 800 miles. Because 300 miles of the actual pipeline lie in Syria, IPC concluded that ⅜ of the total profit should be accredited to Syrian operations and under the 50-50 principle should be shared equally by the Syrian government and IPC. Prices as posted at Basrah and Banias are in dollars per barrel. IPC made its calculations in shillings and pence per ton. In 1955 the posted price differential between Basrah and Banias was twenty-five shillings and five pence per ton. To take care of a possible future increase in tanker operating costs, IPC agreed to add 25 percent to the 25s. 5d. differential, bringing it up to 31s. 10d. a ton. Rounding this off at 32s. and taking ⅜ of the rounded figure gave 12s. per ton to be accredited as the gross profit from its Syrian operations. From this IPC deducted an estimated future operating cost per ton of 4s. 6d. This left a remainder of 7s. 6d. per ton as profits from the Syrian segment of its pipeline. IPC rounded this off, treating it as eight shillings. As Syria's share of the profits in moving one ton of oil over the 300-mile Syrian segment of the line, it took half of eight shillings arriving at a figure of four shillings. It then expressed this profit figure on 100 ton-miles by dividing 4s. by three. This gave a 100 ton-mile profit figure of 1s. 4d. This is the transit fee to which Syria agreed and is incorporated without the above calculation in the 1955 agreement.[4]

The parties also agreed to a loading tariff of 4s. 3d. per ton at the Banias terminal and an estimated cost, including depreciation and administrative expenses, of 2s. 2d. per ton. Deducting the estimated cost from the agreed-on tariff gave an estimated profit from the terminal operations of 2s. 1d. a ton. Thereafter a loading fee of 1s. 1d. per ton was incorporated into the 1955 agreement as Syria's share of the Banias terminal profits.

In compliance with the transit and loading fees as agreed on, IPC has paid the Syrian government, exclusive of the protection fee of £250,000 annually, the following amounts in the indicated years:[5]

	£			£
1955	413,931		1961	8,350,056
1956	4,890,062		1962	8,462,844
1957	2,547,781		1963	9,208,111
1958	5,122,640		1964	9,289,386
1959	6,340,567		1965	9,658,853
1960	7,765,685			

4. Mimeographed statement of IPC, "The IPC-Syria Dispute," pp. 2–3.
5. Mimeographed statement of IPC, "I.P.C. Has Always Paid the Correct Amounts under the 1955 Agreement."

Syria Requests a New Deal

Late in August 1966, the Syrian government requested IPC to enter into immediate negotiations with the government looking toward an increase in the transit royalties under the 1955 agreement. Negotiations began on September 10, 1966, between an IPC delegation headed by IPC's managing director, C. M. Dalley, and a Syrian delegation headed by Syria's finance minister. Negotiations continued through the week of September 10–17, when they were adjourned; they were resumed on October 2 and adjourned on October 10; they were resumed again on October 26 and continued intermittently until November 22. Because the parties had reached no agreement, the government issued a warning on November 16 to the IPC negotiators, setting forth a formula that it was prepared to accept in fulfillment of the profit-sharing principle. On November 23, IPC's negotiating team submitted two memoranda in reply to this warning which, according to the government, "involved in effect the rejection of the formulae set forth in the government's warning."[6] On the same date the government officially terminated negotiations.

The Government Fixes Rates by Decree

On December 1 the Syrian Council of Ministers issued Decision No. 853, raising the transit fee from 4s. to 5s. 10d. for each ton of oil transported through IPC pipelines, the loading fee from 1s. 1d. to 2s. for each ton of oil loaded at the Banias terminal, and levying a new transit fee of 3s. per ton to compensate for alleged IPC underpayments on oil shipments from January 1, 1956, to December 31, 1965. The 3s. fee was to continue until all "accounts are settled with the company."[7]

The prime minister communicated the council's decision to the minister of finance, who then issued Decision No. 2455/W providing for an "executory attachment" of all movable and immovable properties of the Iraq Petroleum Company within Syria as settlement of the sum of £3,743,904 and 10s. 8d. alleged to be owed by IPC for oil shipments through Syria in the first nine months of 1966. The minister of finance's decision also set up

6. *Middle East Economic Survey,* December 9, 1966.

7. The *MEES* translated the full text of the Syrian Council of Ministers' Decision No. 853 of December 1, 1966, and Decision 2455/W of December 7, 1966.

special committees to carry out the executory attachment and thereafter to hand over the attached properties to the company "to ensure continuance of operation."[8]

Apparently the decree uses this language in an effort to place the responsibility on IPC for stopping shipments through the pipeline. But at the same time the government informed the company that it should not export oil from Banias until it paid the £3.7 million. IPC continued to pump oil through the pipeline until December 12. Its storage having been filled, it then discontinued pumping and delivered no oil to the Banias terminal until the settlement of the dispute on March 3. Despite committee instructions to stop doing so, IPC continued to pump oil to the Tripoli (Lebanon) terminal until December 13 when a "series of alleged mechanical faults occurred in the Syrian pipeline station." IPC was not allowed to investigate and pumped no oil over the Syrian pipeline until the dispute was settled.

Such in brief is the chronology of the dispute between IPC and the Syrian government. What apparently began as negotiations over an economic issue soon became a bitter political controversy in which Syria identified itself as the defender of Arab interests against what it described as dishonest, imperialistic, monopolistic exploiters.

Why did this happen?

The Specific Issues in Controversy

A detailed examination of the specific issues in controversy is the first step in answering this question. Both parties agree that it was the intent of the 1955 agreement to grant to Syria a 50-50 share in the "profits" from the operation of the pipeline within Syria's borders. This was set forth in a memorandum submitted by IPC on the day on which the agreement was signed and in the preamble to the draft law providing for the ratification of the agreement.

Although the agreement states that the payments provided for were properly derived from agreed pipeline and terminal tariffs for the transport of oil across Syria (Article 6, paragraph b) and although both parties signed the agreement in 1955, Syria contended in the 1966 negotiations that the rates did not in fact conform to the profit-sharing principle, that

8. *Idem.*

the company through incorrect accounting methods had evaded these principles and had shared with Syria a wrongly calculated sum. The government contended specifically that IPC in calculating the portion of pipeline profits to be accredited to the Syrian segment of the line had wrongly based the entire pipeline profits on an imaginary line 800 miles in length and had wrongly accredited to the 300-mile Syrian segment of the actual pipeline ⅜ of the total profits. Contending that the actual pipeline extending from Kirkuk to Banias was about 500 miles long, 300 miles of which lie in Syria, the government insisted that IPC should have accredited ⅗ of total profits to the Syrian segment of the line. The government also contended that IPC had used inflated pipeline operating costs in calculating its notional profit. The government challenged specifically IPC's use of the 1956 throughput of 26 million tons to arrive at a constant cost factor which it used in subsequent years when throughput was considerably in excess of this figure.

Three Methods by Which Pipeline "Profits" Might Have Been Calculated

The higher posted price for crude at Banias reflects the fact that it bears a lower transportation cost to get to the European market. Basrah oil that competes with it must have a higher transportation cost and hence must sell for less. To calculate the share of the pipeline profits to be accredited to Syria, IPC might have chosen either of three methods: it might have calculated a price for crude at the Syria-Iraq border (which would have been appropriately higher than the Basrah price) and subtracted this from the posted price at Banias, and from the difference deducted the cost of transporting the crude 300 miles, all of the profits so calculated being accredited to the Syrian operation; it might have calculated a price for crude at Kirkuk which would have been a higher price than the Basrah price (because Kirkuk is both geographically and cost-wise nearer the European market), subtracted the Kirkuk price from the Banias posted price, deducted the per barrel cost of transporting crude through the 550-mile pipeline to Banias, and allocated 300/550 of the notional profits to Syria; or it might have taken the difference between the Basrah and Banias prices and subtracted from this the actual per barrel cost of transporting the oil through the Kirkuk-Banias pipeline to get a notional profit,

⅜ of which it would credit to the Syrian segment. This last is in fact what IPC did.

If the Kirkuk, Syrian border, and Basrah prices were so calculated as to reflect their relative advantages of location, the three methods would yield the same notional profit to be accredited to the Syrian segment of the pipeline. As IPC has concluded, "The illogicality of the Government's claim is that the Government is prepared to agree that pipeline profitability should be based on tanker costs from the Arabian Gulf to the Mediterranean but that its own profit should be assessed by reference to the source of oil."[9]

Apparently the gross receipts that IPC accredited to the Syrian segment of the Kirkuk-Banias pipeline were appropriately determined. It is not clear, however, that IPC calculated the profits to be shared with Iraq according to the same principle. In 1955 the Banias price of crude was 129s. 5d. a ton. The gross receipts on its operations in Syria were fixed at 12s. per ton and the loading fee at 4s. 3d. Subtracting the sum of these two items (16s. 3d.) from the Banias posted price gives a Syrian border value of 113s. 2d. The Syrian border value on which IPC was calculating its profits to be shared with Iraq was 110s. 5d. Professor Zuhayr Mikdashi in making these comparisons concludes that "IPC was left . . . with an untaxed profit of 2s. 9d. a ton."[10] Mikdashi is correct. But it should be pointed out that Iraq and IPC had arrived at that border value by agreement in March 1955, before the Syrian agreement of November 1955. The discrimination in favor of Syria was a matter primarily of concern to Iraq rather than to Syria.

The Government Also Challenges IPC's Cost Calculating

Although not disputing the accuracy of the transit and loading-cost figures used in 1955, the government contended that because they were based on an annual throughput of 26 million tons, whereas throughput

9. IPC's undated mimeographed statement released from its Beirut office, "IPC Was Not Responsible for Cessation of Pumping."

10. Zuhayr Mikdashi, "Some Aspects of Pipeline Transport in the Arab World," Third Arab Petroleum Congress, Alexandria, October 16–21, 1961, p. 17. Professor Mikdashi was a member and secretary of the Ministerial Committee on Oil Affairs of the Government of Lebanon in 1956 and took part in the negotiations between Lebanon and the IPC over pipeline royalties.

mounted over the years and was in 1965 more than 40 million tons, the cost figures used in the intervening years were too high and hence the calculated profits to be divided with Syria were too low. The company denied the validity of the government's contention and averred that the average annual costs for the whole period have approximated the 1955 figure despite the increase in annual throughput. To explain this apparent paradox it contends that by destroying segments of the pipeline in 1956 during the Suez crisis, Syria forced IPC to make heavy capital expenditures and hence raised costs; by requiring IPC to keep 3,500 surplus employees on its payrolls, the government kept IPC labor costs high; and IPC incurred large capital expenditures for the construction of a second 30-inch Kirkuk-Mediterranean pipeline to take care of the increase in throughput.

An outsider without access to the relevant cost data cannot pass judgment on the validity of the company's contention although on its face it seems plausible. The company admitted that present costs are well below the average rate of the past ten years and in the course of negotiations agreed to reduce the transit-cost figures used in calculating the profits from 4s. 6d. per ton to 2s. 4d. per ton and to reduce the loading cost figure from 2s. 3d. to 1s. 1d. This would have increased IPC's transit payments to the government from 4s. per ton to 4s. 10d. per ton and its loading fee payments from 1s. 1d. to 1s. 7d. This compares with a transit fee of 5s. 10d. and a loading fee of two shillings as fixed by government decree No. 853.

The Price of Crude for the Homs Refinery and the Labor Surplus Issue

The controversy was aggravated by the additional issues: the price at which IPC delivered oil to Syria's Homs refinery and the retention on IPC payrolls of 3,500 employees for whom there was no work. Under the 1955 agreement IPC was to supply annually 600,000 tons of crude to Syria's Homs refinery at a discount of twelve shillings per ton (22.2 cents a barrel) off the Banias posted prices. Because the posted price at Banias includes terminal costs of 2s. 2d. a terminal fee of 1s. 1d., and port dues of threepence, totaling 3s. 6d. (about 7 cents a barrel), the discount of 22.2 cents is in fact reduced to 15.2 cents. This is less than the prevailing discount IPC's owners grant to third-party buyers to meet the competition of independents. The negotiations were concerned primarily with the larger issues involved in the 50-50 profit-sharing principle and the Homs issue was neglected. If the main issues had been settled, doubtless the Homs refinery issue would have been easily reconciled.

The issue of surplus manpower is more difficult. IPC raised this issue in the course of negotiations. It alleged that it was paying and for some time had been paying wages to 3,500 employees for whom it had no work. The company states the following as facts about this matter. The company hired most of these workers (85 percent) on a temporary basis for construction work starting in 1957. When the construction work was completed in 1960, at the request of the government as a temporary measure, the company put about 700 of the workers on "special leaves" at full pay and they remain on special leave with full pay to this day. Although many of the 700 had worked for the company for less than a year, the "special leavers" brought a case against the company under Syria's social security laws and were awarded annual raises for each of the past five years although it was acknowledged that they were in fact doing no work for the company. The majority of the 3,500 surplus workers, including the "special leavers," draw full pay from the company although "only attending their official place of work to draw their pay; indeed, it is believed that many of them are now employed and drawing salaries elsewhere."[11] The company alleged that it had tried several times to get a settlement of the labor issue. Its most recent offer was to pay each worker the normal severance benefits and the labor law indemnity; to pay additional sums to each worker equal to from 31 to 37 times the worker's basic monthly pay; and to advance to the government £1,000 for each worker to meet the costs of resettling him in other employment.

At a press conference that the Syrian ministers of information, social affairs, labor, and state held on November 25, the ministers explained that the company wanted to solve the labor issue by paying a lump sum to the government for distribution to the workers. The government insisted that the company reach an agreement with the workers themselves. The company alleged that at various meetings with representatives of the workers and the government, the workers demanded lump sums ranging from 80 to 120 months' wages. The company considered such demands exhorbitant.

The company stated that the 3,500 unused laborers represented about 75 percent of its total Syrian labor force and that their wages amounted to approximately £2 million. If the labor problems were solved along the lines suggested by the company, it proposed to share the saving with the

11. Undated, mimeographed statement of IPC.

government by increasing transit and loading fees from the 6s. 5d. (4s. 10d. transit fee plus 1s. 7d. loading fee) which it had offered to 6s. 11d. (5s. 4d. plus 1s. 7d.). Government decree No. 853 fixed the total at 7s. 10d. (5s. 10d. plus 2s.).

The company's offer represented an increase of £3.5 million payment per year over its 1965 payments, or an increase of about 36 percent. The government's rate entailed an increase of £3.7 million for the nine months of 1966 or about 50 percent more than the company's 1965 payment.

With this difference, why could the parties not agree? It was because the government demanded that the annual increase in payments be made retroactive to the beginning of the 1955 agreement. This was to be achieved by an additional payment of 3 shillings per ton, aggregating about £40 million.

The Company's and the Government's Positions

In summary, the company's position was that it had observed meticulously the terms of the 1955 agreement, terms which fairly applied the 50-50 profit-sharing agreement, terms which the government approved and to which it agreed, and terms which over the intervening years reflected accurately pipeline and loading costs. It acknowledged that they no longer did so and that with the amortization of the pipeline investment within the next seven years, costs would decline further. The company made an offer which it contended provided for this contingency.

The government's position was that the agreement was designed to incorporate the 50-50 profit-sharing principle; that the transit and loading fees provided in the agreement transgressed this principle; that IPC used inflated costs in calculating profits; that in dividing profits it allocated too little to the Syrian segment of the pipeline; that the company, having failed to apply the principle properly, was obliged to make restitution and to base its future payments on a proper calculation of costs and a proper division of profits. It demanded an immediate additional payment of £3.7 million for the nine months of 1966 and future payments that would wipe out the company's alleged delinquency.

It is a fact that both parties accepted the cost basis and the distribution basis in 1955 as proper for the calculation of the fixed fees provided in the agreement, and both signed the agreement which certified as to the correctness of the fixed fees.

The Agreement Provides for Arbitration

The 1931 agreement, to which the 1955 agreement is supplementary, placed an obligation on both parties to refer all disputes growing out of the agreement that could not be settled amicably to an arbitration board consisting of two arbitrators, one appointed by each of the parties, and an umpire whom the arbitrators should select. If they could not agree on an umpire, they were obliged to request the president of the Permanent Court of International Justice to appoint one. The decision of the arbitration board was to be binding on both parties.

Stopping the flow of oil through the pipelines involved a hardship on three parties: the Syrian government, IPC, and most severely on Iraq, from which the oil is obtained and which is heavily dependent on oil revenues for its solvency. As the controversy became more bitter and the emergency more acute, IPC requested the government in writing on December 11, 1966, to refer the dispute to arbitration and to appoint its arbitrator within the 30-day period provided for in the agreement. The company appointed as its arbitrator "an international French jurist." Pending arbitration, the company proposed to pay the Syrian government the £3.7 million which it demanded for the nine months of its 1966 operation of its pipelines, to be treated as an advance against any amount which the arbitration board might decide is due the government. The Syrian government did not reply to this offer.

Political Factors Complicate
the Controversy

THE Syria–IPC pipeline dispute developed into a direct confrontation between a state determined to assert its sovereignty and an oil company determined to protect its contractual rights. From Syria's point of view the controversy represented an episode in a broader struggle to free the "Arab nation" from the domination of "Western imperialism" and exploitation by oil "monopolists." As have other Arab states, Syria brought into its conception of the "enemy" Israel and all those countries that had "conspired" to create it. Israel was but a handmaiden of the West in dominating the Middle East.

This, of course, was not a new theme. The unstable political conditions that had plagued Syria throughout the years of her independence created an environment in which such ideas flourished. Mobilization of public sentiment against real or fancied wrongs became an instrument of political security. Monopolistic exploitation by oil companies, Western imperialism, and the Zionist conspiracy afforded an ideal theme for winning Arab support at home and Arab sympathy abroad.

The exercise of Syrian sovereignty in dealing with IPC was an essential step towards the ultimate goal of "Arab oil for the Arabs." Syria for some time had designated itself as leader in this struggle. To prepare itself for leadership in ridding Arab oil of Western exploitation, Syria first set about freeing itself.

Syrian Officals Set Pipeline Controversy in a Broad Political Framework

For some years Syrian officials had been constructing a political framework into which the pipeline controversy could be readily fitted. At a press conference in Damascus on January 13, 1965, General Amin al-Hafiz, head of the Syrian presidency council and prime minister, outlined Syria's oil policy. Disappointed with the slow progress made in developing Syria's oil resources since the discovery of oil in 1959 and disillusioned with the "fruitless talks" with members of the "international cartel," Syria decided to develop its own oil resources, thereby putting into practice the slogan "Arab oil for the Arabs."[1]

As a first step they sought the assistance of Soviet Russia. On February 11, 1965, the Syrian government concluded an agreement with the Soviet Union for deep drilling in northern Syria, where oil had been found. As a second step they took over domestic oil marketing facilities. On March 4, 1965, the government issued a decree nationalizing the country's nine oil-products distributing companies and authorizing compensation with negotiable bonds repayable in fifteen years and bearing 3 percent interest.

On March 8 in a public address on the second anniversary of the revolution that brought the Baathists to power in Syria, General Hafiz, while stating that Syria had liberated its own oil resources from foreign domination, indicated the necessity of liberating the rest of Arab oil, particularly that of the Arabian Gulf area, for use in the battle of Arab destiny. "This Arab oil wealth is the property of all the Arab nations and must be mobilized for the battle."[2]

In an article in the Damascus daily *al-Baath* on March 9 the Syrian minister of industry described Syria's oil policy in these terms:

The oil sector is one of the most important sectors of the economy, and the revolution has dedicated itself to removing oil from the quagmire into which it had sunk under previous regimes. The revolution has mapped out practical steps to develop Syrian oil under direct state control. Revolutionary Syria will thus set a unique example in the Middle East, providing an incentive to the Arab people as a whole and to the people of the Arab oil-producing countries in particular. It will enable them to assume control of their industry, which will provide the springboard for their prosperity and promote the Arab aspiration for unity, liberty, and socialism.

1. *Middle East Economic Survey,* January 22, 1965.
2. *Ibid.,* March 12, 1965.

General Hafiz in his speech of March 8 and again in a speech at Aleppo on April 17 made clear his belief that the Arab world had an obligation to utilize its oil resources in resisting the "Zionist tide" and in liberating Palestine. Oil revenues, he said, must not be monopolized by Arab rulers, for Arab oil is "the rightful property of the Arab people as a whole."[3]

At the conclusion of its Eighth National Congress, held in Damascus on May 4, 1965, the Arab Baath socialist party issued a communique containing the following statement on oil:

> It is the view of the Eighth National Congress that the ultimate objective of any Arab petroleum policy is the nationalization of Arab petroleum, the formation of a unified Arab petroleum organization to undertake all operations pertaining to the production, transportation, refining, and marketing of Arab oil, and the establishment of direct relations on sound commercial bases with the consuming countries; and that this objective is part of the strategy of the Arab revolution which aims to liberate the Arab homeland from the old and new imperialism, to liquidate the reactionary regimes subservient to it, and consequently to realize the national objectives of the Arab nation.
>
> Pending the realization of this concept of Arab petroleum, it is the view of the Congress, particularly after the nationalization of the petroleum industry including product distribution in Syria, that it is necessary to foster popular awareness and to rally popular support around a phased program which aims firstly at the channeling of part of Arab oil revenues into a common fund for development and armament, and secondly at reviewing the relations of the Arab countries with other states—with respect to petroleum affairs—on the basis of the relations of their states with Israel.
>
> . . . .
> It is therefore the view of the Eighth Congress that the question of the development of Arab petroleum is inseparable from the Palestine problem. For Israel, whose economy is organically tied to that of the West, indirectly derives at least a part of its strength from the development of Arab petroleum. It is therefore necessary to kindle the struggle of the Arab masses for the purpose of fulfilling the concept "Arab oil for the Arabs."[4]

On May 12 Dr. Munif al-Razzaz, who had replaced Michel 'Aflaq as secretary general of the Arab Baath socialist party, echoing the official sentiments of the party, made a statement criticizing the misuse of Arab oil revenues by "oil shaikhs" and lamenting the exploitation of Arab oil resources by the West to strengthen its imperialistic sway. And he made clear his notion that oil should be used as a weapon in liberating Palestine.

> For at a time when the Arab states bordering on occupied Palestine are bearing all armament costs in standing up to Zionist conspiracies and paying

3. *Ibid.*, April 23, 1965.
4. *Ibid.*, May 7, 1965.

them with the blood and sweat of their citizens, the owners of the oil in the rear are safely amusing themselves with the squandering of their wealth, contributing nothing more than empty words to the Palestine cause.[5]

On February 23, 1966, the extremist wing of the ruling Baath party executed a coup d'etat which they described as an internal party matter. Dr. Yusuf Zu'ayyin headed the new cabinet as prime minister. Dr. As'ad Taqla replaced Dr. Hisham al-'As as minister of industry, with major responsibility for the country's oil policy. Shortly thereafter the government began an investigation of the IPC-Syria pipeline agreement as a preliminary to negotiations for a better deal.

On August 28, 1966, Prime Minister Zu'ayyin disclosed that a "thorough study" conducted by the Syrian government revealed serious errors in IPC's calculation of the pipeline royalties due the Syrian government resulting in losses of hundreds of millions of Syrian pounds and that his government had invited IPC to enter into immediate negotiations looking towards a correction of the accounts and recovery by Syria of the "huge losses it had sustained."[6]

Syria Warns IPC

Such is the political background of the pipeline negotiations. Available evidence does not indicate that Syria foresaw the breakdown of negotiations. It rather suggests that Syrian officials thought that IPC, confronted with Syria's determination to assert its sovereignty, would bow to Syria's demands. As continued negotiations indicated that IPC was reluctant to do so, Syrian officials shifted the battle from the conference table to the public arena. They publicly charged IPC with procrastination and evasive tactics, monopolistic exploitation, deceit, and cheating. As the breakdown of negotiations became more imminent, the Damascus daily *al-Baath,* organ of the ruling Arab Baath socialist party in Syria, in an editorial on November 10 warned,

there remains no other course for national and progressive forces except that of struggle in all its forms, even if this were to lead to the cutting off of oil supply lines . . . and closing down of oil wells in order to deprive the monopolist, the embezzler, the despot of this oil.

5. *Ibid.,* May 14, 1965.
6. *Ibid.,* September 2, 1966.

[Continued exploitation] makes us call for the shouldering of our historic responsibility by adopting the firm stand needed to bring about our deliverance from the remnants of imperialistic tyranny exemplified by the oil monopolies.

On November 22, the day before Syria broke off negotiations with IPC, Dr. Zu'ayyin in a talk to representatives of the International Confederation of Arab Trade Unions characterized the pipeline controversy as Syria's "battle against the oil companies to recover the usurped rights of the people." Two days earlier the government had threatened to put into effect "measures already decided upon" but not yet announced.

Why in the face of these warnings did IPC not settle on Syria's terms? Although Syria apparently had hoped that it could avoid taking unilateral action, it was clearly prepared to do so. To Syria the issue had become that of the exercise of its political sovereignty—its right to legislate the terms on which IPC would transport oil across its territory. There is no reason to believe that IPC did not realize what Syria might do. Apparently it was prepared to run the risks and, if need be, bear the costs of noncompliance.

The Risk IPC Faced

By refusing to accept Syria's terms, IPC ran the risk of losing temporarily or even permanently access to the Mediterranian through Syria of oil produced from its Kirkuk fields—roughly two thirds of its Iraqi output. This would complicate its controversy with Iraq which had remained unsettled since Qasim expropriated about 99 percent of IPC's original concession. The loss to Iraq of two-thirds of its oil revenues was certain to work a hardship on the Iraqi economy, contribute to the instability of the Iraqi government, and raise anew the spectre of nationalization.

The risks of noncompliance were unpleasant to contemplate; the risks of compliance were even less attractive to IPC. Apparently IPC thought that to accept Syria's demands would have placed in jeopardy the terms of all Middle East oil concessions. IPC's principal owners—BP, Royal Dutch-Shell, Standard of New Jersey, Mobil Oil and Compagnie Française des Pétroles—had extensive interests elsewhere. It was the company's view that to have acknowledged Syria's right to challenge the terms of its concession by unilateral action and to demand higher transit and loading fees retroactive to the date of the concession would create a precedent for any dissatisfied producing country to modify a concession unilaterally and demand unilaterally retroactive payments under the concession's new terms.

IPC's owners apparently decided that to do this would prove disastrous in the long run. It was this retroactive feature of Syria's demands that prevented a compromise settlement.

Syria Launches Propaganda Campaign

Syria was as determined as IPC was adamant. After Syria had fixed transit and loading fees by official decree, to support its action the government launched a propaganda campaign, vigorous, broad, and far reaching.

Premier Zu'ayyin, speaking to the General Federation of Peasants in Syria on the second anniversary of its establishment, declared that the Arab masses in the United Arab Republic, Algeria, Iraq, Lebanon, and other Arab countries stood with Syria in its battle against IPC. He uttered this ringing challenge:

Let the English oil company (IPC) persist in its stubborn attitude . . . and try its strength against 100 million Arabs. It has overlooked the ability of the masses to shatter imperialist fiction and the oil monopolies. It has overlooked the existence of independent companies prepared to market oil outside the framework of the international monopolies. It has also overlooked the existence of the socialist camp as an international economic force and a faithful friend of the freedom movement and the struggle of the masses against imperialism, old and new.[7]

To mobilize and solidify popular support for its action, the Syrian government established a "Higher Petroleum Committee" headed by Prime Minister Zu'ayyin and including 20 members, among them the minister of finance, minister of state for council of ministers affairs, a representative of the Regional Baath Party Command, president of the Syrian General Federation of Labor Unions, president of the General Federation of Peasants, president of the Teachers Association, president of the National Federation of Syrian Students, a representative of the General Federation of Syrian Women, the commander-in-chief of the National Guard, the president of the Engineers Association, the president of the Medical Association, the president of the Lawyers Association, the president of the Agricultural Engineers Association, and representatives of the university and of the petroleum workers.

7. *Ibid.*, December 16, 1966.

At the committee's first meeting on December 27 the prime minister explained that the committee's functions would be continuing and would involve the organization of "guidance seminars" and establishing relations with all popular organizations concerned with the oil problem. At its second meeting on January 2, the committee reviewed recent developments in the dispute and voted to give its unqualified support to the government, to issue a statement addressed to all citizens, and compile a booklet explaining in language for the common man the issues in the controversy.[8]

Subsequently the Baath party organized task forces and sent them throughout the country explaining the government's policy and plans. Meanwhile Deputy Premier Makhus, referring to the Suez crisis, warned that the people were "more inflammable than petroleum" and expressed anew Syria's determination to obtain the people's "full rights" whatever the consequences.[9]

On January 2, speaking to a Lebanese delegation visiting Damascus to express its support for Syria's stand, Prime Minister Zu'ayyin described the struggle in broad terms. He averred that Syria in beginning the "battle of the transit royalties" realized its dimensions and counted on the support of the Arab masses, for "the present battle is part of the larger battle with the monopolistic oil companies, which is, in turn, part of an even larger struggle that the Arab masses and progressive elements are waging against imperialism, Zionism, and reaction."

On January 3, Dr. Nur al-Din al-Atasi, in addressing the graduating class of army cadets at the Homs Military Academy, set the struggle in a similar framework when in characterizing the Syrian revolution as an indivisible part of the revolution of the Arab homeland inspired by its sense of duty to the Arab masses, he vowed that

The battle for petroleum shall be unlimited in scope until the interests of all the Arab people are realized. The battleground is the entire Arab homeland and we shall wage the battle with strength and determination until Arab oil becomes the property of the Arab people.[10]

Syria Seeks Co-operation of Other Arab Countries

To give a sense of reality to such a broad conception of its struggle and to provide a weapon with which to fight it, Syria sought the co-operation

8. *Ibid.*, January 6, 1967.
9. *Ibid.*, December 30, 1966.
10. *Ibid.*, January 6, 1967.

and support of friendly Arab states. Cairo's was promptly forthcoming. On December 20 the United Arab Republic Assembly passed a resolution identifying Syria's struggle with that of the "Arab Nation" as a whole and pledging its full support.[11]

Iraq's co-operation was another matter. Ruling Iraqis, whose idealistic preconceptions and long-run nationalist aspirations were akin to those of their Syrian counterparts and whose attitude toward IPC had been toughened by their own experience, were reluctant to express anything but sympathy for Syria. But the assertion of Syria's sovereignty promised only suffering at best and disaster at worst to Iraq and its people. As long as the Syrian pipeline remained closed, Iraq would suffer greatly through the loss of revenues on which prosperity of the country had come to depend. To nationalize her oil industry would raise more problems than it would solve. IPC's owners with the exception of CFP had other readily available sources of oil and almost unlimited underground reserves with which to supply their customary markets.

The Risks to Iraq of Nationalization

Nationalization would have confronted Iraq immediately with the necessity of finding new outlets and new markets for approximately 40 million tons of oil annually. The Middle East well remembers what happened to Iran when it resorted to nationalization, and although there are many more outlets to world markets since the independents have gained a foothold in them, independents are wary of buying expropriated oil. The Iraq National Oil Company (INOC), although not yet a going concern, had recognized that disposing of its oil was its major problem. In making recommendations for its future development, its officials had only recently made a highly competent study of the problem of finding new markets and had concluded that a joint-venture partnership with established concerns was its best solution.

Nationalization offered no immediate relief to Iraq. Its problem could only be solved by a prompt reopening of IPC's pipelines.

---

11. Kuwait took the part solely of an observer sympathetic to both Syria and Iraq. On December 25, 1966, Kuwait's Council of Ministers passed a resolution affirming (1) support for the Syrian people in their fair demands; (2) opposition to IPC's stopping the pumping of crude oil; and (3) support for any fair solution arrived at by the parties concerned which conforms to the interests of two brotherly peoples and safeguards their legitimate rights. *Ibid.*, December 30, 1966.

Iraq Sends Delegation to Confer with Syrian Government

Apprehensive of the possible consequences of Syria's action to its own material interests, Iraq dispatched to Damascus a delegation headed by Iraq's deputy premier and minister of the interior a few hours before the IPC-Syrian negotiations broke down. After a two-hour conference with Prime Minister Zu'ayyin and a series of meetings with various other Syrian officials on the following day, the delegation announced that discussions had dealt with "the preservation of the rights of a single Arab nation." Despite this show of unity a conflict of Syrian and Iraqi interests soon became apparent. While professing sympathy for and giving verbal support to Syria's "just demands," Iraq made it increasingly clear that it preferred that they be kept within the framework of an IPC-Syrian dispute rather than being pictured as a Syrian struggle in a war for Arab liberation.

Syria Invites Iraq to Nationalize

In its broader complex Syria's position was in effect an invitation to Iraq to nationalize its oil industry. Although the Syrian government never publicly expressed itself in these specific terms, Syrian Deputy Premier Makhus while in Baghdad for talks with Iraqi officials stated in an interview with the *Iraq News Agency,* that "If Iraq decides to sell its oil itself, Syria will not apply the same transit fee as in the agreements with IPC. Such calculations do not apply between brothers."[12] On his return to Damascus he expressed the same idea. "We consider that the oil pipelines and installations in Syria belong to the Iraqi people. . . . We are ready to entrust the operation of these installations to workers and officers so as to ensure, if Iraq desires, the pumping of Iraqi oil to the Mediterranean."[13] On a trip to Paris where he conferred with French officials about the crisis, in an interview with *AFP* he announced a willingness to grant France "general marketing rights for Arab oil." In an interview with a correspondent of the Paris daily *Le Monde,* published on December 20, he stated,

If IPC persists in wanting to transform a simple commercial dispute into a political battle, thereby impoverishing and humiliating the Iraqi people, the

12. *Ibid.,* December 16, 1966.
13. *Idem.*

latter will have no alternative but to place their resources at the service of their own well-being and prosperity. In fact, we are convinced that the major oil companies are today no longer indispensable to the peoples of the area. The present situation is radically different from that which prevailed when Dr. Mossadegh nationalized Iranian oil. New trends have appeared in the world of oil, whose practical effects can be seen in Algeria and Iran. Contrary to what some people think, world markets are no longer the exclusive preserve of the major oil companies."

Tariki Favors Nationalization

Abdullah Tariki, vigorous exponent of Arab nationalism, had already issued an invitation to Iraq to nationalize its oil industry. Speaking softly at first in an interview with the Paris daily *Le Monde* of December 11, he defended Syria's action in taking over IPC's pipeline installations and ventured the thought that nothing prevented Iraq from nationalizing the IPC oilfields. He argued that Iraq was technically in a position to do so; it had the necessary personnel and with Russia's help could readily find customers. He pointed out that Russia's exportable surplus was shrinking and that Russia at present was unable to supply fully its principal external market, East Europe. He prophesied that by 1980 East Europe alone would demand more than 2 million barrels daily. A week later Tariki was surer of himself. What had apparently been an off-the-cuff idea became a more specific proposal. In an article in the Beirut daily *al-Anwar* of December 17 he made an impassioned plea for immediate nationalization of Iraq's oil industry:

The time is ripe and it has become essential for Iraq and Syria to take the plunge together and carry the struggle to a victorious outcome. . . . [It] is no less important as regards the political and economic future of the Arab nation than the battle of the Suez Canal.[14]

He proposed a four-point co-operative program among Arab oil-producing countries to prevent the major oil companies from sabotaging Iraq's nationalization. A week later, admitting that immediate nationalization might lead to a military coup, Tariki urged gradual nationalization beginning with INOC's assuming responsibility for the operation of Syria's pipelines through which it would ship Iraq's royalty oil and such other oil

14. *Ibid.*, December 23, 1966.

as INOC might produce. Such steps Tariki viewed as paving the way for a painless takeover.

Iraqi Baathists Make Similar Plea

Dissident Iraqis, less free to express their ideas forthrightly in their own country, sounded a similar note. The Damascus daily *al-Baath* published on January 29 what was allegedly an official oil policy statement by the "Interim Regional Command of the Baath Party in Iraq." In it the party urged that the Iraqi government immediately assign all oil rights in the areas expropriated under Law 80 to the Iraq National Oil Company; that INOC begin exploration, drilling, and production operations immediately; that INOC in cooperation with the ministry of oil and friendly consumer states seek outlets for Iraqi oil; and that Iraq's government initiate contacts with the "liberated Arab states" to elicit their co-operation in making a unified stand and recruiting all revolutionary forces to face up to any consequence that might ensue.

Other Iraqis Demand Reorganization of the Government

As the crisis deepened, so did the criticism of Iraqi policy. Some prominent Iraqi leaders regarded the oil crisis as symptomatic of the more fundamental political crises that faced the nation. Fayiq al-Samarrai, president of the Iraqi Lawyers' Association and in 1959 Iraqi ambassador to the United Arab Republic, Husain Jamil, former secretary of the National Democratic party (now dissolved) and in 1950 minister of justice, and Muhammad al-Durrah, a retired army officer active in politics before the 1958 revolution, expressed their views in a memorandum which they submitted to President 'Arif and which the Iraqi newspaper *Saut al-'Arab* published on January 6. The oil crisis, they contended, reflected a political vacuum which had plagued the country for the past eight years. The government's lack of a positive oil policy reflected its lack of a solid popular base on which to lean. What was needed to cope with the oil situation was a government representing the will of the people. The memorandum called for a return to parliamentary government. The first step required was the establishment of a constituent assembly, the adoption of a permanent constitution, the removal of restrictions on the freedom of the press, the election of a parliament upon which should rest the responsibility for all decisions vital to the country's interest including the development of an oil policy.

Samarrai in a press statement on the same day, in urging that Iraqi authorities begin a study of pipeline projects to connect Kirkuk with Basrah on the Persian Gulf, which would enable Iraq to dispense with the Syrian pipeline, warned that Iraq had not faced such a serious crisis even under Qasim's regime when Communist influence had reached its peak and when Syria, then in union with Egypt, was openly hostile to the Iraqi regime.

Government Officials Bring Pressure to Settle the Dispute

As the internal political situation became more unstable, Iraqi government officials began to exert pressure first on IPC and eventually on Syria to get the oil flowing again.

As early as December 15 Premier Talib was quoted by *Saut al-'Arab* as having indicated that the Iraq government did not consider itself a party to the dispute and as having informed IPC that it would expect to receive its full revenues on the basis of "regular annual rate" of exports. On the following day President 'Arif stated in an interview in the Baghdad daily *al-Thaurah al-Aribiyah* that the prime minister had submitted a memorandum to IPC urging that the dispute between the company and Syria should not be permitted to reduce Iraq's oil revenues.

In an interview with the Baghdad daily *al-Jumhuriyah* on January 8 Prime Minister Talib discussed at length the government's attitude on the oil crisis. Despite the different views on ultimate oil policy expressed by various critics, the prime minister recognized a consensus on the importance of immediate resumption of the flow of oil. He said,

This is a point I would like to impress upon our brothers in Syria and Arab public opinion in general.

In the present circumstances there is only one solution to the problem that would ensure an immediate resumption of pumping, namely, a direct agreement between the Syrian Government and the company, and the Iraqi Government is making every possible effort to bring this about.

Shortly thereafter the Iraqi government issued an official warning to IPC that within a week from January 21 the company must either settle the dispute with Syria in such a way as to permit the resumption of the flow of Iraqi oil to the East Mediterranean or undertake to pay Iraq its full revenues for the lost production from the northern oil fields.

President 'Arif Visits Nasser

The expiration of the ultimatum brought no action either by IPC or Iraq. It was followed by President 'Arif's five-day state visit to Cairo where he discussed with Nasser the oil crisis along with other problems of mutual interest. A joint communique issued at the end of the visit stated somewhat equivocally:

> The UAR side heard the Iraqi side's point of view on the oil crisis and supported the Iraqi Government's stand. It reaffirmed the necessity of the flow of Iraqi oil in order to safeguard Iraq's vital interests and its national economy. The two sides also backed the legitimate demands of the Syrian Arab Republic in this connection.[15]

In a speech that Nasser delivered about two weeks later on the anniversary of the Syrian-Egyptian union, he stated with equal equivocation Egypt's position in the oil crisis.

> We support the Syrian people in their struggle against imperialism and reaction, and we support the Syrian Government in its national, progressive stand for securing its rights from the oil company, in the same way that we support the government and people of Iraq.[16]

Following the Cairo conference and after four days of talks with IPC's Managing Director Dalley, Iraqi Premier Talib headed a delegation to Damascus in another effort to break the IPC-Syrian deadlock. They returned empty handed.

Syrian Intransigence Abates

Despite this fact there is evidence that Syria's intransigence was abating. In a statement to the press following the talks with the Iraqi delegation,

15. *Ibid.*, February 10, 1967.
16. *Ibid.*, February 24, 1967.

Syria's Prime Minister Zu'ayyin seemed to narrow the conflict to its initial dimensions when he said,

The matter is quite simple. We have not prevented the flow of oil from North Iraq to the Mediterranean, but have only applied the laws of our country when we impounded the company's property pending payment of sums due from it to the state treasury. The door has always been open—and is still open—for the company to pay the sums due. Naturally, loading operations at the terminal and the flow of oil will be resumed as soon as the company fulfills its obligations towards the state treasury.[17]

Ten days later to the surprise of seasoned observers IPC and Syria announced a settlement of the controversy. Meanwhile the controversy had gone full circle. Beginning as a financial issue regarding the amounts due Syria under the pipeline agreement, it had broadened under Syria's guidance into a full-scale political battle designed to bring about the nationalization of Iraq's oil industry. It was settled, as it had begun, as a controversy over the amount of pipeline royalties owed by IPC to Syria on terms close to those that IPC had offered at an earlier stage of the struggle.

Why Syria Shifted Its Position

Why did Syria shift its position? With the facts now known a precise answer to this question is not possible. However, two developments during the controversy throw some light on Syria's changing mood. It became evident to Syria that Iraq was unwilling to become the sacrificial lamb for the salvation of the Arab world. Although the closing of the pipelines had shut off approximately two thirds of Iraq's oil production from world markets, nationalization would have meant at least temporarily loss of a market for all of its oil and a disappearance of all governmental oil revenues so essential to the political security of the present regime.

More important, it had become evident to both Syria and Iraq that IPC could readily endure the losses that closing the pipelines entailed. From the world's surplus of oil-producing capacity the IPC owners could easily obtain the oil to meet their commitments in world markets. In truth, during the course of the controversy IPC announced that it had already supplied its oil deficiency from other sources.

tankers have been rerouted to establish a new supply pattern which could well be permanent and which has ensured that the sharcholders of the Iraq Petro-

17. *Idem.*

leum Company have oil supplies available to carry on their business without interruption. The Company and its shareholders have therefore no immediate need of this Kirkuk oil.[18]

On January 23 *Platt's Oilgram,* reporting on the new arrangements, gave detailed "informed guesses" on where IPC owners were getting the oil they needed. According to this information most of it was coming from other Persian Gulf countries in which the various IPC owners had concession rights. Saudi Arabia, Kuwait, and Iran were the chief beneficiaries.

With Iraq desperately concerned with the short-run problem of getting the oil flowing but uncommitted to nationalization and IPC already having made arrangements to meet its needs, Syria's broader struggle for "Arab liberation" seemed destined to defeat. Convinced of this, governmental officials once more came to grips with the immediate issues in the controversy.

How much pressure neighboring Arab states brought to bear on Syria is not known and may never be known, but some Iraqi sources say that Iraq is indebted to the UAR and Algeria for persuading Syria to reach an agreement with IPC.

The Settlement

On March 2, IPC and Syria formally signed and ratified a new agreement. By 4 P.M. of the same day oil began flowing from the Kirkuk field to the Banias and Tripoli terminals. As a part of the agreement Syria rescinded its decree of December 1 imposing the new schedule of transit and loading fees and formally returned the impounded properties to IPC. The new agreement provided for transit fees of 5s. 10d. for each ton of oil transported through the pipelines and for a terminal loading fee of 2s. for each ton loaded, making a total charge of 7s. 10d. These were the rates that Syria had incorporated in its December decree. Gone, however, was the 3 shillings-per-ton charge to compensate for the alleged inadequate back payments. The new agreement is supplemental to the agreements of 1931 and 1955. Under the new agreement IPC will continue to pay Syria the $250,000 annual protection fee. Presumably also the profit-sharing principle is implicitly recognized in the new agreement though not mentioned.

18. IPC's *Offer to the Syrian Government,* p. 8. Mimeographed material released by IPC.

The new agreement contains no provisions for any change of rates based on changes in transportation costs. The arbitration clause of the 1931 agreement remains binding on both parties. Whether it would be observed in another crisis is doubtful. Under the new schedule of fees, assuming an annual throughput of about 43 million tons, the pre-shutdown rate, IPC will pay the Syrian government about £5 million, approximately 50 percent more a year than under the old agreement.

Because of IPC's increased cost of oil shipped by pipeline to the Mediterranean, however, Banias and Tripoli are now less profitable loading points than they were. They remain, of course, the only outlet for oil from Iraq's northern fields, but IPC's long-run plans apparently contemplate an expansion in the southern fields. It would not be surprising if time should witness a decrease in the relative importance of the northern output. If such should happen, it would probably not involve any decrease in oil shipped by way of Syria, but it might mean a slowdown in the rate of growth in Syrian revenues.

The agreement made no provision for a reduction in the price IPC charged for oil delivered to Syria's Homs refinery. This was never a critical issue in the dispute, however, and that it was dropped probably reflects the fact that Syria would soon be in a position to supply the Homs refinery from its own production.

Under the new agreement IPC and Syria agreed to review the accounts for the period from January 1, 1956, to December 31, 1966, "on the basis of all the contents of the 1955 Supplement Convention and all its attachments." This raises anew the issue of retroactive payments under an agreement that does not provide for them. Since IPC has demonstrated its unalterable opposition to making retroactive payments on accounts that it insists it has meticulously kept in accordance with the provisions of the 1955 agreement, no agreement on back payments is likely to emerge from the investigations. If such is the case, the only recourse that Syria will have aside from unilateral action of the sort that grew out of the original dispute is to submit the issue to arbitration.

The IPC-Iraq Dispute Reconsidered

It will be recalled that before the Syrian-IPC controversy had reached a critical stage Prime Minister Talib had publicly rejected the Wattari agreement and, in doing so, had stated that his government had formulated an oil policy in harmony with the spirit of Law 80 and that it would be implemented by INOC. Opposition to the Wattari agreement stemmed largely from its provision to return to IPC the expropriated area, the North Rumaila field, in which it had discovered oil in commercial quantities but which it had not yet developed.

On January 8, 1967, in his interview with the Baghdad daily *al-Jumhuriyah* the prime minister referred to the provisions of Law 11 governing the allocation of acreage to INOC, indicated that in accordance with the law INOC had applied for acreage, that the nine-man permanent advisory committee recently created by the government had made recommendations on the matter, that he had suggested to the advisory committee that the allocation should be made by law rather than by a decision of the Council of Ministers, and that the oil ministry had in fact prepared a draft law.

Following an extraordinary meeting of the Cabinet on January 21, the Council of Ministers issued a statement on the oil controversy which Radio Baghdad broadcast. The statement referred to the government's decision to request a guarantee from IPC that Iraq's oil rights would not be adversely affected by any delay in settling the pipeline controversy and emphasized that Iraq was not prepared to enter into any new commitments with IPC to solve the crisis.

In a speech before the opening session of the Eighth Conference of Iraqi Engineers in Baghdad on January 23 President 'Arif accused IPC of using the dispute to bring pressure on Iraq "to relinquish some of its legitimate rights."[19] He assured them that the government would sacrifice none of the peoples' rights. Whether he was implying that IPC had brought pressure on the government to approve the return to IPC of its North Rumaila field that Law 80 had expropriated is not entirely clear.

The following facts are clearer: Syria at this juncture was standing fast on the action it had taken which stopped the flow of oil from Iraq's northern field; IPC had made other arrangements, perhaps temporary, to obtain oil from other sources to make good its deficiency; with the pipelines closed, Iraq was losing approximately two thirds of its oil revenues; the stoppage of the pipelines might prove permanent; if so, IPC would be under pressure to develop within Iraq other production to insure that Iraq's loss of revenue would not become permanent; the Rumaila field was in reality an extension of Iraq's other southern fields, which accounted for about one third of Iraq's total production. The surest way for IPC to

19. *MEES*, January 27, 1967.

expand significantly and promptly Iraq's southern production would be by regaining the Rumaila field. For this the Wattari agreement provided.

This leads us back to where we began. Whether IPC brought pressure on the government to ratify the Wattari agreement is not clear. But these facts afford circumstantial evidence that it did. Meanwhile, fast-moving developments killed the Wattari agreement for all time.

14

The Israeli-Arab War
and Iraqi Oil Policy

THE Syria-IPC pipeline dispute may be regarded as a battle in a war waged by militant Arabs to gain control of the oil industry in the Arab world. So considered, Prime Minister Zu'ayyin's boast that the "revolution" had achieved a victory over IPC sounds like political rationalization; his expression of gratitude for the support that the UAR, Lebanon, and Algeria had given him, sounds like the euphemistic acknowledgment of behind-the-scenes pressure designed to force Syria to accept a settlement that IPC would approve. But the war went on.

IPC-Syrian Pipeline Dispute Embitters Iraqis

Syria's seizure of leadership in the Arab struggle and its disregard for the immediate consequences to Iraq left Iraqi leaders bitter. But they expressed their bitterness in public criticism not of Syria, but of IPC. In his speech before the opening session of the Eighth Conference of Iraqi Engineers in Baghdad on January 23, 1967, President 'Arif accused IPC of using the dispute to exploit Iraq mercilessly in the excercise of its legitimate rights. "It utilizes all opportunities to ensnare the developing countries in order to obtain the largest amount of benefit for itself."[1] He urged Iraqis to

1. *Middle East Economic Survey*, January 27, 1967.

be patient, close ranks, be prepared for austerity, and stand firm in the face of adversity. He assured his audience of the government's determination to sacrifice none of the people's rights.

The IPC-Syrian pipeline was closed during most of the first quarter of 1967. When the flow of oil was resumed, a controversy immediately arose over the tax revenue and royalties due Iraq. The government demanded payments based on the amount of oil that would have been shipped had no interruption occurred. More specifically, it demanded first-quarter payments totalling £40 million roughly equivalent to the revenue it had received in the first quarter of 1966. IPC, insisting that the pipeline closure resulted from a *force majeure,* proposed to pay taxes and royalties only on the oil actually shipped. The Iraqi government, contending that IPC's dispute was with Syria alone, rested on Article 7 of the 1955 agreement, which provided that company obligations should not be reduced "for reasons of commercial expediency or convenience." The parties compromised on a settlement in which each professed to find support for its principles. IPC announced that to assist the government it had agreed to advance £14 million to be repaid out of future oil revenues. The government announced that the £14 million would accrue from increased oil output. In the government's view, it was not a loan but a settlement for the decrease in production occasioned by the pipeline closure.

What to Do with North Rumaila?

Behind the negotiations over revenues lay a deeper dispute: What to do with IPC's discovered but undeveloped North Rumaila field, expropriated under Law 80? It will be recalled that the Wattari agreement, in addition to providing for the return of North Rumaila, also provided for the organization of a joint-venture partnership with IPC for the exploration and exploitation of a 32,000 square kilometer concession within the expropriated area but outside North Rumaila. This would have left the government free to work out other joint ventures covering most of the expropriated areas with whomever it could arrange profitable deals. Among those primarily responsible for negotiating this arrangement were a group of young pragmatic realists consisting of some of Iraq's most dedicated, experienced, and competent engineers and economists. As a part of the negotiating process they had examined with technical skill and analysed with professional competence the role that INOC might play in developing Iraq's oil reserves and its domestic economy. They presented the results of their study in a 5,000-word paper to the First Conference of Arab Economists, held in Baghdad November 1 through 7, 1965.

After considering the alternatives facing INOC—developing the expropriated areas alone, or in cooperation with foreign oil companies—they concluded that the surest, most efficient, and most economical procedure was to enter into joint ventures with foreign companies. For IPC's acknowledging the validity of the expropriation and for entering a joint venture with INOC they were willing for INOC to return North Rumaila to IPC.

The Conflict of Purpose

For ten years following Qasim's revolution, oil remained one of Iraq's major political and economic problems. It supplied the government with its major revenue. It influenced most major political decisions. It served as a scapegoat by which politicians could divert the people from a consideration of their misery and provided them with an instrument by which they believed they could achieve their economic salvation. While promising a haven of economic security, it proved to be a turbulent sea of political instability. In most of the six years since Qasim's overthrow, two forces have struggled to shape governmental oil policies. INOC's management, anxious to get on with the job of developing Iraq's oil resources and concerned primarily with the economic consequences of its program, used its influence to convince the government that it should ratify the Wattari agreement.

Those more definitely committed to political ideals advocated that INOC independently develop the Rumaila field. The Baathists and leaders of the Arab National Movement were the principal spokesmen within Iraq for independent development. Abdullah Tariki from his office in Beirut endeavored to mobilize public sentiment in support of independent development throughout the Arab world. His was a popular plea consistent with the Arab slogan "Arab oil for the Arabs." Those at the head of an insecure government were hesitant to turn their backs on it. They were also reluctant to reject the advice of their own oil experts. This was their dilemma.

President 'Arif, anxious to maintain a consensus, remained indecisive.

On May 10, 1967, he reorganized his cabinet.[2] Indicative of his insecurity, his new cabinet was the largest in Iraq's recent history. It included four deputy premiers and was composed for the most part of moderates. No hard-core Baathists nor Arab Nationalists were included. But outside the cabinet they were ready to attack any oil policy that in their judgment failed to take account of Iraq's national interest. President 'Arif's somewhat ambiguous statement on oil was apparently designed to reassure them. He pledged that the government would work seriously to develop Iraq's national resources and to implement the national oil company in doing so and "thereby make up what we have lost as a result of the domination of the oil companies operating in Iraq."

Israeli-Arab War Shapes Iraqi Policy

This sounded as though it might herald the death knell of the Wattari agreement. The Israeli-Arab war removed all doubt about it. On the eve of hostilities the Arab states indicated that they were prepared to use all the political and economic weapons at their command in a war against Israel. To determine how they could effectively use their oil they announced a meeting of delegates for June 4 in Baghdad. Iraq promptly indicated its determination to cut off oil supplies to any state supporting Israeli "aggression" and to terminate the concession rights of any company not complying with its decision. Iraq was the first Arab country to stop all oil exports and the last to resume shipments after the war.

On July 10 President 'Arif announced the formation of another cabinet. In it he surrendered his premiership, appointing as his successor Lieutenant General Tahir Yahya. Premier Yahya characterized his government as a government of "war and reconstruction."[3] In the "war" he proposed to use oil as a weapon in the hands of INOC. In the new cabinet Adib al-Jadir, who as a bitter opponent of the Wattari-IPC agreement had urged that INOC be the instrument for developing the expropriated oil properties, became the minister of industry.

In a policy statement on July 28 Premier Yahya elucidated further the government's intentions with regard to INOC. The new government, he

2. *Ibid.,* May 12, 1967.
3. *Ibid.,* July 14, 1967.

said, proposed to carry out the provisions of Law 80 (the Qasim expropriation decree) by allocating territories to the Iraq National Oil Company, by
precluding the granting of new concessions, and by breathing life into
INOC to enable it "to build up a national oil sector and engage in oil-pro-
duction and marketing operations."[4]

Law 97

Law 97, issued nine days later, gave more precise meaning to the
premier's policy statement. It represents a sharp and presumably final
break with the traditional concession system. It bars the restoration to IPC
of North Rumaila (Article 2). It assigns to INOC the exclusive right to
develop oil in all Iraq territory except that which IPC was permitted to
retain under Law 80 (Article 1). Although it permits INOC to develop any
area in association with others, it specifically prohibits its doing so through
"a concession or the like" (Article 3).[5]

Law 97 did not specifically bar INOC from developing the North
Rumaila field in association with IPC. In truth, INOC officials who had
helped negotiate the 1965 agreement were still hopeful that they could
negotiate a settlement of all issues in controversy with IPC, and within the
framework of Law 97 work out an arrangement for the joint development
of the North Rumaila field. This accomplished, they anticipated a prompt
increase in Iraq's oil output and a continuing increase in governmental
revenues. It soon became evident that bitterness against the West had
reached an intensity that made difficult if not impossible any new association with the Anglo-American concessionaire. The opponents of reconciliation, distrustful of the management of INOC, demanded its reorganization.
They got what they wanted in Law 123. Published in the Official Gazette
on September 21, 1967, it provided for a complete reconstruction of the
Iraq National Oil Company.[6] The law broadens the scope of INOC,
enlarges its responsibilities, narrows its authority, and brings it more se-
curely and directly under the control of the government.

4. *Ibid.*, August 4, 1967.
5. *Ibid.*, August 11, 1967, gives a complete translation of Law 97.
6. *Ibid.*, September 29, 1967, publishes a translation of the law and the accompanying memorandum.

Law 123 Reorganizes INOC

Law 11, which originally established INOC as an instrument for developing a national oil industry, placed authority primarily in the hands of a managing director, a post initially held by Ghanim al-'Uqali, an engineer by training who by long association with oil problems had gained recognition as one of the Arab world's leading oil technocrats. Law 123 in effect abolishes the post of managing director and concentrates immediate authority in the president and chairman of the board, a post filled by official decree on October 3 by Adib al-Jadir, leading critic of former Minister Wattari's oil policy. The minister of oil nominates the members of INOC's board of directors subject to the approval of the Council of Ministers and the president (Article 13). To insure that the company will adhere to the general oil policy of the state it is attached to the ministry of oil. Although the president of the company with the approval of the prime minister may attend discussions of the Council of Ministers on matters relating to oil policy and its implementation (Article 16), should the oil ministry and the company disagree, the Council of Ministers makes the decision (Article 16).

The new law contemplates that INOC shall not only serve as an instrument for developing a national oil industry but as a base for developing Iraq's industry in general. It is expected to work within a general program

for the development and expansion of oil exploitation in Iraq in all phases of the petroleum industry including the production of crude oil and petrochemicals and refining, export and marketing operations; and to undertake all domestic and foreign operations required to promote the growth of the national income and achieve a selfsufficient and balanced economy.

In short, INOC by developing an independent national petroleum industry is to establish the basis for the country's industrialization.[7]

To aid in achieving these broad ends, INOC is directed to "take steps to raise the educational, technical, and social standards of its personnel and those of its owned companies" by establishing vocational centers to im-

7. *Idem,* explanatory memorandum.

prove workers' capacities and skills, by organizing educational seminars, by introducing work incentives through a bonus system, by establishing educational and technical institutes, laboratories and research centers, and by sending educational and technical missions abroad (Article 19).

The INOC-ERAP Agreement

Armed with the authority to develop the expropriated areas alone or in co-operation with others, INOC looked to the co-operation of industrial countries that had aided or abetted the Arabs in their war with Israel and that possessed the necessary capital and technical skills. They found France and Russia anxious to fill the gap created by the alienation of the Anglo-American oil companies and their countries.

The Iraq News Agency disclosed that on October 23 the French ambassador in Baghdad delivered a letter from President De Gaulle to President 'Arif in which he expressed "the French Government's complete readiness to establish the closest possible ties with the Iraqi Republic in all fields in which French-Iraqi cooperation could be fruitful for both countries." He characterized previous agreements entered into as providing "a firm basis for promoting co-operation between the two countries in a manner commensurate with the importance that France attaches to economic and social development in the Arab world particularly in the present circumstances."[8]

Within a month a more specific agreement had been worked out. On November 23 the Iraq government announced a twenty-year contract between INOC and France's state-owned group of oil companies, Entreprise de Récherches et d'Activités Pétrolières (ERAP), for the development of a portion of the expropriated acreage. This represents the first concrete step in INOC's developing an oil industry owned and managed by it. In its basic characteristics it is quite similar to Iran's contract with ERAP signed in the summer of 1966 and characterized by Chairman Eghbal of NIOC as "revolutionary." (See Chapter Seven.)

The arrangement provides that ERAP shall explore an area of 10,800 square kilometers under a contract with INOC. ERAP will release one half of the acreage in the third year, one half of the remaining area in the fifth

8. *Ibid.*, November 3, 1967.

year, and at the end of the sixth year all the remaining acreage except that on which it has discovered producing fields.[9]

Of the discovered reserves ERAP will hand over to INOC 50 percent to be developed or held by INOC as a national reserve. ERAP will develop the remaining 50 percent in co-operation with INOC. ERAP is obligated to provide as a loan without interest all exploration costs, repayable only if oil is developed in commercial quantities, in fifteen annual installments or at the rate of ten cents a barrel, whichever is greater. ERAP will advance interest-bearing funds for the development costs and INOC will repay them in the form of crude oil over a five-year period. This will be compensated in part by ERAP's payment to INOC in annual installments an unrecoverable bonus totaling $15 million over the life of the contract. Five years after exports begin ERAP will hand over the entire venture to INOC, which will both own and operate all the equipment including pipelines and export terminals.

When production begins in the area jointly operated by ERAP and INOC, ERAP will buy 30 percent of the output, 18 percent at a price equal to the cost of production plus a royalty of 13.5 percent plus one half the difference between their sum and the posted price of crude in the Persian Gulf. In short, on 18 percent of the oil produced INOC will get what the Organization of Exporting Countries has long sought but not yet fully obtained from the Middle East traditional concessionaires.

On the other hand, ERAP will buy 12 percent of this output at a bargain price equal only to the cost of production plus a 13.5 percent royalty. On this 12 percent Iraq will not receive the 50 percent tax or profit split which traditional concessionaires pay on their whole output.

INOC as owner is free to sell the remaining 70 percent of production at such prices as it can command. If it cannot market this at satisfactory prices, ERAP is obliged to market for INOC's account at such prices as it can command five million tons a year receiving one half a cent a barrel for services rendered, and an equal amount on which it will receive one and a half cents a barrel. What INOC derives from this 70 percent will obviously depend on the market price of oil, and all the profits from its sale will belong to INOC. This compares with an expensed royalty of 12.5 percent and a 50 percent tax based on posted prices (not market prices) that the traditional concessionaire now pays.

On February 3, 1968, shortly before President 'Arif left for a state visit

9. *Ibid.*, January 12, 1968, published the text of the agreement.

to France, a special session of the Iraqi Cabinet under his chairmanship approved the agreement, and at an official ceremony al-Jadir, as chairman of INOC, and Jean Blancard, representing ERAP, signed the contract. Law 5 of 1968 ratified it and the *Iraqi Official Gazette* for February 5 published it.

Significance of the INOC-ERAP Agreement

The INOC-ERAP agreement represents a repudiation of the traditional Middle East concession policy. As such it evoked wide acclaim throughout the Arab world. To many articulate Arabs it represented Iraq's declaration of independence, freedom from a bondage imposed on them by imperialistic oil companies of the Western world. The Lebanese economist, Dr. Nicholas Sarkis, in the Beirut journal *Le Commerce du Levant* for November 29, 1967, referred to Laws 97 and 123 and the ERAP-INOC agreement as evidence of the determination of an Arab country disillusioned by the bitter experience of the concession system to recapture by legislation its rights and to escape from a tyranny imposed on it by a foreign power. He said,

The major oil companies are now driven to admit the *fait accompli* and to reconsider their relation with Iraq. Apart from the immense advantage that this new situation entails for Iraq, the facts prove again that in this field only firmness pays and that the improvement in the system of oil exploitation depends first and last on the will of the Arab countries to shoulder their responsibilities and take in hand the development of their resources.

Abdullah Tariki, reportedly a consultant in negotiating the INOC-ERAP agreement, in a statement to the Baghdad press on November 29 praised INOC for its initial step in becoming a fully integrated oil company operating in world markets. He expressed the opinion that "Whatever else can be said of the INOC-ERAP agreement, it is a point of departure for a better life for the Arab people of Iraq."

Much of the Iraq press, swayed perhaps by the influence the government exerts on its freedom, voiced similar sentiments. *Al-Jumhuriyah* (Baghdad, November 24, 1968) avowed that by concluding the ERAP agreement "Iraq has turned a new leaf in the history of the revolution and has become master of its land and the wealth contained therein." It expressed its appreciation for the "noble stand adopted by France *vis-à-vis* the Arab nation in its transient predicament" and for the evidence that "the

Arab people know how to return favors to a friend and how to treat their enemies." *Al-Arab* (Baghdad, November 25) characterized the agreement as "a new type of co-operation . . . a fair partnership free from injustice and fraud" and expressed the hope that it will exert a salutary influence on the "monopolistic oil companies" persuading them "to alter many of their ways which have for long been the object of severe disputes in more than one oil-producing country."[10]

Iraqi Experts Challenge the INOC-ERAP Agreement

With the promulgation of Law 97 and Law 123 and the signing of the ERAP and Soviet agreements, Iraq launched an oil-development program on which, barring a revolution, there is no turning back against its broad objectives and the political independence that the government manifested in adopting it. No organized Arab voice of dissent has been raised. But whatever may be its long-run material benefits and however great its spiritual rewards, two competent Iraqi engineers, former members of INOC's management who resigned after the promulgation of Law 123, challenge the ERAP agreement as an instrument for increasing Iraq's oil revenues. In short, they regard it as a bad business deal. In two memoranda addressed to President 'Arif they analyze in detail its several provisions and they conclude that it offers oil to France at bargain prices and provides the government a smaller profit than does the old type of concession.

As former members of INOC's management with access to the geological and engineering records that IPC released to it after the promulgation of Law 80, they can speak with some authority. They allege that the acreage to be developed under the ERAP contract can be considered among the world's best and that it contains fifteen geological structures, as revealed in a preliminary seismic survey; they estimate that the acreage contains more than fifty billion barrels of oil (more than one and one half times the United States' unmined reserves).

Applying production costs based on IPC's experience and taking account of the output division provided in the ERAP contract and of the payments ERAP is obligated to make to INOC, they conclude

that the French group will be incurring a lower per barrel cost for its purchase entitlement from INOC than the corresponding lifting costs attributed

---

10. *Ibid., Supplement* for December 8, 1967, presents a roundup of press comment on the Iraq-ERAP deal.

to BPC under a hopelessly outmoded concession—an arrangement which would not have survived were it not for the sanctity of contracts concluded by previous governments under pressure of bygone circumstances, and for the economic power of the companies operating in Iraq . . . and their monopolistic control of world markets.[11]

They calculate that ERAP's annual return on its investment will be somewhere between 200 and 330 percent. They conclude,

So long as we denounce the imperialistic oil concessions which permitted the monopolistic companies to realize an annual return on investment of 66 percent and are doing our utmost to modify and lower the rate of return, whether by expensing royalties, eliminating market allowances, or raising taxes, surely we have the right to question the reason for allowing the French group, ERAP, such a high rate of return.[12]

The Validity of the Experts' Criticism

The uncertainties with regard to investment, output costs, selling prices, and other imponderables are too great to accept the conclusion of the experts as accurate and final. If we apply their assumptions and calculations to the national reserves, that is, if we assume the same production costs and market prices that the experts apply to the 50 percent of the reserves developed in common, INOC's per barrel revenue on the whole output well exceeds that which the conventional concessionaires now pay. The most doubtful assumption is with regard to the future market price of oil. If INOC becomes an important producer of oil, the price it can command is likely to be affected adversely by two factors: the increasing flow of oil into international markets and the adverse conditions that INOC must face in selling it in competition with the fully integrated worldwide operations of the well-established international oil companies.

But it should be observed that the engineers were not alone in their criticism. The Baghdad *al-Taakhi* for November 26, although recognizing that the government by exercising its sovereign rights had broken the "monopolistic fetters" by which it had hitherto been bound, expressed regret that the government had not subjected the draft agreement with

11. This quotation is from the first memorandum, signed by Tariq Shafiq on behalf of the 'Uqaili-Shafiq petroleum consultants firm in Baghdad. *MEES*, December 1, 1967, published a translation of the memorandum.

12. This quotation is from the second memorandum signed by Ghanim al-'Uqaili, a summary of which the *MEES*, January 19, 1968, published.

ERAP to public criticism before signing it. It lamented the government's failure to enlist the advice of Iraqi oil experts, university professors, and representatives of various political groupings. In its following issues *al-Taakhi,* apparently referring to the reconstruction of INOC under Law 127, became more specific in its criticism. It alleged that when Iraq most needed the help of oil experts it had dispensed with their services. It charged that the government in its oil policy had been more concerned with the political affiliations of its advisors than with their skills and competence.

In the November 30 issue of the Baghdad daily *al-Nasir* the Iraqi economist Zayid Muhammad criticized the ERAP contract more boldly, charging that it was simply a disguised concession agreement.

In order to quiet further dissent, the government by decrees published in the *Official Gazette* for December 3 abolished Iraq's sixteen privately published newspapers and replaced them with five nationalized dailies edited by state officials.

This drove some critics abroad. Mr. Mahmud al-Durrah, a retired army officer active in local politics, in an interview with the Beirut daily *al-Hayah* (January 9, 1968) lamented the muzzling of public opinion by the nationalization of the press and insisted that no agreement could be valid without a return to parliamentary government.

INOC-ERAP Agreement A Political Move

Any conclusion as to whether Iraq chose the most profitable method of developing a portion of its oil lands is beside the point. To evaluate the course upon which Iraq launched, it must be viewed in historical perspective.

The British mandate over Iraq brought political dissension culminating in bitterness and eventually in revolution. Although IPC through its application of modern technology and capital has developed an oil industry profitable to it and Iraq alike, its concession covering virtually the whole of Iraq has been a constant source of controversy and conflict. During the mandate's life the oil companies came to be regarded as instruments of Western colonialism and imperialism.

The creation of Israel and the wars of 1948, 1956, and 1967, with their unfortunate consequences for the Arabs, have exacerbated their feelings toward the West and heightened their hostility toward the oil companies.

Arab leaders have identified their oil problems with the "Arab struggle against imperialism" and their struggle against imperialism with their struggle against Israel. Adib al-Jadir expressed this identity of relationship in a statement published in *al-Thaurah* on December 31. The oil question, he said, cannot be separated from the Arab struggle against imperialism, and the United States is the demon in both.

the shameful position taken by the U.S. at the UN during the war and afterwards, its flagrant aggression against the Vietnamese people, and its economic pressure on India on account of its understanding attitude toward Arab causes—all these are a part of a pattern emanating from the hysteria which has seized America of late.

The "oil monopolies" in exerting pressure to stifle the aims of an eager and energetic people are reflecting the will of the United States. He proclaimed that the law expropriating IPC's concession (Law 80), the law abolishing the concession system (Law 97), the law reorganizing INOC to make it a more effective instrument to develop Iraq's oil lands and its economy (Law 127), all are steps in extricating Iraq from the straightjacket which the oil monopoly imposed on it. They are steps essential to the fruition of the Revolution of July 14, 1958. The ERAP and the Soviet agreements represent "the first stone in the edifice of a real national oil industry. . . . It is not mere vanity to say that these two steps represent a point of departure of which every Iraqi with his country's interest at heart can be proud."

Despite the dissent and criticism that the ERAP deal evoked, it is probably fair to say that Jadir reflects the consensus of most literate, articulate Iraqis.

IPC Protests Law 97

The recently proclaimed statutes forbade the return of North Rumaila to IPC. They forbad the disposing of North Rumaila reserves under a conventional concession. They did not forbid INOC's working out an arrangement with IPC within the framework of Law 97, but they made such an agreement extremely unlikely. On August 10, 1967, IPC issued a formal protest against the enactment of Law 97. It referred to its continuous endeavor to resolve its differences with the government. It called attention specifically to the 1965 agreement that it had worked out providing for a joint venture with INOC. It alleged that governmental approval of this agreement would have resulted in an expansion in oil operations with increasing quantities of Iraquian oil moving to world markets and increasing oil revenues to the Iraq government. It averred that the agreement would have enabled Iraq to offer new companies undisputed developmental rights to large areas relinquished by IPC. It expressed regret that by enacting Law 97 the government had rejected negotiations as a means of settling their differences with IPC and characterized this as a "further breach of international law." Finally, it announced that its shareholders would take "every step available . . . to prevent the future exercise by other parties of the company's rights."

In September the British government sent a note of protest to the Iraq government through the Swedish Embassy echoing the sentiments expressed by IPC. The government alleged that by failing to ratify the 1965 agreement and by promulgating Law 97, the Iraq government had "rejected companies' efforts to settle outstanding issues by negotiation" and in doing so had violated "the principles of International Law."

While expressing confidence that the government of Iraq would fulfill its obligation to resolve the issues by arbitration, at the same time

the British Government reservers all the rights of the British Government and the rights of the companies acquired through their concessions. The British Government also wishes to make it clear to the Government of Iraq that it is determined, for the purpose of safeguarding the rights of companies registered in the United Kingdom, to assist them in reaffirming their rights.

Following these protests IPC dispatched negotiators to explore further the possibility of settling all issues in dispute with the Iraq government, including the return of North Rumaila. But by this time the point of no return apparently had been reached. In a full-page statement published in the Beirut daily *al-Anwar* on February 23, 1968, the Iraqi minister of oil, 'Abd al-Sattar ali al-Husain, reproduced the texts of the British government's protest and related circumstances leading to Iraq's irreconcilable position. He alleged that IPC during the June Israeli-Arab war had refused to honor his government's request for oil when needed to meet a temporary emergency. About this, he said,

For us, this was the last straw, and we realized the importance of breaking the monopoly. We began to study the problem till we had the whole picture before us. We went to the Council of Ministers and put forward our views to President 'Arif. We were told that what we were planning was very dangerous indeed—perhaps a more fateful step than the nationalization of the Suez Canal

and its aftermath of war. Taking any steps against the monopolistic oil companies, it was said, would be followed by landings of foreign troops. The question
put to us was whether we were ready to face all eventualities. Then the
President stood up and addressed us as follows: "We are ready to make to do
with bread and onions, so long as our Arab dignity is upheld. Our rights must
be restored to us and nobody can intimidate us in pursuit of our aim." With
these words, President 'Arif announced the birth of Law 97 which was the
subject of a British Government protest and considered by the oil companies as
a sentence of death.

Iraq Turns to Russia

While INOC officials were negotiating with ERAP for a limited contract within the framework of Law 97 providing for the co-operative
development of a relatively small portion of the expropriated area, they
were trying to formulate a definite program for the exploration and development of the remaining vast area made available by Law 80 and particularly for the development of the North Rumaila field. To achieve this, they
looked to Russia. As early as September 1967 the Iraqi oil minister invited
Soviet representatives to Baghdad to discuss arrangements looking to this
end. On December 24, 1967, Chairman Adib al-Jadir of INOC and
Semyon Skachkov, who headed a sixteen-man Soviet delegation on a
four-weeks visit to Iraq, signed a letter of intent providing for Soviet
assistance in Iraq's development program.

It was not until the summer of 1969 that this letter of intent was
translated into a definite program designed to permit INOC to realize its
full potentialities as a national oil company and to bring the expropriated
properties into continuing development. This was done by two agreements:
the one between the governments of Iraq and the Union of Soviet Socialist
Republics; the other between INOC and the Soviet Machinoexport Organization. The agreement between the two governments, signed in Moscow on
July 4, 1969, obliges the Russians to "prepare and put into operation" the
oil fields of North Rumaila with an immediate objective of producing 5
million tons annually (about 100,000 barrels daily) to be increased to 18
million tons (about 365,000 barrels daily); to "draw up a program for the
development, survey and design work to prepare and put into operation"
five designated fields in South Iraq; "to provide technical assistance in
preparing and putting into operation the Ratawi oil field"; to conduct a
geological and geophysical drilling survey necessary to explore for oil and
gas in various undesignated areas; to lay a pipeline from the oil-producing
areas to the Fao port on the Persian Gulf, and to provide the necessary

equipment and installations in accordance with design instructions provided by Iraq.[13]

To finance this program Russia agreed to provide 60 million rubles ($70 million) bearing interest at 2.5 percent and repayable in annual installments beginning in 1973. The loan and interest are repayable in crude oil delivered by Iraq to the Soviet Union at "realized prices prevailing in the free international market." The agreement contemplates that the North Rumaila field will be ready for operating by the first quarter of 1972.

The agreement between INOC and the Soviet Machinoexport Organization, signed in Baghdad on June 21, 1969, designated the latter as the instrument by which the terms of the broader agreement will be realized. The preamble to the agreement acknowledged INOC's decision to "undertake the direct national development" of the areas assigned to it (Law 97 assigned to INOC the exclusive right to develop oil in all of Iraq except that area which Law 80 permitted IPC to retain) including fields where the presence of oil had been established and covering all phases of the oil industry from prospecting and exploration to drilling, producing, gathering, storage and transportation; and it recognized INOC's desire to obtain technical assistance from a competent party in realizing these objectives.

Machinoexport agreed to provide the machinery, the equipment, all materials, and the technical assistance necessary to develop INOC as a fully integrated national oil company. As a start toward this end Machinoexport agreed to provide ten drilling rigs and five geophysical and geological survey teams and all ancillary services needed. It will also provide short-term credit up to $72 million to cover the costs.

The Soviet-Iraqi agreement constitutes the most significant development in the recent history of the Middle East oil industry. It is significant in four respects. First, it marks the culmination of a program conceived in Qasim's revolution whereby a national oil industry was to be built on lands reclaimed by expropriation, with no provision for compensation. Second, in doing so, Iraq obtained the technical and financial aid of Russia, who by her help increased greatly her influence in the Middle East. It marks Russia's first foothold in an important Middle East oil-producing country. Third, it will be instrumental in establishing INOC as a fully integrated national oil company with an immediate market for its output. Fourth, while it guarantees INOC an immediate market for its oil, it does so with no guaranteed price. INOC will sell oil in world markets for what it will bring. It will add to the downward pressure on prices.

13. *MEES Supplement, August 8, 1969, published the two agreements.*

15

Saudi Arabia and Aramco Settle
Some Issues

ABDULLAH TARIKI more than any other single individual is responsible for having focused attention on those terms and interpretations of the Aramco-Saudi agreement which the government considered in conflict with the country's interest. With a B.S. degree in geology from the University of Cairo (1945), an M.A. degree in petroleum geology from the University of Texas (1947), a year's experience in Texaco's training program, and six years' experience as head of the inspection office of Saudi Arabia's Bureau of Mines, he had apparently qualified himself by training and experience for the post of Director General of Petroleum and Mineral Resources to which he was appointed in 1954. His bright mind and technical training soon gave him an unusual insight into the controversial issues inherent in the concession agreement. His dedication to public service, his ambition, boldness, and loquacity soon made him the leading Arab spokesman on these issues. Men of such qualities are rare in a desert culture where the battle for survival develops many rugged individualists but few public servants. Although Tariki's experience, training, and personality drove him upward in the hierarchy of a government whose chief problems more and more centered around oil from which the royal family derived its affluence and upon which the country's prosperity had come to depend, their influence in qualifying him for the position that he was soon to occupy was not wholly salutary. His experience as a minor government official in Dhahran in dealing with Aramco had left him embittered. Strong preconceptions, a fiery temperament, a determined will nurtured in an intellectual environment hostile to "Western imperialism and colonialism" made Tariki more effective as an agitator than as an administrator. His explosive and pugnacious temperament is reflected in a dramatic paper he presented before the Second Arab Petroleum Congress in 1960 to an audience of several hundred oil executives, oil technicians, journalists assembled from all over the world, and government officials from the several Middle East countries, in which he alleged that the oil companies by arbitrary and discriminatory pricing of Middle East oil had pocketed $5,474,290,133, one half of which under proper pricing would have gone to the Middle East producing countries.[1]

Tariki's elevation to the post of minister of petroleum and mineral resources in the cabinet shakeup that resulted in the resignation in December 1960 of the king's council of ministers and the temporary retirement from governmental affairs of Crown Prince Faisal gave Tariki a more responsible platform from which to agitate for changes in the concession agreement and its interpretation, and his new responsibilities seemed to temper somewhat the fervor of his public pronouncements and to focus them on specific issues.

Tariki States His Views on the Oil Agreements

In an interview with the Mecca daily *al-Nadwah* almost a year before his elevation to cabinet rank, he characterized the early agreements between the several Middle East countries and the concessionaires as "grossly unfair to the host countries," enabling the concessionaires to "control the economy of the host state, drain its resources and impede its development." That they had initially been acceptable to the several Middle East countries, he contended, was understandable only in the light of the special circumstances of the times, the lack of sophistication of the governmental negotiators, and the urgency of the countries' needs. He pointed to the Japanese-Saudi offshore agreement, in the framing of which he had played a major role, as providing a "fruitful partnership" between the concessionaire and the government.[2]

1. Abdullah Tariki, "The Pricing of Crude Oil and Refined Products," Second Arab Petroleum Congress, Beirut, October 17–22, 1960. Tariki departed from his prepared text to make this accusation.
2. *Middle East Economic Survey*, January 8, 1960.

Shortly after his appointment as minister of petroleum and mineral resources. Tariki directed his criticisms specifically towards the Aramco and the Trans-Arabian Pipeline Company (Tapline) agreements and urged the importance of bringing them into accord with his conception of Saudi Arabia's needs.

Our people and government now expect more from life, and our government is determined to improve the lot of its people and to provide a better life for them.
This can only be accomplished through an increase in revenue and by emulating other nations in making maximum use of natural resources. For this reason the oil agreements should be altered to conform to present-day conditions.[3]

Changes are necessary, he averred, not only for Saudi Arabia's benefit, but to insure that Aramco and Tapline continue their "work in an atmosphere of friendship, understanding and cooperation."

In an interview with the Cairo economic fortnightly *al-Ahram al-Iqtisadi* on February 15, 1961, Tariki was more specific in stating his ultimate aims for remaking the Aramco concessions. He advocated the conversion of Aramco into a fully integrated company, producing, transporting, refining and marketing petroleum products and sharing profits with Saudi Arabia in all of its operations; greater Arab participation in all phases of the oil industry both inside and outside the Middle East; and the eventual "Arabization" of Middle East oil so that Arab resources would come under Arab control.

In an interview with the Baghdad daily *al-Zaman* on April 17, 1961, Tariki was more specific in his statement of the immediate issues between his government and Aramco. Among these he listed participation by the government in the profits derived from the operations of Tapline; Aramco's relinquishment of the unutilized concession areas; Aramco's practice of granting marketing discounts in disposing of its oil to its parent companies.

Two months later, in an interview with two of Saudi Arabia's leading newspapers, the Jiddah daily *al-Bilad* and the Mecca daily *al-Nadwah,* Tariki made what was currently regarded as his most important policy statement since his promotion to cabinet rank. In addition to outlining the work of the newly created Supreme Planning Board, of which he was a member, and commenting on some broad questions of governmental pol-

3. Interview with Jiddah *al-Bilad* as reported by the Riyadh weekly, *al-Yamameh,* January 15, 1961, and translated by the *MEES,* January 27, 1961.

icy, he again directed his attention to matters in dispute between his government and Aramco. Contending that foreign capital was realizing an annual return of 60 percent on its capital investment in Saudi Arabia as compared with only a 10 percent return in the United States, he urged amendment of the oil agreements to limit capital to a "reasonable return sufficient to encourage its continued operation" in Saudi Arabia and to insure that the balance of the profits went to "the people." Specifically he advocated a revision of the profit-sharing agreement; relinquishment of all undeveloped portions of the concession, including all oil fields not being actively exploited; a governmental voice in Aramco's accounting procedure; prohibition of the flaring of any natural gas produced in association with crude oil, subjecting Tapline's operations to the profit-sharing agreement; and the development of a petrochemical industry.

Tariki on the Tapline Issue

With respect to governmental participation in the profits derived from Tapline's operation, he pointed out that when Aramco delivers crude to its parent companies at the Mediterranean terminal of Tapline, it charges the prices posted in the Persian Gulf rather than the prices which the parent companies post at the Mediterranean port of delivery. These prices are higher than the Persian Gulf prices by approximately the tanker cost of delivery by way of the long water route through the Arabian and Red Seas and the Suez Canal to the Eastern shores of the Mediterranean. By this practice, Tariki contended, Aramco shifted the profits of pipeline transportation to the parent companies and thus avoided sharing them with the government. Regarding this as grossly unfair, Tariki insisted that the profits from Tapline's operations be shared 50-50 with the Government and that Aramco pay the Government $180 million in settlement of past losses. In concluding his interview, Tariki professed the government's desire to maintain good relations with the oil companies and its intention of settling its disputes in an amicable and legitimate manner, "watching over the companies' interests with the same enthusiasm that it shows for the interests of its own people."[4]

To Tariki's charge that Aramco owed the government $180 million in income taxes on the sale of crude oil at the Mediterranean port, Aramco, in

4. MEES, June 23, 1961, Supplement gives a summary translation of the interview.

a press release reviewing briefly its program for aiding Saudi Arabia's development, emphasizing its constant interest in the country's welfare and acknowledging its "utmost desire to find a solution for any differences in accordance with a spirit of co-operation and good will," stated,

On this issue the Government and Aramco hold honest differences of opinion since Aramco believes it owes the Government nothing in this connection. In view of this, Aramco wishes to assure the gracious Saudi Arab people that it stands ready to discuss or arbitrate, if necessary, any point of difference with the Saudi Arab Government.

To Aramco's expression of willingness to arbitrate, Tariki replied in an interview with the Saudi weekly *al-Khalij al-'Arabi,* "Tax problems concerning the oil companies in our country must be settled in accordance with the Income Tax Law promulgated by Royal Decree, to which Aramco is subject."[5]

In an interview with the Cairo economic journal *al-Ahram al-Iqtisadi* in October, Tariki lapsed into his earlier belligerency. He vowed that his country would not tolerate Aramco's "iniquitous exploitation" much longer and would request the Organization of Petroleum Exporting Countries' (OPEC) endorsement of Saudi Arabia's stand in its dispute with Aramco.[6]

About two weeks later Radio Mecca characterized Saudi Arabia's oil policy in somewhat more conciliatory language:

Under the leadership of His Majesty the King, Saudi Arabia is employing all possible means to derive maximum benefit from its oil resources; and to this end it is cooperating with all other states which are working along the same lines. Our method of achieving this aim is to base our course of action on the principles of equality and justice within the framework of voluntary association with the oil companies.[7]

Tariki Resigns

On March 15, 1962, from Riyadh the government announced a new cabinet with King Saud as prime minister and Crown Prince Faisal as deputy prime minister. Tariki, who as administrator of Saudi Arabia's petroleum affairs since 1954 had been a zealous and persistent protagonist

5. *Ibid.,* September 8, 1961.
6. *Ibid.,* October 20, 1961. For the organization and functioning of OPEC see Chapter Seventeen.
7. *Ibid.,* November 10, 1961.

in the protection of his country's oil interests as he saw them, was not in the new cabinet. Neither the king nor the deputy prime minister made any public explanation of the omission of Tariki's name from the new government. The Beirut economic bi-weekly *Le Commerce du Levant,* one of the few non-Egyptian papers to comment on it, editorialized that Saudi Arabia, about to orient its policy so as to come to terms with Aramco and Tapline, chose to be represented by a less intransigent negotiator.[8] If by "coming to terms" *Le Commerce* meant to imply that Saudi Arabia was preparing to settle the issues by surrendering its claims, events have proved that it was in error.

On leaving the ministry Abdullah Tariki went into voluntary exile in Beirut where he has since conducted a consulting office on Middle East oil affairs and functioned as self-appointed spokesman in the protection of Middle East oil interests. Tariki's departure from the government was not an isolated event; it was an incident in a movement reflecting dissatisfaction within the royal family with the king's profligate expenditures and the way in which he was running his kingdom, which eventually resulted in Crown Prince Faisal's assuming the throne and Saud's leaving the country. The king was responding too slowly to forces both within and without his kingdom pressing for improvements in the lot of his people. Tariki, viewing his country's oil resources as the most effective means of promoting Saudi Arabia's economic progress, was moving too rapidly. He looked anxiously to the day when his country's oil industry would be nationalized and Arabized, and he rarely missed an opportunity to express his views on the subject.

Yamani Becomes Oil Minister

Sheik Ahmad Zaki Yamani succeeded Tariki as minister of petroleum and mineral resources. He brought to the ministry a calmer, more hard-headed, common-sense spirit, equally dedicated to using oil for promoting his country's welfare but differing from Tariki in his methods. With the expertise and judicial attitude of a well-trained law student, the breadth that foreign study engenders, and the aspirations that an educated Arab is heir to, he was well qualified for his task. Yamani received his law degree from the University of Cairo in 1951 when he was only twenty-one years

8. *Le Commerce du Levant,* March 21, 1962.

old. After two years of service with the Saudi Arabian government, he went to the United States for graduate study and obtained the degree of Master of Comparative Jurisprudence from New York University in 1955. Thereafter he did additional graduate work at Harvard. On returning to Saudi Arabia, he engaged for a short time in private law practice. In 1958 the government employed him as legal advisor to the council of ministers and in 1960 made him a member of the council and minister of state. He was only thirty-two when he became minister of petroleum and mineral resources. He soon let it be known that he had not assumed office to settle outstanding issues with the oil companies at his country's cost. In his first oil policy statement, carried in the April 1962 issue of the ministry's monthly bulletin, *Akhbar al-Bitrul wa al-Ma'adin,* he made it clear that he recognized oil's basic importance to the social and economic development of his country; that he was aware of the "fabulous profits" that the oil companies had made; that he appreciated the fact that three groups have a vital interest in oil—the governments, the oil companies and the consumers; that he believed that these interests could be reconciled if the oil companies would show good faith and realize that times have changed. They must recognize, he stated, that those who tamper with the march of history do themselves a disservice and run the risk of touching off a reaction that will wipe out their gains. He concluded,

The oil companies, who have harvested a fortune from our treasure and lived among us for so long that they have become a part of us, would, with a little dispassionate thinking, realize that tug-of-war tactics will not lead to a satisfactory solution of outstanding problems; they would realize that good judgment, far sightedness, and a recognition of the realities of the day are the best ways to arrive at fair solutions to these problems, thereby promoting stability and removing all bitterness from our relationship. When reason and good faith prevail there is no reason why disagreement should arise between partners.[9]

OPEC Resolutions

At its fourth conference, held in Geneva, April 5–8 and June 4–8, 1962, OPEC defined three grievances that its member countries had against the oil companies and adopted resolutions designed to adjust them. It resolved that the member countries should immediately enter into negotiations with the oil companies to ensure that they re-establish posted prices

9. *MEES,* April 27, 1962.

no lower than those prevailing before August 1960 and to work out a formula under which royalty payments should be fixed at a uniform rate which member companies consider equitable and be treated as costs rather than as credits against income-tax liability. It further resolved that the member countries affected take measures to eliminate any contribution to the marketing expenses of the companies concerned.[10]

Shortly afterward OPEC Secretary General Fuad Rouhani announced that OPEC had designated Iran and Saudi Arabia to initiate negotiations with the oil companies to achieve these objectives.

Saudi Government and Aramco Undertake Negotiations

On July 21, 1962, the Saudi Arabian government began negotiations on these issues and on certain other issues peculiar to Saudi Arabia and Aramco. Oil Minister Yamani captained a governmental negotiating team consisting of twelve members experienced in the legal, administrative or accounting aspects of Saudi Arabian oil problems. Robert Brougham, Aramco's senior vice-president, headed Aramco's six-man team, which included Aramco's general counsel, the general manager of its public relations, Tapline's president and Tapline's legal adviser and accountant.

Negotiations, which proceeded intermittently over an eight-month period, apparently were conducted in an atmosphere of mutual good will. When the meetings were temporarily suspended on July 26, Brougham in a public statement said,

It was agreed by both sides that the company, which is the only source of so much of the government's income, must emerge from the negotiations with arrangements which will permit it to meet its own responsibilities as well as those of the Saudi Arabian Government and the people.

With an atmosphere of reason and understanding on both sides we will, God willing, bring the discussions to a mutually satisfactory conclusion.

Commenting on Brougham's statement, a spokesman for the ministry of petroleum remarked that it would be difficult for the company to "survive in an atmosphere alien to the host country," and he expressed optimism with regard to the outcome of negotiations.

Later Crown Prince Faisal, when questioned in New York City by

10. *Resolutions of the Fourth Conference of the Organization of Petroleum Exporting Countries,* Geneva, *Resolutions* IV.32, IV.33, and IV.35.

reporters about his government's relations with foreign oil companies, stated,

Our policy has always been to cooperate with foreign companies, while at the same time upholding the country's rights in respect to its material resources. At present our relations with the foreign companies are excellent and we do not expect these relations to undergo any change, though there are a number of issues of secondary importance currently under negotiation.[11]

These negotiations, the crown prince said, were proceeding in an amicable atmosphere, but he indicated that the companies must take into consideration the producing countries' interests.

Although OPEC had authorized Saudi negotiators to consider the three issues of concern to all Middle East countries (oil prices, treatment and amount of royalties, and the eliminating of marketing discounts), they made no significant progress on the pricing and royalty issues. As negotiations proceeded, they were directed primarily to the marketing-discount issue and three other issues of direct concern only to Saudi Arabia and Aramco. These were the relinquishment of Aramco's unexploited concession areas, the sharing of Tapline profits, and company accounting practice with respect to certain costs.

Negotiations continued intermittently during a period of eight months. In the latter part of March 1963 the negotiations reached an accord on relinquishment of concession areas, and on the government's claims against any profits derived from Tapline operations.

Relinquishment of Concession Areas

The California Standard Oil Company's original concession granted the company exclusive rights to an area covering the eastern portion of Saudi Arabia embracing approximately 360,000 square miles, and preferential rights to approximately 177,000 square miles adjoining the exclusive area on the west.[12] The 1939 supplemental agreement provided for an extension

11. *MEES*, October 26, 1962.

12. Articles 2 and 3 of the concession agreement of 1933 define the areas but do not indicate their size. *Middle East Basic Oil Laws and Concession Contracts, Petroleum Legislation*, P. O. Box 1591, Grand Central Station, New York 17, New York, Vol. I, 1959, p. A-4. Staff Report of the Federal Trade Commission, submitted to the subcommittee on monopoly of the Select Committee on Small Business, U.S. Senate, *The International Petroleum Cartel* (Washington, D.C.: Government Printing Office, 1952), p. 114, gives the figures on the size of the concession.

of the exclusive area.[13] At its greatest extent the exclusive area embraced 495,827 square miles and the preferential area, 177,037 square miles, for a total of 672,864 square miles, an area equal roughly to the combined areas of Texas, New Mexico, Arizona, and California.

As previously indicated, the original concession granted the company discretion on relinquishment. Article 9 provided that

Within ninety days after the commencement of drilling, the Company shall relinquish to the Government such portion of the exclusive area as the Company at that time may decide not to explore further, or otherwise use in connection with this enterprise. Similarly from time to time during the life of this contract, the Company shall relinquish to the Government such portions of the exclusive area as the Company may then decide not to explore or prospect further, or to use otherwise in connection with the enterprise.

Article 7 of the Supplemental Agreement signed May 31, 1939, extending the area of the original concession, relieved the company of any obligation to relinquish any part of the exclusive area for a period of ten years. Thereafter the company was given the same discretionary rights as it obtained in the original agreement.

In an agreement covering Aramco's offshore rights signed on October 10, 1948, Aramco surrendered its preferential rights to all lands west of longitude 46 degrees east of Greenwich and agreed to a relinquishment program covering its exclusive area providing for the periodic relinquishment of 33,000 square miles at six specified dates extending to July 21, 1970. The government agreed that this program, providing for a total relinguishment of 198,000 square miles in the exclusive onshore area, constituted "Aramco's relinquishment obligations." By 1962 Aramco had surrendered its preferential rights covering an area of 135,200 square miles and had relinquished 143,521 square miles of its exclusive area.[14]

Aramco's program of relinquishment, to which the government had agreed, had hardly gotten under way before the government expressed dissatisfaction with it. The government was also dissatisfied with the rate at which Aramco was developing oil fields which it had discovered. Accordingly, it included these two matters in a list of grievances submitted on February 13, 1952, to Aramco for discussion.

13. Article 4, Article 5, and attached schedule of Supplement Agreement, May 1939. *Petroleum Legislation,* Vol. I, pp. A-21, A-22, A-28.

14. Saudi-Aramco Tapline Agreements, March 26, 1963. *MEES* published these agreements as a supplement to its April 5, 1963, issue. On p. 24 it gives the details of past, present, and future relinquishment.

In response Aramco agreed to "continue its study of the development of individual fields and to study a revision of the relinquishment program agreed to in October, 1948," and within two months to begin a serious discussion of these matters with the government in an effort to reach a final decision on them as quickly as possible.[15]

Accord Reached on Relinquishment

No progress had been made on this issue when negotiations opened in 1962. As indicated above, the negotiators reached an accord in March 1963. According to the agreement, Aramco obliged itself to relinquish immediately all rights in the unrelinquished portion of the preferential area and to 227,306 square miles of its exclusive area. This left Aramco with exclusive rights on only 125,000 square miles (25 percent of its original exclusive area and only 18.6 percent of its maximum area). Of the 125,000 square miles on which it still held exclusive rights Aramco agreed to relinquish 20,000 square miles in each of the following years: 1967, 1972, 1977, and 1982. Of the 45,000 square miles it will retain after 1982, it agreed to relinquish 15,000 square miles in 1987 and 10,000 square miles in 1993. From that year until the expiration of its concession, it will retain exclusive rights to 20,000 square miles. When account is taken of the fact that the original concession agreement left relinquishment to the discretion of the company and that under the 1948 agreement covering Aramco's off-shore rights, the government accepted as final a relinquishment program that by 1970 would have left Aramco with preferential rights to approximately 43,000 square miles and exclusive rights to approximately 353,306 square miles, the relinquishment accord of March 1963 is likely to impress un-biased observers as extremely liberal and as calculated to end the relin-quishment issue for all time. Certainly it reflects Aramco's desire to main-tain amicable relations with the Saudi government.

In passing judgment on the finality of the 1963 settlement, one would do well to bear in mind that Aramco selects the areas to be relinquished; that Aramco has already had more than a quarter of a century in which to explore the entire area and evaluate its potentialities for oil production; and that under the 1963 accord it has successive intervals of five years within which to locate the least promising areas for relinquishment. Since oil is

15. *Petroleum Legislation,* Vol. I, p. A-88.

where you find it, all this does not preclude the possibility of Aramco's relinquishing lands that may prove highly productive, but it reduces the likelihood. On the other hand, a provision of the original agreement may restrain the company from a single-minded endeavor to surrender only the least promising areas. Article 26 obliges the company to "supply the Government with copies of all topographical maps and geological reports (as finally made and approved by the Company) relating to the exploration and exploitation of the area covered" by the contract.

As might have been expected, OPEC at its sixteenth conference set as one of the major goals for its members acceleration of relinquishments with the government's exercise of a voice in the areas to be relinquished. (See Chapter Twenty-two.)

On the Tapline Issue

To understand the Saudi government's demand that the 50-50 profit-sharing agreement of 1950 be made applicable to Tapline's operation, it will be helpful to review briefly the relevant provisions of the several agreements under which Aramco and Tapline operate.

Article 1 of the original concession agreement, defining the activities in which the company may engage, includes the right to "carry away and export petroleum" and petroleum products. Article 22 recognizes the company's right to use all means and facilities necessary to carry out the activities enumerated in Article 1, including the construction and use of facilities in "connection with the transportation . . . or exportation of petroleum or its derivatives." Article 32 forbids the company to assign its rights to anyone without the government's consent, but recognizes the company's right, on notifying the government, to assign its rights and obligations to "a corporation it may organize exclusively for the purpose of this enterprise" and to create such other corporations as it may consider necessary. Any such corporation shall "be subject to the terms and conditions of" the agreement.

On July 11, 1947, the Saudi Arabian government "in consideration of the fact that the execution of the undertaking will increase the exportation of oil," signed an agreement with the Trans-Arabian Pipe Line Company, a Delaware corporation organized by Aramco's owners, granting it the right to construct and operate a pipeline and all the necessary appurtenances thereto from Saudi Arabia to a terminal port on the Mediterranean for transporting petroleum and petroleum products produced by or for Aramco.[16] In return for this right Tapline agreed (Article 2) to give its pipeline to Saudi Arabia when its concession expired (1999), to reimburse the government for all reasonable expenses it incurred in providing the necessary service in connection with the construction of the pipeline,[17] and after fifteen years from the official publication of the agreement to pay the government a "reasonable transit fee on the transportation of petroleum and petroleum products" by pipeline equal to the highest transit fee that Tapline pays to neighboring countries over whose territory the pipeline will pass and to the highest fee charged by any Middle East country on any other pipeline in the area. (Paragraph 2 of accompanying letter.)

Before Aramco signed the Tapline agreement, its owners, Standard of California and Texaco, had arranged to sell a 30 percent interest in Aramco and Tapline to Standard of New Jersey and a 10 percent interest to Socony Vacuum. Simultaneously with Standard of New Jersey's and Socony's purchase of an interest in Aramco and Tapline, the owners entered into a so-called offtake agreement, providing that each owner would buy crude oil and refined products from Aramco in proportion to its ownership at prices determined by Aramco's Board of Directors "in accordance with the principle that seller (Aramco) should be run for its own benefit as a separate entity."[18]

It was on a basis of the above facts that the Saudi government after the signing of the 50-50 profit-sharing agreement, contended that profits derived from Tapline's operations should go to Aramco and be shared in accordance with the income tax laws of November 1950 and December 1952, to which Aramco had subscribed.

Aramco Owners Post Crude Prices at Mediterranean Ports

When oil began to move over Tapline to the Mediterranean, it did so at much lower cost than oil moving in ocean tankers from the Persian Gulf to the Mediterranean by way of the Suez Canal. Taking account of this fact, Aramco's owners posted prices at Tapline's Mediterranean terminal

16. *Ibid.,* Vol. I, pp. B-1, B-47, Convention between Saudi-Arab Government and Trans-Arabian Pipe Line Company and Related Documents.

17. Paragraph 1 of letter from William J. Lenahan to the Saudi Minister of Finance, the signers of the agreement. *Petroleum Legislation,* Vol. I, p. B-23.

18. *The International Petroleum Cartel,* pp. 125–126.

roughly equivalent to the prices they had posted at Ras Tanura plus the cost of ocean transportation. On the other hand, Aramco sold the oil to its owners at a lower price, based on the Ras Tanura price plus the cost of pipeline transportation. In effect, Aramco was treating Tapline as a facility organized for the benefit of its owners rather than as a facility organized for its own benefit, and accordingly passing on to its owners the profits that Aramco would otherwise have realized. By doing so, Aramco avoided sharing these profits with the government.

Without the 1950 fifty-fifty profit-sharing agreement this would have been of no concern to Saudi Arabia. But with it, the matter was quite different. The government regarded this practice as a device for avoiding the tax provisions of the 1950 income-tax decrees. It accordingly protested.

When the controversy came before the negotiators, two related issues confronted them. The basic issue was whether Aramco owed the government 50 percent of the profits derived from Tapline's operations, which had previously gone to its owners. A related but scarcely less important issue was how Aramco should treat transit fees in calculating profits to be shared. Should they be regarded as Tapline costs or as additional payments to the government?

The settlement of these issues is set forth in the agreement between Aramco and the government concluded on March 24, 1963, covering other issues as well, and a supplemental agreement between Tapline and Aramco concluded on the same date. These agreements, like many legal documents, are complex and abstruse, and some of those not parties to them have misconstrued them. Stripped of unessential complexities, the agreements provide that for the period between October 1953 and December 31, 1962, for the purpose of calculating profits to be shared with Saudi Arabia, Aramco will increase the prices at which it sold oil to its owners at Sidon by the difference between its Ras Tanura price plus Tapline's cost (including all payments that Tapline made to the governments of Syria, Jordan, and Lebanon) and its owners' posted price. Aramco agreed to pay to the government 50 percent of its increased profits thus calculated (Paragraphs 1 and 2). Beginning Jan. 1, 1963, Aramco has charged its owners for oil delivered at Sidon their posted price for such oil, less any discounts allowed by the government (Paragraph 6a). In calculating its profits thus derived from Tapline, which thereafter were to be shared 50–50 with Saudi Arabia in accordance with the 1950 agreement, it deducted all payments to Tapline covering its costs including transit fees paid to the various countries through which Tapline passes.

If the Tapline transit fees due the government, which worked out at 9.7 cents a barrel, exceed the 50 percent profit which Aramco is obliged to pay the government, Tapline will pay the government the excess (Paragraph 3 of supplemental agreement). If transit fees are less than the government's 50 percent share in Aramco profits derived from Tapline operations, Tapline pays the government nothing; but Aramco pays the government its 50 percent share. In brief, Aramco guarantees the government a transit fee of 9.7 cents a barrel (equal to that paid any other country) or 50 percent of Tapline's profits, whichever is greater.

Under this settlement the government apparently got all that the concession agreement entitled it to. The concession agreement clearly authorized Aramco to construct and use pipelines for transporting and exporting petroleum or to assign its rights to do so to a separate corporation organized for this purpose. It did not specify where or to what ports such pipelines would be built. When Aramco's owners organized Tapline its proposed pipeline was to extend northwestward from Aramco's oil field across Saudi Arabia, Syria, and Lebanon. When Tapline agreed to pay Saudi Arabia the transit fee equal to that paid by any other country, Aramco was not sharing its profits with the government and Saudi Arabia accepted the transit-fee arrangement. Because after the profit-sharing agreement the benefit of Tapline's lower cost went to its owners, unshared by Saudi Arabia, Saudi Arabian protest seems reasonable. But to exact a fee from Aramco for the use of facilities it had been authorized to construct, which are essential if it is to maximize its earnings to be shared with the government, would seem to be double taxation of an onerous sort. It is to penalize Aramco for its foresight and efficiency. Moreover, and this apparently was decisive, the 1950 profit-sharing agreement specifically provided that all payments including royalties shall constitute a part of the 50 percent due the government.

The amount due the government for the period from October 1953 to December 1962 was to be determined by an audit committee established under the agreement. By Royal Decree (No. 48) issued on March 13, 1963, the government announced that the revenue received from Aramco in settlement of this claim would be allocated to an economic development fund to be used for special development projects recommended by the Supreme Planning Board and approved by the council of ministers. The fund was to be enriched by an estimated $100 million as a result of Aramco's payment of the back revenues and thereafter by the annual payments which will depend on Tapline's throughput.

In commenting on this settlement, Saudi Oil Minister Ahmad Zaki Yamani stated that as early as 1956 a leading U.S. certified accountant, Francis X. Prior, had advised the government that the differential between the Ras Tanura price plus pipeline transportation cost and the posted price at the pipeline's Mediterranean terminal was subject to the Saudi Arabian income-tax laws. Thereafter as the government pressed its claim it had the advice of other leading international tax and accounting experts. Secure in their conviction that their claim was legally justified, the government negotiators made it clear at the outset of negotiations in July 1962 that they would accept no compromise on the applicability of the Saudi income-tax law. According to the minister, the "tenacity and patience" of Saudi negotiators finally persuaded Aramco of the justice of their claim and "the company recognized what was right, for which it is duly thanked."[19]

Aramco's Cost Accounting Practices

On cost accounting, three sets of issues confronted the negotiators: deduction of marketing expenses; the treatment of exploration and intangible developmental costs; and the treatment of Aramco's U.S. office expenses, its public relations and its governmental relations expenses, and its charitable and philanthropic donations.

As long as Saudi Arabia's oil receipts were on a tonnage basis, the government's direct interest was in the amount of oil sold. After 1950, when Aramco agreed to submit to an income tax to yield the Saudi government, including its royalty receipts, 50 percent of Aramco's profits, the government was directly concerned not only with the amount of oil sold but in Aramco's selling price and its costing methods. Although Aramco, in contrast to IPC, is a profit-making enterprise, it sells oil primarily to its owners—Socal, Texaco, Jersey, and Socony. Each is entitled but not obliged to buy Aramco's crude in proportion to its ownership—30, 30, 30 and 10 percent respectively. Aramco as a business unit and Saudi Arabia as a political taxing unit were both interested in Aramco's making as much money as possible. Aramco's and the government's interest in maximizing Aramco's net revenues were identical; not so with Aramco's owners and the government. Income to Aramco was cost to its owners.

19. *MEES*, April 5, 1963.

Until 1955 Aramco in selling oil to its owners had granted them a discount of 18½ percent from the posted price. After the profit-sharing agreement of 1950, Aramco used the discounted price in calculating profits to be shared with the government. Granting a discount of 18½ percent was consistent with the government's interest only if it so stimulated sales as to increase the profits to be shared. There is no evidence that it did so.

For the most part Aramco's owners sold Aramco's oil to their affiliated refineries. They sold it at the posted not the discounted price. They continued to do so until quite recently. When the European Cooperation Administration was financing sales of crude to West Europe under the Marshall Plan, Aramco owners billed ECA at their posted prices, despite the fact that they were buying oil from Aramco at a discount of 18½ percent. With all major refiners paying identical prices for their crude there is little reason to believe that they bought more of Aramco's oil than they otherwise would have. The 18½ percent discount might well have induced Aramco owners to increase output of Aramco crude at the expense of other sources of supply. But the evidence does not support this hypothesis. From 1949 to 1955 Saudi Arabian production of crude increased by less than 100 percent, while Kuwait's increased by more than 300 percent and Iraq's by approximately 700 percent. Of the Middle East countries, only Iran showed a less rapid rate of increase than Saudi Arabia, and here nationalization had brought production to an abrupt halt in 1952. The Middle East's rapid increase in output no doubt reflected its lower cost of production, and although Saudi Arabia shared in this advantage, no evidence is available to indicate that Aramco's 18½ percent discount was of itself a significant factor.

Although evidence does not support the proposition that the discounts increased Aramco sales, they undoubtedly increased the profits of its owners. When the government learned that Aramco was calculating profits on a basis of the discounted prices at which it sold to its owners, it protested. Under a settlement reached in 1955 Aramco agreed to discontinue the 18½ percent discount and to pay the government $70 million as a quit claim for all price matters then in controversy.

Discounts to Cover Marketing Expenses

After abandoning the 18½ percent discount, Aramco with the acquiescence of the government continued to grant its owners a discount to cover the cost of marketing Aramco oil. Initially this was set at 2 percent. In 1956 it was renegotiated and replaced by a deduction based on "actual audited expenses incurred in marketing." This amounted to about 4.2 cents a barrel.

It will be recalled that before 1955 the Iraq Petroleum Company had been granting a discount of approximately 20 percent in selling oil to its parent companies and that when the government protested, IPC reduced the discount to 2 percent. It justified the 2 percent discount as necessary to cover the cost of selling Iraqi oil. Later the consortium included in its costs a marketing allowance of 1 percent of posted prices.

At its fourth conference, held in Geneva from April 5–8 and June 4–8, 1962, OPEC had passed a resolution recognizing that neither its members "nor the companies operating in their countries participate in the world-wide marketing operations of the Oil Companies" and "that the bulk of crude oil produced by the Operating Companies is marketed through their parents or parents' affiliates with no brokerage charges being incurred." OPEC accordingly recommended "that the Member Countries . . . take measures to eliminate any contribution to the marketing expenses of the Companies concerned." In the spring of 1963 Saudi Arabia and Aramco negotiated the issue raised by OPEC's resolution.

Granting discounts to cover marketing expenses raised two questions: Were expenses actually encountered? If so, by whom? As previously pointed out, Aramco sold its oil to its owners at prices they established—subject, however, to the marketing discounts in controversy. The buyers of crude oil are customarily independent refiners or corporate subsidiaries of integrated companies who buy for affiliated refiners. It has long been the practice for the buyers to post the prices at which they will buy oil at the well. Obviously the buyer or his intermediary encounters some expenses in buying crude oil, in such matters as transfer of title, keeping accounts, seeking out sources of supply, providing facilities for the acceptance of the oil. These expenses tend to vary with the number of sellers, increasing as the sources of supply multiply. The buyer or other functionary intervening between the seller and the refinery takes account of these expenses in announcing the price at which he is willing to buy. When competing buyers have access to an oil field, several may post a price at which they will purchase, but the prices they post are customarily identical.

Some buyers may at times pay a premium over the posted prices; at other times they may buy at a discount. Whether they pay more or less than the posted price depends upon the urgency of their needs and the abundance of supply. But nowhere except in the Middle East has it been the practice for the buyer to discount the posted price to cover the expenses of buying. And here the justification for doing so appears to be particularly weak. The buyers of Aramco crude are the owners of Aramco. Aramco plans its production in response to the anticipated needs of its owners. In recent years its owners have provided Aramco with a capital budget which enables it to maintain a production capacity somewhat in excess of actual daily production. The owners, who are entitled to buy Aramco output in proportion to their ownership, customarily state their anticipated needs so as to insure that each will share in proportion to its ownership if Aramco's full capacity should be utilized. Except as it may decrease the percentage of total output that the buyer is entitled to nominate in the future, his failure to take all the oil nominated carries no penalty. A buyer who takes more than he nominates is permitted to increase his percentage of total output in future nominations.

Clearly this close relationship between buyer and seller, with buyers vying with each other for a larger share in allowable production, involves no marketing expenses for Aramco. And except when the owner sells his oil to someone outside his own organization, it involves little or no expense to the buyer. The owners have customarily delivered the oil to their refineries for whose accounts they have bought and they have customarily delivered it at the posted price plus the cost of getting the oil to the refineries. When the owners have sold oil to a nonaffiliated customer, in recent years they have generally sold it at a discount from their posted price. When they have done so with governmental approval, they have paid Aramco only the actual price received less the marketing discount.

If the above analysis is logical, it would seem to follow that on sales by Aramco owners to their affiliates who pay the posted price plus the cost of getting the oil to their refineries, Aramco should be charged with no selling expenses. On sales to nonaffiliates Aramco should bear the entire cost of getting its oil on the market and the government should share in this burden. In practice that would mean a marketing discount to cover actual marketing costs only on sales to nonaffiliates.

On April 24, 1963, Aramco and Saudi Arabia, as a result of their negotiations, signed a new agreement covering marketing allowance. It provided that marketing allowance that hitherto had amounted to about 4.2 cents a barrel be reduced to one half cent a barrel retroactive to January 1, 1962. The additional income to the government of about 1.85 cents a barrel added substantially to the government's annual oil revenue, amounting to more than $10 million for 1962 alone.[20]

This settlement represents a compromise between Aramco's practices and OPEC principles, but apparently it is acceptable to OPEC. At any rate the Saudi Arabian daily *al-Bilad* on April 25, 1963, reported that "the agreement had been cleared with OPEC beforehand and had received the unanimous approval of its members."

Negotiations Settle Other Accounting Issues

Two minor accounting issues confronting the negotiators were Aramco's treatment of its New York office expenditures, its public relations' expenditures, and its philanthropic gifts, which before the settlement Aramco had treated as costs in calculating its income taxes payable to the Saudi Arabian government; exploration and intangible developmental expenditures, which Aramco had treated as costs in the year in which they were incurred.

The oil ministry challenged both practices. On the treatment of its New York office and public relations' expenditures and its gifts to philanthropy, most of which were to Middle East institutions, the negotiators reached a mutually acceptable compromise. On the treatment of exploration and intangible developmental costs, Aramco acceded to the government's demands. It agreed that thereafter it would treat these as capital expenditures to be amortized over a number of years, exploration costs at the rate of 5 percent a year, intangible developmental costs at the rate of 10 percent a year. This settlement has the effect of increasing Aramco's taxable income, and hence the government's revenue. Assuming regular annual expenditures for these items, the increase will be greater in the year in which the practice is inaugurated, but it will decline as the amortization period of the first capitalization is completed.

In an interview in Washington in the spring of 1963 Oil Minister Yamani stated that the tax arrears due the government under the settlement of the several claims that it had made against Aramco, most important of which was the treatment of Tapline earnings, would total $160 million.

20. *Ibid.,* May 10, 1963.

16

Saudi Arabia's Program of Self-Help

In his dealings with Aramco Oil Minister Yamani has done well for his country, but it is his program outside the framework of the Aramco concession that has been most significant to Saudi Arabia's economic development and to the growth of a Saudi Arabian oil industry independent of Aramco.

In this program he has tried to harness private enterprise with public enterprise as teammates in his country's economic progress. The instrument through which he has operated primarily has been the General Petroleum and Mineral Organization (PETROMIN), established by Royal Decree No. 25 on Nov. 22, 1962. Mr. Yamani is chairman of its Board of Directors. Its governor is Dr. Abdulhady Hassan Taher, who holds the Ph.D. in business administration from the University of California, Berkeley, and is the author of *Income Determination in the International Petroleum Industry*. He is an articulate gentleman with a fine sense of humor and an understanding of the technical aspects of the oil industry. Other original members of PETROMIN's board included the governor of the Saudi Arabian Monetary Agency; the deputy minister of petroleum and mineral resources; the deputy minister of mineral affairs; the deputy minister of finance and national economy; the deputy minister of commerce and industry; and two representatives of the business community.

While the proposal to create PETROMIN was before the council of ministers, Yamani indicated that he viewed the contemplated organization as the operating arm for the ministry. It could undertake those tasks that for administrative or policy reasons the ministry could not perform. PE-TROMIN would provide a framework within which the private sector could take part in financing various oil and mining projects. Shortly after its creation Dr. Taher elucidated further the purpose of PETROMIN. He stated that it would engage in three types of activity: the development of a petroleum and natural gas based petrochemical industry; the development of mining industries; and co-operation with private firms in exploring for, producing, refining, and marketing petroleum and its products.

The Ministry Initiated Surveys and Grants Concession

The areas relinquished by Aramco and the uncommitted areas, particularly along the Red Sea, constituted a promising region for oil exploration activities. The ministry, either directly or through PETROMIN, promptly inaugurated a series of surveys to determine the oil potentialities of these areas. At the same time it made known that it was interested in granting new oil concessions and extended an open invitation to oil companies to participate.

As a more positive step in creating an active interest among potential concessionaires, the oil ministry awarded a contract to the Robert H. Ray Geophysical Company of Houston, Texas, to conduct seismic surveys in three zones of the Red Sea coastal area and made an initial appropriation of more than a half million dollars to finance the survey scheduled to begin in April 1963. Later the council of ministers authorized the oil ministry to award the Ray Company a new contract to conduct a survey in the "preferential area" to which Aramco had relinquished its rights. Early in 1964 Oil Minister Yamani announced that the Ray survey in the preferential area had revealed a twenty kilometer–long structure favorable for the occurrence of oil. On June 28, 1964, in a radio broadcast from Jiddah, Minister Yamani announced that the Ray survey in the Red Sea area had revealed a number of extensive structures lying under relatively shallow water.

Meanwhile several foreign concerns had expressed an interest in concessions and the government through the oil ministry had been quietly conducting negotiations with France's state-owned Régie Autonome des Pétroles (RAP). The negotiations were successful. On April 4, 1965, Yamani, on behalf of the Saudi Arabian government, and André Martin, on behalf of Société Auxiliaire de la Régie Autonome des Pétroles (AUXI-RAP) with which RAP has since merged, signed an agreement granting AUXIRAP exclusive exploration and production rights covering areas in the Red Sea zone, the precise limits of which were to be subsequently agreed upon in a separate document.[1]

Innovations in the AUXIRAP Agreement

In providing for the exploitation of the concession by a joint-venture enterprise, the agreement follows the general pattern of recent Middle East joint ventures. It introduced some important innovations, however, and represents another step in the effort of Middle East oil-producing countries to assert their sovereignty over their oil resources.

Worth noting is the fact that the concession grant differs in legal form from that of the Iranian joint-venture concessions. It consists of two documents instead of one. First is the "Agreement" which grants to AUXI-RAP a "license" to explore for oil in the concession area for a preliminary period and, if and when oil is discovered in commercial quantities, an "Exploitation Concession Lease." It also sets forth the rights and privileges the government grants AUXIRAP and the obligations each assumes. The second document is a "Contract" between AUXIRAP and PETROMIN which specifies the details regarding the establishment, management, and operation of a joint company to be established if and when AUXIRAP discovers oil in commercial quantities.

The agreement's most important innovations are with regard to taxation, royalties, integrated operations, and arbitration. Less important but none the less worth noting are innovations covering the use of gas and the amortization of bonuses.

As this study has emphasized, under the original concessions the Middle East countries surrendered in perpetuity their power to tax the incomes of their concessionaires however great they might become. In the 1950s, when the so-called 50-50 profit-sharing agreements were incorporated into the concessions, the producing countries obligingly enacted corporation income tax laws in order that the companies might credit the payments to

1. *The Middle East Economic Survey*, April 5, 1965, published the agreement and the contract in a supplement to a special issue.

producing countries against their home country corporate income tax obligations. The 50-50 profit-sharing agreements were not obtained by the unilateral exercise of the country's sovereignty but through negotiations with their concessionaires.

The Saudi Arabia-AUXIRAP agreement goes a long way in asserting the government's sovereign right to tax. Article 22 provides that the

Concessionaire shall pay on all its integrated operations covered by this Agreement both inside and outside Saudi Arabia, including the sale of crude oil, refining, transportation, and marketing, the income taxes imposed by the relevant Saudi Laws and Decrees *now existing or which may later exist* at any time during the life of this Agreement. [Italics supplied.]

The recent Iranian joint-venture agreement introduces the same principle but compromises it by the proviso that the concessionaire's taxes shall not be less favorable to the concessionaire than the taxes paid by other companies which together produce 50 percent of Iranian crude. Because the consortium is likely to account for more than 50 percent of Iran's crude production throughout the life of the consortium, this proviso makes the new concessionaire's tax obligations virtually identical with the consortium's, that is, the joint-venture agreements limit the government's right to tax in the same manner as does the consortium agreement.

The Saudi-AUXIRAP agreement is not devoid of compromise on the tax issue however. It is understood, although the agreement does not say so, that the tax laws immediately applicable to AUXIRAP will be Royal Decree No. 17/2/28/3321 of November 2, 1950, as modified by Royal Decree No. 7/2/28/576 of October 19, 1956. This law is applicable to all corporations, domestic as well as foreign, except those under the special hydrocarbon income tax law which provides the basis for the 50-50 profit-sharing agreement. At the time it was enacted Aramco, for which the law was designed, was Saudi Arabia's sole concessionaire. The general corporate income tax law now provides for a tax rate of 40 percent of profits for those companies making yearly profits in excess of a million Saudi rials. Because it is applicable to domestic as well as foreign corporations, internal political pressures may make it difficult to raise the rates. The government's reliance on Aramco as the major source of its revenue will work towards the same end. The agreement, while preserving Saudi Arabia's sovereignty on taxes, would seem to give AUXIRAP preferential treatment.

AUXIRAP Agreement Provides for Integrated Operations

The agreement provides that the joint company which will operate the concession lease once AUXIRAP discovers oil in commercial quantities shall carry out all those operations of an integrated oil enterprise, including production, refining, transportation, and marketing (Article 26) and that it shall locate a refinery in Saudi Arabia (Article 28)—goals for which Tariki had vigorously agitated. In c.i.f. sales of crude oil the agreement obliges the joint company either to use its own tankers for transporting the oil or to give preference to tankers owned by Saudi Arabian subjects provided they offer rates equal to free market rates (Article 26, paragraph 2).

The joint company's obligation to engage in refining is conditional, however. It is allowed fifteen years in which to build a refinery, and it need never do so if conditions make it unlikely to prove profitable or place an undue financial burden on the company (Article 26, paragraph 1).

In short, the agreement seems to leave it to the discretion of the joint company as to whether to go into the refinery business. Article 28, however, sets forth guidelines for the company in the use of its discretion. It provides that when the joint company's crude production averages 100,000 barrels daily for ninety days, the company

as an application of Article 26.1, shall commence and prosecute to completion as rapidly as reasonably practicable and in no more than three years, the construction *in Saudi Arabia* [italics supplied] of a refinery or refineries having a minimum daily throughput capacity of not less than thirty thousand barrels . . . unless it may be contemplated that a refinery of such a capacity would result in a loss for the Company, in which case the Company and the Government shall agree on the proper capacity to be adopted.

Arbitration of Disputes

It would seem not unlikely that the provision with regard to the company's obligation to engage in refining might lead to disagreement. The validity of this observation may be affected by consideration of the composition of the decision-making body. The ownership of the company for which the agreement provides is divided between PETROMIN, holding Class A shares only and representing 40 percent of the company's capital stock, and AUXIRAP, with Class B shares representing 60 percent; the profits of the company are to be shared in similar ratio. Although PE-

TROMIN has only a 40 percent ownership in the company, however, it has equal voting power in its management and has equal representation on its board.

The company's executive management consists of an administrative executive manager, who at the outset shall be a Saudi Arabian designated by PETROMIN and a technical executive manager, who at the outset shall be a Frenchman designated by AUXIRAP (Article 6 of the contract).

If disputes should arise, both the agreement and the contract provide a novel way of settling them, one calculated to insure their being settled by arbitrators in whom both parties have confidence. Article 63 of the agreement provides that "any doubt, difference or dispute" between the government and AUXIRAP regarding the "interpretation or performance" of the agreement shall be referred to a committee of two experts, one appointed by the government and the other by the concessionaire. If the experts cannot agree on a settlement, the dispute shall be referred to the Board of Concession Appeals provided for in Article 50 of the Saudi Arabian Mining Code. Article 50 provides that this board shall consist of "not less than three and not more than five members to be chosen without regard to their nationality from eminent and highly reputable jurists and judges who are experienced in international law and in problems relating to concessions."

In language similar to that of the agreement, the contract provides that disagreements between PETROMIN and AUXIRAP concerning the interpretation or performance of the contract be referred first to three arbitrators, one chosen by each of the parties and a third chosen by the other two arbitrators. In the event of failure of either party to designate an arbitrator or of the two to designate a third, the Board of Concession Appeals provided in Article 50 of the Mining Code shall designate the "missing arbitrator" from a list set up and agreed to by PETROMIN and AUXIRAP (Article 13 of the contract). The arbitrators must sit in Saudi Arabia or in some other place that the parties may agree upon.[2]

Royalty Provisions

In its royalty provisions the agreement goes further in fostering the interests of the country than any other Middle East agreement. It provides

2. The agreement also provides that "If and when an international court is created for the settlement of disputes arising [out] of Middle East concessions, then the parties shall examine together the possibility of substituting that court for the Board of Concession Appeals." Article 63.

that AUXIRAP pay royalties of 20 percent of the crude produced from the concession area, graduated downward on a basis of daily crude production. With an average daily production of more than 80,000 barrels, the royalty will be 20 percent; with a daily production of from 60,000 to 80,000 barrels, the royalty will be 17 percent; with production of less than 60,000 barrels, it will be 15 percent. Royalties are to be calculated on the price the company posts for crude. They are to be counted as a cost of production by the joint company and not as a part payment on AUXIRAP taxes. These royalties compare with the usual 12.5 percent of the older concession agreements and with no royalties in Iran's recent joint-venture agreements under which NIOC derives its profits from its ownership and the government its revenue from the 50-50 income tax profit-sharing principle. PETROMIN as a part owner will bear its share of the royalty cost.

Waste of Gas

Two other minor innovations in the Saudi-AUXIRAP agreement are its provisions against the wastage of gas and for the nonamortization of the bonus that AUXIRAP agrees to pay to get the concession. Article 16 of the agreement provides that "gas produced as a result of operations under this Agreement shall be conserved to the maximum possible extent." More specifically, it places on AUXIRAP "the constant duty . . . to make economical use of such gas through sale, reinjection or other commercial or economic employment, development or disposition, except where the Government agrees that such gas is insufficient for such purposes."

Concessionaires in the Middle East have been slow to make use of gas released in the production of oil except as fuel in their operations. The flaring of gas, amounting to hundreds of millions of cubic feet daily, has been a characteristic over the years of the major concessionaires' operations in the Middle East. Its justification has been that no market has existed for the commercial sale of gas. The thickness, porosity, and saturation of the producing horizons and the wide well-spacing in the Middle East have all been conducive to high daily production and long life and have delayed the use of gas for repressuring fields. Waste of gas is one of the practices against which articulate Arabs have proclaimed long and loud. Decisions with regard to its use have been the exclusive province of the concessionaires. The Saudi Arabia-AUXIRAP agreement places on AUXIRAP the obligation not to waste gas and gives the government a voice in determining what constitutes waste.

The Treatment of Bonuses

To obtain the agreement AUXIRAP obliged itself to pay the government a bonus of five and a half million dollars—a half million to be paid within one month of the effective date of the agreement, one million within two months after the granting of the concession, and four million when crude production averages 70,000 barrels daily for ninety consecutive days. As compared with the bonuses which recent concessionaires paid Iran under their joint-venture agreements, this bonus is small. When account is taken of the fact that AUXIRAP will be operating in wildcat territory remote from proven acreage, it looks large. But the novel feature is that it is never recoverable. The Iranian joint ventures permit the amortization of bonuses after commercial production.

The other obligations of AUXIRAP are similar to those assumed by the concessionaires in the Iran joint-venture agreements. The joint company is a profit organization, however, following the pattern of AGIP rather than the joint agencies of the other NIOC venture agreements. As with the Iranian joint ventures, AUXIRAP as concessionaire will meet all costs of exploring for oil, obliging itself to spend not less than $5 million during the first two years. If commercial production results, the expenditures will be treated as a part of AUXIRAP's contribution to the capital of the joint company rather than as amortizable operating costs.

The provisions on production and pricing are designed to permit a flow of the joint company's oil into world markets without disturbing international prices. The partners in the venture will buy the joint company's crude output in proportion to their ownership at f.o.b. prices fixed by the board of directors at a level "fairly comparable to the level of prices on the international market" (Article 11(1) of contract). If PETROMIN is unable to sell all its share of oil, AUXIRAP is obliged to buy it at the price it realizes on resale (subject to a commission graduated according to the relationship of the realized price to the price determined by the board) (Article 11(2)e of contract). AUXIRAP must have PETROMIN's approval to sell any of PETROMIN's share to independent third parties at a discount of more than 2.5 percent from the price set by the company's board. "Sales at prices lower than 10 percent below the board's price should not be proposed" (Article 11(2)b of contract). If AUXIRAP can't sell PETROMIN's share except at more than a 10 percent discount, PETROMIN "shall decide on the solution to be adopted." The contract places similar restrictions (with different allowable discounts) on AUXI-

RAP's selling PETROMIN's share at discount prices to AUXIRAP affiliates (Article 11(2)c of contract). The agreement places no restriction on
the prices at which PETROMIN or AUXIRAP sells its own share of oil,
except that each is required to buy it at prices fixed by the company's
board.

Under the agreement the government derives its revenue from three
sources: royalties paid either in cash or oil; PETROMIN's share of the
joint company's profits; and taxes imposed on AUXIRAP's net earnings
from the venture. These are expected to yield the government well over 70
percent of total net profits.[3] Whether the venture proves profitable at all
and, if profitable, what the government's aggregate revenue will be depend,
of course, on whether oil is found in commercial quantities, the richness of
the reserves, the cost of production, and other factors that make oil
production a bonanza or a bust.

Other Companies Become Interested

The signing of the AUXIRAP agreement created a flurry in international oil circles and a number of international companies promptly expressed an interest in Saudi acreage. Reports emanating from the oil
ministry during the next several months indicated that at least sixteen
companies including ENI, Standard of Indiana, Gulf Oil, British Petroleum, Continental, Pure Oil, Union of California, and Phillips Petroleum
were conducting negotiations with the government at various stages.

For two years none was carried to completion. It began to look as
though the oil companies were reluctant to subscribe to the terms upon
which the oil ministry was insisting. Perhaps most of them considered the
price too high to pay for the privilege of hunting for oil in wildcat territory.
Perhaps the fact that six oil groups had recently paid $192 million to the
National Iranian Oil Company for such a privilege and that only two had
thus far succeeded in finding oil deterred them.

3. Variables are too numerous to permit a precise calculation. Royalties as percent
of total profits will vary with average daily production. How PETROMIN disposes of
its oil and at what prices will affect significantly the profits of the government. On the
assumption that government royalties of 20 percent will be paid and that PE-
TROMIN will dispose of its entire share of oil whether through direct sale or through
AUXIRAP at the prices determined by the company's Board (an unlikely contingency), the government would realize approximately 74 percent of net profits. The
calculation is as follows: Royalties at 20 percent treated as costs will equal 10 percent

But not all of them. On December 21, 1967, Oil Minister Yamani announced the signing of a concession agreement with an Italian corporation, AGIP-Saudi Arabia, S.p.A., a subsidiary of Italy's ENI state-owned corporation, and on December 23 an agreement with a United States group composed of Sinclair and Natomas. The former is a complex document consisting of a contract between PETROMIN and ENI's subsidiary, AGIP; a contract between Saudi Arabia and PETROMIN; and a petrochemical contract between PETROMIN and ENI's subsidiary ANIC.

The Saudi Arabia-PETROMIN Contract

The contract between Saudi Arabia and PETROMIN grants PETROMIN a six-year oil exploration and prospecting license covering 77,382 square kilometers in the Rub 'al-Khali (Empty Quarter) desert area of southeastern Saudi Arabia, adjacent to or surrounded by areas previously relinquished by Aramco.[4] Under the terms of the contract, PETROMIN assigns all its rights and obligations to AGIP, which will be responsible for the exploratory work. If and when oil is discovered in commercial quantities, the government will grant a 30-year exploration concession to PETROMIN which will assign it to AGIP, retaining a 30–50 percent share depending on the concession's productivity. On the oil produced the concessionaire will pay the government taxes based on posted prices at the rate of 50 percent (or such higher rate as may be declared by Saudi laws or decrees) and an annual royalty of from 12½ to 14 percent of the oil produced, depending on the rate of annual production.

Two significant features of the concession agreement are those providing for integrated operations and those governing the selling price of oil. On integration the agreement provides that when production shall average 200,000 barrels daily for ninety days the parties shall undertake a study of the feasibility of establishing operations in refining and marketing. If the study indicates that integrated operations will yield profits on the investment, calculated by the discounted cash flow method, equal to or higher

---

of PETROMIN's share in the net profit of the joint venture; PETROMIN's ownership entitles it to 40 percent of the company's net profits; and AUXIRAP's tax obligation under Royal Decree of November 2, 1950, as modified by the decree of October 1956, will yield the government 40 percent of AUXIRAP's 60 percent share in net profits.

4. *MEES* published the full text of the Saudi-PETROMIN-ENI agreements in a supplement to its March 8, 1968, issue.

than the profits derived from the sale of oil to third parties, AGIP and PETROMIN will organize a separate company to conduct integrated operations.

The provisions governing the sale of oil are somewhat complicated. Their significant feature is that PETROMIN may require AGIP to sell all or any portion of its share of production not at posted prices but at the "highest competitive price obtainable in a free market." This is to be determined by AGIP's selling a specified amount of oil to third parties "acting in good faith" and with "due diligence" to obtain the best price possible. On a basis of its experience AGIP is required to determine for each yearly period the "expected third-party price." If AGIP does not determine the expected third-party price within one month of the beginning of any yearly period, the matter shall be referred to arbitration. If the arbitration shall hold that AGIP has not determined the expected third-party price because of failure to comply with the "due diligence" clause governing the sale to third parties, PETROMIN may demand that production be curtailed to an "economic level," defined as that level of production that will result in AGIP's receiving a return after income taxes of 10 percent per annum on the cumulated net worth from date of contract.

The purpose of this requirement is twofold: to assure the sale of oil and to give some protection to the price at which it is sold. But the protection to price is apt to prove elusive. What this contract would seem to mean is that oil produced on the concession will be available at market prices. Although PETROMIN will pay taxes based on posted prices, it will sell at market prices. What one arm of the government gains by this arrangement, another will lose.

Neither the safeguards to price stabilization in this or in the AUXIRAP contract are apt to prove adequate to check the pressure of crude oil supply on market price.

PART III

FROM INDIVIDUAL TO COLLECTIVE
ACTION

17

OPEC: An Experiment in International Collective Bargaining

WHILE the Middle East countries were bargaining individually with their concessionaires to obtain better terms, at the same time they presented a united front to improve their lot. The instrument through which they worked is the Organization of Petroleum Exporting Countries (OPEC). At the outset OPEC's primary concern was the stabilization of oil prices. International co-operation to stabilize the price of commodities on which the prosperity of producing countries largely depends is not new, but among such endeavors OPEC is unique.

Other programs have usually involved the co-operation of the producer companies and the producing countries, and more recently, the consuming countries as well. OPEC represents an exercise in self-help by the leading oil exporting countries of the world without the co-operation or indeed the sympathy of the producing companies or the consuming countries. Moreover, it has not confined its efforts to stabilizing crude oil prices but has served as an instrument through which oil-exporting countries may present a united front in obtaining better terms from their concessionaires and in settling grievances with them. In effect, the governments of the OPEC countries have sought by collective bargaining to gain benefits in the utilization of their most important source of national wealth which they have not been able to obtain through the exercise of their sovereignty. But they have sought more than that. Dr. Muhamad A. Mughraby in his penetrating pioneer study of Middle East oil concessions and legal change[1] analyzes OPEC as an instrument by which the oil countries are endeavoring to reestablish economic sovereignty over their oil resources. His analysis places OPEC in a broader legal framework than that in which OPEC has customarily worked. This study has emphasized that in granting the original concessions for the exploitation of their oil resources, Middle East countries surrendered important sovereign rights to their concessionaires and it recognizes as does Mughraby that recent negotiations between the governments and concessionaires represent an effort to re-establish and assert their sovereignty. But OPEC was not created for this purpose. As stated above, it conceived of itself as a collective-bargaining agency to improve concession terms for its members. But it also conceived of itself as an agency for establishing group authority over oil-field operations that the members regarded as essential to realizing the benefits that they sought. The exercise of national sovereignty on the part of member countries might well aggravate some of the ills that have befallen the industry and from which the countries have suffered. One OPEC goal has been to substitute monopoly power obtained by co-ordinated activity among the producing countries for the monopoly power that its members accuse their concessionaires of exercising. OPEC's activities and goals throughout most of its history resemble those of a modern industrial trade union dealing with closely knit corporate enterprises or of a cartel of producing countries rather than those of individual countries in search of their surrendered sovereignty.

The immediate occasion for creating OPEC was Esso Export Corporation's announcement in August 1960 of a reduction in its posted prices for Middle East crude oil and the downward readjustments which this occasioned in the posted prices of all major Middle East producers. But its roots go deeper. They originated in the humus of distrust and suspicion laid down by the abrasive impact of western technology and a business culture on economically underdeveloped countries wholly dissimilar in their political and social institutions and their history and traditions. Once planted, they thrived under a blanket of hostility kept warm by the clash of a corporate quest for profits with the interests of underdeveloped countries as conceived by their politicians and their people.

1. Muhamad A. Mughraby, *Permanent Sovereignty over Oil Resources* (Beirut: Middle East Research and Publishing Center, 1966).

Venezuela Takes the Lead

Venezuela, whose representatives sparked the movement that culminated in OPEC, had officially protested to the British government the Middle East price cuts inaugurated by British Petroleum in February 1959, following smaller cuts by Shell and other major buyers of Venezuelan crude —cuts that destroyed the customary relationship between Middle East and Caribbean and Gulf Coast prices and gave an advantage to Middle East oil in world markets.[2]

In a memorandum addressed to the British ambassador in Caracas in March 1959, the Venezuelan government charged that the British Petroleum Company had taken "advantage of the readjustment which followed the Suez crisis and . . . the economic recession on the American continent . . . to break the logical tie between the international petroleum price structure and prices in the United States," the world's largest producer and consumer of petroleum. "Such an attempt to break the unity of markets and prices," the protest contended, "is obviously contrary to the common interest of the producing countries as well as the long-run interest of the consuming countries."[3]

On May 25, 1959, the Venezuelan government in a note to the American ambassador protested the mandatory restriction of United States imports of petroleum, pointing out that an indiscriminate limitation would work to the disadvantage of Venezuela and urging continuing consultation designed to maintain a stable market for western hemisphere oil and Venezuelan oil in particular.

Alfonzo Confers with Arab Oil Officials

A month earlier Dr. Juan Perez Alfonzo, Venezuelan minister of mines and hydrocarbons, while attending as a visitor the First Arab Petroleum Congress in Cairo, conferred with representatives of Iran, the United Arab

2. *Venezuela and OPEC,* Publication of the General Secretariat of the President of the Republic (Caracas: Imprenta Nacional, 1961). English edition prepared by the Division of Petroleum Economics, Ministry of Mines and Hydrocarbons, pp. 99–101. This is a collection of documents, speeches, and press comments relating to the creation of OPEC. Hereafter cited as *Venezuela and OPEC.*

3. *Ibid.,* p. 100.

Republic, Saudi Arabia, Kuwait, and Iraq on problems of mutual interest to the leading oil-exporting countries. There he struck a responsive chord. These representatives, like the Venezuelan officials, had been greatly disturbed by the February price cuts and welcomed collective action in settling common grievances with the oil companies.

After informal discussion the conferees agreed among other things on the desirability of their governments' jointly establishing a consultative commission to meet annually and discuss and seek a solution to problems of mutual interest; the desirability of individually establishing an organization "to co-ordinate from a national point of view, the conservation, production and exploitation of petroleum"; the importance to the Middle East countries of obtaining a revision of their concessions, designed to increase each government's share in net profits to 60 percent and thereby place them on a parity with Venezuela's take through its income tax laws; and the importance of the oil companies' consulting with the several interested governments before initiating changes in posted prices. What the group had agreed to informally, the Petroleum Congress echoed formally.

The delegates apparently left the congress with a determination to transform discussion into action. At any rate, Abdullah Tariki, who by this time had assumed the role of spokesman for the Middle East on oil matters, outlined an international proration program before the annual meeting of the Texas Independent Producers' Association and Royalty Owners. A few days later in Cairo Dr. Alfonzo and Tariki issued a declaration favoring an international petroleum agreement embracing Venezuela, Kuwait, Saudi Arabia, Iraq, and Iran.

The Baghdad Conference Gives Birth to OPEC

The several ideas remained in the discussion stage until the price cuts of August 1960. The resentment of the Middle East countries was vigorous, prompt, and decisive. Within less than a month a conference in Baghdad of representatives of Iraq, Iran, Kuwait, Saudi Arabia, and Venezuela gave birth to OPEC.

Although born in an atmosphere of hostility and mistrust and actuated by a resolute determination to protect and better the interests of its members through collective action, OPEC conceived its role in broad and sometimes pretentious terms, formulated its purposes with an imaginative but not always practical reach, and pursued its specific objectives with moderation and patience. At its first conference it called attention to the developmental programs on which its members had launched, and it emphasized the role that oil, a wasting asset, necessarily played in these programs. It stressed the importance of oil revenues in maintaining balanced budgets within the member countries. It recognized that oil as a primary energy resource was indispensable in maintaining and improving living standards in all the nations of the world. Asserting that fluctuations in oil prices were disturbing to the economies of both producing and consuming countries, it professed responsibility in insuring their stability. Its members, it avowed, could no longer remain indifferent to price changes initiated unilaterally by the international oil companies. They had an obligation to producing and consuming countries alike to participate in price-making decisions.[4]

Consistent with the responsibilities which OPEC had assumed, its members pledged themselves to demand that the oil companies maintain stable prices and dedicated themselves to the restoration by "all means available" of the pre-cut price levels and to making certain that the oil companies would consult with them whenever they believed price changes were necessary.[5] While emphasizing their obligations to the producing and consuming countries, they also acknowledged responsibility to their concessionaires on whose know-how and capital their countries' welfare had become so dependent. To fulfill these several obligations—"a steady income to the producing countries, an efficient, economic and regular supply" of oil to consuming countries, and a "fair return on their capital to those investing in the petroleum industry"—they proposed to formulate after investigation and study an output control program.[6] To insure continuous, unified effort in pursuing these goals, each member vowed to reject any special benefit that any company might offer to a single member in an effort to persuade him to break ranks.

Having acknowledged their responsibilities and outlined their goals at the Baghdad meeting, the members adjourned to enlist from their several governments official approval of the action they had taken. They promptly got it.

---

4. *Resolutions Adopted at the Conferences of the Organization of Petroleum Exporting Countries,* First Conference, Baghdad, September 10, 1960, Preamble.
5. *Ibid.,* I.1. (Roman numerals indicate the number of the conference and Arabic numerals indicate the number of the resolution.)
6. *Idem.*

OPEC's Organization

At its next two meetings OPEC created the administrative machinery and adopted the procedures essential to the achievement of its objectives. These reflected the intent of its members to pursue their goals with deliberation and caution. They invested supreme authority in the conference, composed of one delegate from each member country and meeting at least twice a year. The conference formulates general policies and makes decisions on reports and recommendations submitted by its board of governors. Conference action requires the unanimous approval of the founding members.

Recognizing that its effectiveness would depend on the comprehensiveness of its coverage, OPEC provided that any substantial net exporter of crude petroleum might become a member with the unanimous approval of the founders. By 1968 its membership had expanded to include Qatar, Libya, Indonesia, and Abu Dhabi. In 1967, member countries produced 44.4 percent of the world's oil, contained 71.6 percent of estimated world reserves, and accounted for 84.7 percent of world exports.

OPEC vested management of its affairs in the board of governors which it appoints—one governor on the nomination of each founding country, one nominated by all other members collectively. The chairman of the board of governors, appointed by the conference for a period of one year, acts also as secretary-general and chief executive officer of the organization. The secretary-general appoints all staff members and with the approval of the board of governors all departmental heads. To insure the international character of the organization the secretary-general in making staff appointments is obliged to give as far as possible "adequate geographical distribution" among member countries. All those accepting positions in the secretariat become international civil servants pledged to discharge their duties and regulate their conduct solely in the interest of the organization without regard to national origins.

OPEC's Internal Structure

In shaping its structure and outlining its functions, the organization reveals its apparent intent to formulate its general policies, articulate its specific demands, publicize its recommendations, and implement its decisions with knowledge and understanding based on research and investigations. To aid the secretary-general in the discharge of his duties the organization established five departments—administration, economics, enforcement, public relations, and technical.

The administration department is not limited in its responsibilities to the internal affairs of the organization—general services, personnel matters, budgets, accounting, and the like—but is obliged to study and review continually the general administrative policies and industrial relations methods of the international oil industry as they may affect the organization.

The economics department is obligated to conduct continuous research on the economics of world oil, to make such special studies as the interests of the producing countries may require, and to alert the secretary-general to any development in the industry which may affect the welfare of the member countries.

The technical department parallels the economics department in duties and purposes. Its subject matter is different. It is designed to keep the secretariat abreast of developments in geological and geophysical methods, production and engineering techniques, and the like as they bear on oil discovery, reserves, and potential production.

The enforcement department is designed to explore legal problems and developments that have a bearing on the aims of the organization and their realization. It apparently reflects the organization's desire to achieve its objectives within the framework of international law.

The public relations department has responsibility for developing a public relations program designed to keep the public informed on OPEC goals and methods.[7] For this purpose it is obligated to maintain and expand an information center, including a library and competent translators.

Its structure completed, OPEC voted itself an initial budget of £150,000, selected Geneva, Switzerland, as its headquarters, and appointed Fuad Rouhani of Iran as its first board chairman and secretary-general. Its selection of Rouhani was consistent with its apparent intent to follow informed, prudent, and persistent policies in pursuit of its objectives. Rouhani is a cultured, articulate, urbane gentleman used to the ways of the

7. The public relations department was later abandoned as a separate department and attached to the secretary-general's office. In June 1965, OPEC signed an agreement with the Austrian government providing for the transfer of the headquarters from Geneva to Vienna.

West, with a courtesy adapted alike to the drawing room and the conference table. Experienced in diplomacy, unimpeachable in integrity, and with
a quiet determination that belies his gracious manner, he seemed remarkably equipped to represent OPEC in its effort to negotiate better terms with
the concessionaires. His colleagues manifested their confidence in him by
voting him a two-year appointment despite a one-year limitation prescribed
in OPEC's bylaws.

OPEC Considers Prices and Proration

Its policies outlined, its structure completed, its budget provided, its
secretary-general designated, OPEC was ready for business. To support
members' contentions that their share of oil industry profits was not
"commensurate with the contribution" of the host countries to the industry
nor with their "social and economic needs," OPEC's members instructed
the board of governors "to prepare a comprehensive study based on
objective research and consultation with experts" of concessionaires' investments and rate of return in the oil industry compared with those of
other enterprises in other countries.[8] As a step toward stabilization of oil
output and country revenues, they instructed the board of governors to
make a detailed study including, if necessary, a study of "international
proration" designed "to arrive at a just pricing formula."[9] Apparently
anxious to secure the sympathetic understanding and co-operation of the
United States, they suggested consultation on import quotas with "friendly
countries" that had imposed restrictions on oil movements,[10] and they
expressed the hope that the international oil companies would manifest "a
spirit of understanding" in discussions then underway with some member
countries.[11]

In OPEC's first conference (Baghdad, September 1960) it had stated
as an immediate objective the restoration of prices to the precut level.
Reflecting its desire to pursue this goal with caution and understanding, at
its second conference (Caracas, January 1961) OPEC instructed each of
its members to report to the board of governors within two months its
precise position on the price issue in the light of the specific terms of the

8. *Ibid.*, II.11.
9. *Ibid.*, II.13.
10. *Ibid.*, II.15.
11. *Ibid.*, II.16.

concession contracts and subsequent agreements or correspondence modifying or interpreting these provisions and of the provisions governing the settling of disputes. The conference instructed the board of governors to submit the members' reports to competent legal advisors for study and recommendations.[12]

Fourth Conference Sets Three Specific Tasks

Despite its preoccupation with the restoration of prices, two years after its birth in Baghdad OPEC had made no progress in settling the issue. Indeed, its members had manifested reluctance to utilize the organization's collective strength in grappling with the issue and such steps as the individual countries had taken had proven futile. The fourth conference (Geneva, spring and summer 1962) reflected a more determined tone and a more restricted aim. The conference set its members three specific tasks: restoration of prices, the expensing of royalties, and the elimination of market allowances.[13]

OPEC was eventually to attain its objectives on the royalty issue and on marketing allowances. On the price issue it was doomed to fail. It could scarcely have been otherwise. Its goal with respect to prices reflected its hopes rather than a sense of reality. Although OPEC had not based its demand for price increases on the elementary thesis that Abdullah Tariki developed with dramatic effect at the Second Arab Petroleum Congress in Beirut in 1960 in alleging that the oil companies had conspired to reduce prices and by doing so had robbed the Middle East oil countries of more than $2.5 billion, it has been slow to free itself entirely of the notion that oil monopolists have reduced prices in contravention of the best interests of the Middle East oil countries. In arguing for higher prices, OPEC has seemed at times more concerned with equity than with economic principles. This is reflected in its fourth conference resolution on prices. Without having explored the intricate statistical and administrative problems involved or the role that market forces exerted on oil prices, it instructed member countries to formulate a rational price structure linking crude oil prices to an index of prices of goods which the member countries imported.

More significant, it instructed its members to begin immediate negotia-

12. *Ibid.*, II.12.
13. *Ibid.*, IV.32,33,34.

tions with the oil companies "or any other body deemed appropriate" to insure member countries being paid "on the basis of posted prices" prevailing before August 1960. This is something different from the restoration of the precut prices and lends itself better to collective bargaining. Before the price cuts the oil companies had been selling oil for some time at a discount from posted prices although they had continued to calculate their tax payments on a basis of posted prices. When they lowered the posted prices, they reduced the basis for calculating their income tax payments to the member countries. OPEC had in effect instructed its member countries to negotiate an agreement with the oil companies to restore the oil prices for determining their tax obligations. This issue is better adapted to collective than to individual bargaining. Were the companies collectively to acquiesce in this, it would mean in effect that they increase the countries' tax take at the expense of company profits.

While this was a more realistic immediate aim, it was one that the members were unlikely to achieve, especially through their individual bargaining efforts. Any company that alone acquiesced in such a demand would place itself at a competitive disadvantage in world markets. Moreover, in view of the current world surplus of crude oil, it was scarcely to be expected that the oil companies collectively could be persuaded to surrender their administrative discretion on pricing decisions. It has not proven otherwise. In the course of negotiations over the next two years the pricing problem was apparently lost sight of.

Not so with the expensing of royalties and the elimination of marketing allowances. As a basis for member negotiation OPEC published three "Explanatory Memoranda" setting forth the logic of their demands. These it sent to the oil companies and to the governments of the consuming countries. It also requested Saudi Arabia and Iran to take the initiative in negotiating with their respective concessionaire companies on all these issues.

Saudi Arabia Begins Negotiations with Aramco

As previously indicated, the Saudi Arabian government began negotiations with Aramco in the summer of 1962. These negotiations were concerned with other issues peculiar to the Aramco concession—the most important of which were the relinquishment of concession areas and the sharing of Tapline earnings—and apparently the royalty and price issues played an inconsequential role. In the spring of 1962 Aramco and Saudi Arabia signed agreements covering relinquishment of concession areas, sharing of Tapline profits, and the granting of marketing allowances. As previously indicated, the market allowances settlement, which reduced marketing deductions from about 4.2 cents to ½ cent a barrel, was a compromise, satisfactory, however, to OPEC, whose members gave it unanimous approval. Iran and the consortium subsequently reached a similar settlement.

Settling the market-allowance issue was relatively simple. Discounts to meet the cost of marketing crude oil sold by the producing companies to their parents at prices which the buyers determined appeared to have little justification in logic or fact and the cost of complying with the governments' demands was relatively slight.

OPEC Begins Negotiations on Oil Royalties

The expensing of royalties was something else. The negotiations on this issue, which lasted from 1962 to 1965, have been characterized by OPEC as among the "longest, toughest, and most revealing in the history of the international oil industry."[14] During this period OPEC utilized all of the agencies of negotiation at its command. At the outset it requested Saudi Arabia and Iran to conduct negotiations directly with their concessionaires. Later it commissioned Rouhani to conduct negotiations on behalf of all the countries affected. Still later it replaced Rouhani with a three-man negotiating committee and eventually it referred the issue back to the several individual countries. The bargaining process had run full cycle.

Oil Companies Reluctant to Recognize OPEC as a Collective-Bargaining Agency

Throughout the negotiations the oil companies persistently rejected the principle of collective bargaining. Although two or more of them were joint

14. Organization of Petroleum Exporting Countries, "OPEC and the Principle of Negotiation," paper presented at the Fifth Arab Petroleum Congress, March 16–23, 1965, p. 8.

owners of concessionaire companies in each of the Middle East exporting countries and were accustomed to using their collective power in negotiating with individual countries, they were reluctant to recognize and accord a similar joint interest to the countries with which they dealt. So strong and obvious was their opposition to joint action by the host countries that the London *Times,* although recognizing the difficulties of collective action among countries whose interests were diverse, commented editorially in its August 3, 1963, issue:

> This [opposition] could be represented as yet another instance of the [oil companies'] attempts to slow down the whole process of negotiation and to behave in an out-of-date imperialistic way. To some extent the industry would have only itself to blame if this happened. Some of its more prominent members have not hesitated to treat O.P.E.C. as an inevitably hostile body—if indeed, they are prepared to recognize its existence at all. Yet the Governments have a perfect right to set up a body to look after their common interests. Moreover, these interests are to a large extent shared with the companies. All make money out of oil. If producing Governments sometimes forget this, the industry should show its greater wisdom by being less haughty about O.P.E.C.

In a similar vein the *Economist* commented in its August 10, 1963, issue:

> The oil companies have always been extremely anxious to avoid negotiating direct with OPEC as an entity in itself, because they have not wanted to magnify the importance of what they like to think of as simply a forum for discussion and an advisory information bureau for oil governments.

Consistent with this attitude, when OPEC members authorized Rouhani to negotiate on their behalf, a step characterized by OPEC as the first "ever taken by producing countries to institute collective bargaining with the major oil companies"[15] and by the *Middle East Economic Survey* as a "turning point in the royalty battle,"[16] the oil companies informed the press that their three-man committee (J. M. Pattinson of British Petroleum, H. W. Page of Jersey Standard, and G. L. Parkhurst of Standard Oil of California—all from the companies owning the Iranium Consortium) would be negotiating purely on behalf of the consortium with Rouhani as a representative of the Iranian government. When OPEC's three-man committee designated Geneva as the place to begin negotiations on behalf of Iraq, Qatar, and Kuwait, to which countries their concessionaires had as

15. *Ibid.,* p. 11.
16. *Middle East Economic Survey,* August 28, 1964, Supplement.

yet submitted no offers, the oil company representatives first demurred; and when they eventually acquiesced, representatives of IPC, Qatar Petroleum Company, and Kuwait Oil Company successively met separately with the committee. Yet, as the *Middle East Economic Survey* observed, "if the fact of collective bargaining was precluded by legal obstacles, the substance was certainly there."[17]

Except for its duration the bargaining did not differ radically from industrywide industrial union bargaining in American industry.

Obstacles to Agreement

Although several international oil companies are joint owners of each of the concessions, ownership among the several countries is not identical, and the interests of various companies do not always coincide. The companies differ in the degree of self-sufficiency in crude production, in the richness and variety of their crude reserves, in alternative sources of supply, in the balance between their crude production, refining facilities, and marketing operations, in their overall profits, and in the character and pliability of their top executives. Although all have been shaped in the framework of a profit economy, are committed by training and experience to business goals, and are limited in their independence of decision making by the forces of the markets, they differ in their individual judgments as to what the traffic will bear and their sense of social and corporate responsibility. Consensus is not always easy.

Similarly, despite their commitment to collective action through OPEC, the member countries differ in the richness of their resources, in their financial needs, their political stability, their sense of brinkmanship, their attitude towards the West, their willingness to compromise their long-run demands in return for immediate advantages which the companies might offer, and their commitment to unilateral action if no agreement is reached. And both the oil company and the country participants in the bargaining process were representatives of their principals, whose approval was necessary to make effective any agreement that the negotiators might reach. These were obstacles to a mutually satisfactory agreement. To the extent that they bargained on a rational basis, each group had to consider the alternatives to a peaceful settlement.

17. *Idem.*

Process of Collective Bargaining

OPEC has described the process of negotiation as falling into exploratory talks in late 1962 between Iran and Saudi Arabia on the one hand and the consortium and Aramco on the other, and five stages of active negotiations: stage one, in which Rouhani negotiated with consortium representatives in London between September 16 and October 31, 1963; stage two, in which Rouhani continued negotiations in New York between November 4 and 8, 1963; stage three, in which the member countries consulted informally in Beirut and formally at the OPEC's fifth conference in Riyadh during December 1963; stage four, in which OPEC's three-man team negotiated either directly in Geneva with company representatives or considered written offers which the major companies had submitted; and stage five, in which negotiations culminated in final talks between OPEC and company representatives in London on November 12, 1964. During the course of these negotiations the companies shifted their position from a complete and positive rejection of the principle of expensing royalties to an acceptance of the principle and the offer of specific terms which OPEC characterized as meeting their minimum demands and which OPEC referred for action to its member countries.

The Logic of OPEC's Demands

It will be helpful in understanding the progress of negotiations and in evaluating their outcome to state more precisely what OPEC was asking for and why it was asking.

Under the institutional arrangements that have evolved in contemporary societies, oil is considered an attribute of the land under which it is found. The owner of the land owns the oil. In the United States this is true whether the land is owned privately or publicly. Since oil is a wasting asset, the owner of it is entitled to remuneration by anyone who exploits it. As it is depleted he receives compensation which tends to conserve the capital values inherent in the land. Whether regarded as a rent or compensation for depleting the land of its original value, or as a payment for the intrinsic value of the raw material in the land, it is something quite different from the tax which the exploiter may be required to pay the state on the profits he derives from his enterprise. In calculating his taxable earnings, the entrepreneur appropriately regards royalty payments as costs encountered in the productive process. This distinction is clear in Western countries. Oil companies in the United States may pay bonuses and annual rentals to landowners for the privilege of searching for oil, but when they find it, they customarily pay the landowner a royalty on every barrel of oil produced. In private transactions custom has established this payment at one eighth of the value of the oil at the wellhead.

The practice is no different where the state is the landowner, although the amount of the royalty demanded and gotten on unusually promising land may be larger. If an individual or company contracts with a government, whether state or federal, for the privilege of hunting for oil, when he finds it he pays an agreed royalty to the state. In his accounting he treats this as a cost to be deducted along with other costs from gross revenues to determine net earnings, and any income tax he may be required to pay will be calculated on his net earnings. This is also the practice to which the oil companies conform in Venezuela.

Oil Companies Treat Royalties as Costs Except in Calculating Governmental Taxes

That this practice was not followed in the Middle East concessions when OPEC began its negotiations is to be explained in terms of historical circumstances. Most of the concessions originally provided for a royalty payment of four shillings a ton. As concession agreements were modified, the royalty payments came to be specified on a per barrel basis, customarily at 12½ percent of the posted price of crude. When the so-called 50-50 profit-sharing was introduced as a result of the negotiations of the 1950s and income tax laws were passed to give it an institutional basis, the laws and the concession contracts provided that the royalty payments, which the companies had hitherto considered as a cost, be treated as a payment to the government in calculating its share of the profits. This did not change the accounting procedure of the oil companies in calculating their costs. Though the companies treated royalty payments as income in calculating the host countries' share of profits, they treated them as costs in determining their tax obligation, if any, to the country of their origin. By merging royalty payments with earnings for tax purposes, royalties as payments for the "intrinsic value"[18] of the raw material produced lost their identity.

Rouhani, in his London talks with representatives of the Iranian Consortium in October 1963 characterized this anomaly as follows:

either the Consortium Member Companies are paying income tax at the full rate prescribed by law, but no royalty ["stated payment"]; or they are effectively paying royalty but their income tax payments amounted to about 41% of income not 50%.

And he argued that

In the first case it would have to be recognized that the Consortium agreement (in the case of Iran) in effect lacks a royalty clause, or, in language more germane to the juridical nature of the transaction, provides no consideration for a contract of sale and purchase. In the second case, one can see no reason why prosperous companies should enjoy a substantial tax reduction in a country which has such a vast need for development financing and which is faced every year with a crisis arising out of a serious budget deficit. It should not be forgotten that the Oil Companies enjoy another immunity as compared with the general public; that from the payment of customs duties and indirect taxes.[19]

Rouhani was fully aware, of course, of the fact that the consortium concession agreements, as well as the agreements in other Middle East countries, which had been renegotiated in the 1950s, provided for just this practice. In truth, Rouhani, as a representative of the Iranian government, had participated only a little more than a decade earlier in negotiating this arrangement.[20] But he believed with the other Middle East spokesmen that regardless of the legal validity of the oil companies' position on royalties, the time had come for a change.

18. This is OPEC's language used in its *Explanatory Memoranda on the OPEC Resolutions,* Fourth Conference, Geneva, April–June 1962.

19. Fuad Rouhani, *Memorandum Concerning the Expensing of Royalties,* presented by the secretary general of OPEC to oil company representatives during the London talks, October 1963, p. 3 (unpublished). Quoted in Mughraby, pp. 141–142.

20. When I asked Rouhani in 1963 why he had not held out for the expensing of royalties in the Iranian negotiations with the oil companies following Iran's nationalization of the industry a decade earlier, he said that at that time neither he nor his fellow negotiators were aware of customary practices in treating oil royalties. Rouhani has learned a lot about the international oil industry in the intervening years.

OPEC's Specific Demands and the Companies' Initial Offer

Resolution 33, adopted at OPEC's fourth conference in Geneva in the spring of 1962, set as its goals the fixing of royalty payments at a "uniform" and "equitable" rate and their treatment as company costs of production rather than as credit against its income-tax liability. When negotiations began on the royalty issue, payments in the Middle East countries were calculated at 12.5 percent of the posted price of oil except in Saudi Arabia where a flat rate of 22 cents a barrel prevailed. Although available evidence does not indicate that OPEC at any time demanded a specific royalty rate in excess of that being paid, in its explanatory statement defending its royalty demand it called attention to rates as high as 25 percent paid on petroleum production on government lands in the Western world and to the 20 percent royalties provided for in recent agreements between Saudi Arabia and Kuwait with the Japanese Trading Company. After reviewing customary rates in other countries, it concluded "that a minimum of 20% would be a just and equitable rate" in the Middle East.[21]

Expensing of royalties with no increase in rates would have increased governmental income from oil production by about 15 percent or approximately 11 cents a barrel. This presumably represented a minimum target at which OPEC originally aimed.

On October 16, 1963, approximately one month after Rouhani began negotiations with representatives of the consortium, the companies made their initial concession to the principle of expensing royalties. In doing so, they expressed a willingness to expense royalties in return for recognition by the member countries that Middle East oil was selling in commercial transactions at a discount below the posted prices which were used as a basis for calculating income taxes. Specifically, they agreed to expense royalties at the rate of 12.5 percent provided the countries would allow a 12.5 percent discount from the posted prices in calculating their income taxes. Such a settlement would have brought no financial gain to the governments. Moreover, the companies conditioned their offer on OPEC's acceptance of it in satisfaction of all of its resolutions—including the price

21. OPEC, *Explanatory Memoranda on the OPEC Resolutions,* p. 14.

adjustment that it had set as its goal—and on the understanding that OPEC countries would make no more favorable adjustment with any other operating company. In the language of OPEC "under this . . . arrangement the major oil companies wanted a quit-claim on all OPEC demands in return for their acceptance of the mere principle of expensing royalties without the slightest financial benefit on a higher per-barrel revenue."[22] OPEC rejected this offer.

Companies Improve Their Offer

In their second offer, made in New York early in November 1963, the companies agreed to expense royalties at 12.5 percent in return for the governments' agreeing to allow a discount of 8.5 percent off posted prices in calculating company tax liabilities. They accompanied the offer with the same conditions originally submitted and the added stipulation that the countries concerned should impose no restriction on the production or movement of oil.

OPEC characterized the conditions as unacceptable and the amount of royalty adjustment as inadequate. While negotiations were in progress, OPEC members, who were in frequent consultation on the matter, had considered from time to time what retaliatory unilateral action they might take if an acceptable offer were not forthcoming from the oil companies. OPEC did not disclose officially the measures contemplated, but they are believed to have included *inter alia* a tax on tanker earnings, legislation providing for expensing of royalties, and in the event of noncompliance by the companies, restrictions on production or exports and price control.

In response to pressure from OPEC, company negotiators improved their offer from time to time. The record discloses that OPEC made only one specific counter offer. In response to a written offer which company negotiators made to Iraq, Kuwait, and Qatar at Geneva in the spring of 1964 (similar in its essentials to offers they had previously made to Iran, Saudi Arabia, and Libya) OPEC countered with the proposal that the companies expense half the royalties (about 5.5 cents) immediately and the remainder over a specified number of years. In rejecting this offer, company negotiators early in July in Geneva countered with a proposal that they expense royalties but that governments permit them to calculate

22. "OPEC and the Principle of Negotiation," p. 12.

government revenues on a basis of price discounts of 8.5 percent in 1964, 7.5 percent in 1965, and 6.5 percent in 1966, according to an escalation formula based on the gravity of the crude. Subject to future market conditions, the companies also agreed to consider a further reduction in allowable discounts. This offer, which would increase government revenues by 4.5 to 5 cents a barrel, fell far short of complete expensing of royalties, which would yield the countries an additional revenue of about 11 cents a barrel, but it acknowledged the principle of expensing royalties and in its financial results came close to OPEC's proposal that half of the royalties (5.5 cents) be expensed immediately. The companies, however, still insisted that the countries relinquish all their other demands including those relating to prices.

OPEC Defines Minimum Requirements of an Acceptable Offer

At its sixth conference (Geneva, July 6–14, 1964) OPEC defined for its members, but did not publish the minimum requirements, particularly "the nonfinancial ones," of an acceptable offer. Despite the fact that the companies' revised offer of July 1964 failed to meet in all respects the minimum requirements of an acceptable offer, OPEC decided at its sixth conference that "subject to certain improvements" the offer provided "a suitable basis for full implementation" of its royalty demands and instructed its members to explore with their respective concessionaires the "possibility of reaching a final agreement."[23] Meanwhile OPEC members were to keep each other informed on the progress of negotiations and to report at a special meeting to be held in Beirut on September 14, 1964. Failing a satisfactory settlement, the members were to request the secretary-general to convene a special conference to review the situation and make recommendations looking to a final solution.

Although OPEC has never revealed the details of its minimum acceptable offer, the *Middle East Economic Survey,* which maintained frequent and close contact with OPEC members, reported on August 28, 1964, that OPEC would submit before mid-September on a "take-it-or-leave-it" basis a specific counter proposal containing three demands: that all expensing of royalties be gradually reached within a definite time limit, that the royalty

23. *MEES Supplement,* August 28, 1964.

agreement limit in no way unsettled issues, and that benefits offered to any country be accorded to all OPEC members.

Whether or not the *Middle East Economic Survey* was correctly informed of OPEC's plans, they were not fully realized. At their Beirut consultation OPEC representatives concluded that the oil companies should further clarify their July offer and resumed negotiations with them in London. After OPEC representatives had conferred for four days with oil company representatives, they cabled OPEC's secretary-general on October 19 that their minimum requirements had been met and requested him to convene a special consultation meeting in Geneva to review the company offer. According to OPEC, on the evening of the same day that its representatives were preparing to leave London the companies "communicated to them clauses re-establishing some of the unacceptable conditions." Ten days later the parties resumed negotiations, concluding their talks on November 12. On November 16 the companies submitted another offer in writing to OPEC members.

OPEC Members Disagree and Refer Settlement Back to Individual Countries

Whether the companies brought their final offer within the limits of OPEC minimum requirements is not clear, not even to OPEC members. A resolution adopted at OPEC's seventh conference held in Djakarta from November 23 to November 28, 1964, notes that Iran, Kuwait, Libya, Qatar, and Saudi Arabia contended that *it did* and expressed their intention of accepting it. Hence they argued it was not a proper topic for review at the Djakarta conference. On the other hand, the resolution notes that the Iraq delegation had rejected the company's offer because it had not met minimum requirements to which OPEC members had agreed and because the "nonfinancial provisions of the . . . offer constitute a clear infringement on its sovereignty and restrict its freedom" in achieving the "higher objectives of the Organization."

The delegations of Venezuela and Indonesia, not involved directly in the royalty settlement because the contracts with their concessionaires already provided for expensing royalties, supported Iraq's position, contending not only that the company offer fell short of the minimum requirements, but that these requirements could not be modified without a new conference resolution. Accordingly they prepared a resolution recommending the rejection of the offer.

Confronted with this division, which apparently had been preceded by increasing bitterness among the delegates throughout their long negotiations, the conference voted to leave acceptance or rejection of the royalty-offers issue to the individual member countries. As a balm to the breach, the members dedicated themselves anew to the "struggle for the realization of OPEC's objectives" and declared their "solidarity to each other" and reaffirmed their "determination to strive for steady and equitable petroleum prices."[24]

After OPEC washed its hands of the royalty issue, the company's offer of November 16 became a subject of further negotiation between the companies and the governments. After four weeks of negotiations the Iranian government, represented by its finance minister, Amir Abbas Hoveyda, and the consortium, represented by its three-man negotiating team (H. W. Page of Jersey Standard, G. I. Parkhurst of California Standard, and J. M. Pattinson of British Petroleum) reached an agreement. The finance minister submitted it to parliament for ratification on January 9, 1965. By January 26, 1965, Saudi Arabia and Qatar had ratified. Libya did not reach a settlement of the issue until December 1965. Kuwait, after acrimonious consideration, ratified in May 1967. Iraq had not ratified by the end of 1968.

The oil companies originally stipulated that if four companies had ratified the royalty offer by January 26, 1965, the settlement would be made retroactive to January 1, 1964. By January 26, 1965, only Iran, Saudi Arabia, and Qatar had ratified. Since Kuwait's finance minister, an OPEC member, had approved the agreement before the deadline, the oil companies conceded that in spirit, if not in fact, four companies had approved; they accorded Saudi Arabia, Iran, and Qatar the financial benefits of the agreement as applicable to 1964.

Kuwait Submits Agreement to Its Assembly

On January 19, 1965, the Kuwait government submitted the royalty agreement for ratification to the National Assembly. Whether anticipating

24. *OPEC Resolutions,* VII.49,50.

no opposition or, if any developed, hoping to railroad the agreement through the assembly, the speaker of the house had neglected to supply assemblymen with copies of the agreement although they had been available for two months. In urging speedy approval, the speaker pointed out that Kuwait would forfeit the 1964 payments (approximately £10 million) if the assembly failed to ratify by January 26, 1965. In his defense of the agreement Assistant Undersecretary Ahmad al-Sayyid Omar reviewed briefly OPEC's struggle to settle the royalty issue and its decision to leave final action to the individual member countries. In complimenting OPEC on its work, he implied that only by favorable action could OPEC's solidarity be maintained. He assured the assembly that the agreement contained no secret promises nor any provisions detrimental to the country's welfare.

It was perhaps to be anticipated that the assembly would quickly rubber-stamp the agreement. Here was a country inexperienced in the ways of democratic government, accustomed to the rule of a tribal leader and his advisors, with the world's richest oil reserves, with oil revenues far in excess of its needs, with the world's largest per capita income, with foreign investments running into the hundreds of millions of dollars, and assets so large that it was rapidly assuming the role of banker to the Middle East. Why should Kuwait haggle over a few paltry millions?

But haggle it did. Opposition to the agreement was prompt and vigorous and grew in intensity. Led immediately by an Arab Nationalist bloc of twelve out of a total of fifty elected deputies on whose motion the assembly referred the agreement to a special parliamentary committee for study, the opposition came to embrace practically the whole of the assembly, encompassing even the government spokesman.

Basis of Assembly Opposition

Assembly opposition stemmed from its Arab Nationalist members. To what extent they were engaged in a domestic power play or to what extent they saw in the agreement a forfeiture of Kuwait's basic interests is not clear to an outsider—or perhaps even to an insider—but at the outset in opposing the agreement, they appealed to the fears and suspicions that reflect the uninformed but deep-seated hostility of Arab Nationalists toward "colonialism" and "exploitation" by foreign oil companies. As the agreement was subjected to intensive study by the parliamentary committee, to careful scrutiny by experts whose advice the committee had sought, to debates within the Assembly extending over a year and a half, and to critical review by a nascent nationalistic press, its opponents, who grew more numerous and voluble, found a more specific basis for their opposition. Three provisions of the agreement—the quit-claim provision, the arbitration provision, and the most-favored-company provision—though easily justified from the point of view of the oil companies, lent themselves to exploitation by patriotic Kuwaitis suspicious of the companies and easily mobilized to protect the interests of a young nation scarcely out of its swaddling clothes.

The quit-claim clause provided that the royalty-expensing agreement constituted a full and final settlement of all of the government's outstanding financial claims against the concessionaire. Because Kuwait, as well as the other Middle East oil countries, had never acknowledged the validity of the 1959 and 1960 price cuts, this would have meant foregoing all claims for tax losses incurred by reason of these cuts. Faisal Mazidi, a prominent oil official, chairman of the Kuwait Fertilizer Company and government-appointed director of the Kuwait Oil Company, in a report to the assembly's financial and legislative committee alleged that these amounted during the six years 1959–64 to $449.5 million.

The arbitration clause provided for the arbitration of disputes growing out of the agreement under the auspices of the International Court at the Hague, the president of which was authorized to name an arbitrator should either of the parties fail to name one within the prescribed period. Opponents of the agreement regarded this as a denial of state sovereignty and in conflict with the provisions of the 1955 Kuwaiti income tax decree. Article 13 of the decree initially provided that disputes growing out of the decree be adjudicated by Kuwaiti courts. Under the British protectorate these were created by the British political agent in Kuwait. When Kuwait became an independent nation, these became truly national in character.

The most-favored-company claim provided that Kuwait extend to the Kuwait Oil Company fiscal treatment as favorable as that granted to any present or future concessionaire. Critics of the agreement contended that this imposed an intolerable limitation on the government's freedom of action, particularly with reference to the Kuwait National Petroleum Company, a quasi-governmental enterprise recently organized to exploit Kuwait's unleased oil lands.

Critics Widen Their Attacks

As the debate over ratification grew in intensity, criticism was directed not only at the specific provisions of the royalty-expensing agreement but at the basic concession agreement as well. Critics professed to be disturbed over the fact that the proposed agreement perpetuated what they regarded as inequities in the basic concession under which the oil companies operated. The agreements, old and new, placed serious limitations on the government's taxing power.

They were also disturbed by the failure of the Kuwait Oil Company to meet the production quota experimentally and unilaterally established by OPEC for the year 1965–66 and by an actual decline of 2.4 percent in Kuwait's annual rate of production during the first quarter of 1966. To the Kuwaitis it appeared that the Kuwait Oil Company was penalizing Kuwait for its failure to accept the agreement.

As negotiations between the government and the oil company were renewed, dissatisfaction with their progress expressed itself in official and unofficial statements of the government's intention to take whatever measures were necessary to protect Kuwait's rights. Hints of restrictive legislation and, as a last resort, of nationalization emanated from government sources. By the summer of 1966 the government had more clearly defined its intentions. At a press conference early in July Undersecretary Ahmad al-Sayyid Omar revealed that government officials were conducting studies preparatory to drafting a petroleum law which would not merely include "an agreement on expensing royalties" but would deal with the tax problem as well and regulate relations between the government and oil companies in all their aspects. The secretary indicated that the proposed law would be referred to the council of ministers for study before being submitted to the national assembly for debate at its full session.

At the expiration of its four-year term the ruler dissolved the national assembly on January 3, 1967, and called for an election. The electors went to the polls on January 25. The Arab Movement, which had furnished the core of the opposition, suffered a severe defeat, capturing only four of the fifty seats. Whether it was a contrived defeat is uncertain, but thirty-eight candidates, including six successful ones, protested the election. They were joined by the Kuwait Lawyers Association, Kuwait Newspaper Association, the National Union of Kuwaiti Students, the Kuwaiti Graduates Association, the Kuwaiti Women Graduates Association, and the Executive Council of the Federation of Government Workers and Employees Unions. The protests were of no avail.

On May 2 the newly elected assembly formally ratified a revised version of the royalty expensing agreement. In doing so it had the advice of two oil experts, Dr. Nadim al-Pachachi, formerly with the Iraq oil administration, and Abdullah Tariki. The agreement was similar in its main provisions to that accepted by Saudi Arabia, Iran, Libya, and Qatar. It had been modified, however, in form if not in substance, to meet the major criticisms that had been directed against it. The quit-claims provision was so worded as to recognize Kuwait's right to challenge the level of prices as fixed after 1966; the Kuwait Oil Company agreed to submit to the Kuwait national courts any dispute relating to Kuwait's income tax legislation; and the most-favored-nation clause was so worded as to exclude the Kuwait National Petroleum Company from its provisions. Payments by the KOC amounting to KD 33 million were made retroactive to January 1, 1966.[25]

Libya Grapples with the Royalty Issue

As previously noted, Libya did not reach a settlement of the royalty issue until January 1966. Two factors contributed to this delay. Many companies held Libyan concessions, but only the major internationals had participated in OPEC's negotiations and had offered to settle the royalty problem. Moreover, Libyan concessionaires were calculating their tax obligations on a basis of realized rather than posted prices, and this complicated the companies' offer.

Libya soon made it apparent that it envisaged the royalty problem as part of the larger problem of the relations between the government and its concessionaires and the terms under which it had opened its lands to exploitation by foreign oil companies. The government was apparently determined to get better terms under new concessions than it had obtained under old and to place the old and new on a common basis by amending its petroleum laws.

25. *MEES*, May 5, 1967.

Libya's 1955 Petroleum Law

In 1955 Libya enacted its first general petroleum law, defining the terms under which concessions would be granted. At that time, although numerous companies were definitely interested in Libya as a promising source of oil, none had drilled a well nor even obtained a concession. The major companies, anxious to invest capital in the exploration and development of Libya's potential oil resources under favorable terms, had offered their services in drafting a law. And Libyan officials, untutored in such matters and anxious to attract capital for prompt and rapid development, were glad to utilize the experience of the oil companies in shaping a law that would insure the attainment of a common goal. The 1955 law reflected the aims of those who shaped it. It provided that profits from concessions be subjected to a 50 percent tax and that a royalty of 12½ percent be treated as a partial payment of the tax on profits. These had become standard provisions of the older Middle East concessions.

But concessions under the Libyan law differed in some significant respects from what might be termed the standard concessions. Their terms were more favorable to the concessionaires in the concepts of both cost and income. In calculating costs, Libyan concessionaires could deduct a depreciation charge of 20 percent on all physical assets acquired before the date of production and could choose between a 20 percent amortization of all preproduction expenditures and a depletion allowance of 25 percent of gross income. And unlike the older Middle East concessions, which with certain exceptions and allowances tied income to posted prices, the Libyan concessions tied income to realized, as distinct from posted, prices.

Both the cost and price provisions of the concessions were attractive to all potential producers, but the price provisions were particularly attractive to independent companies anxious to obtain a foothold or a larger share in rapidly expanding foreign markets and realizing that to do so they might be compelled to sell their oil below the customary posted prices.

Companies Respond Quickly to Opportunities in Libya

Oil companies were quick to respond to the attractive terms of the 1955 law. Within three months of its passage fourteen different concessionaires, including six of the seven major international companies, together had obtained forty-seven concessions. By 1961 it had become apparent to both the companies and the government that Libyan oil development was to be no flash-in-the-pan. Esso International was the first to discover oil in large quantities, followed shortly by the Oasis group of independent companies—Marathon, Continental, and Amerada. With exportation on a considerable scale imminent, Libya, realizing that its oil development was likely to yield it far less revenue than it had hoped for and far less than a comparable development yielded other Middle East countries, took steps to recapture some of the advantages it had granted to the early comers. It enacted two amendments to its 1955 law.

The 1961 Amendments and Their Failure

A July 1961 amendment abolished the depletion allowance and endeavored to place a floor under income by tying per barrel taxable income to posted prices less "marketing expenses" as defined by its regulations. "Marketing expenses" as conceived by government officials apparently referred to the marketing allowance which customarily did not exceed 2 percent in most other Middle East countries. Some concessionaires were persuaded to adjust their grants to the new legislation, but some insisted that "marketing expense" be defined as including all rebates that the concessionaire deemed necessary to sell his oil. The government accepted this interpretation and to further satisfy the recalcitrants enacted the November 1961 amendment stipulating that no amendment or repeal of regulations would affect existing contractual rights of concession holders without their consent.

The new laws failed to accomplish all that the government had hoped for. As Libyan output rapidly increased, some disturbing anomalies appeared. In 1964 Esso International sold most of its oil to affiliates at a posted price of $2.21–$2.22 a barrel and paid taxes to the government averaging about 90 cents a barrel on an average realized price of about $2.16. In contrast, the Oasis group sold its oil at an average price of $1.55 (a discount of 67 cents a barrel) and paid to the government an average of less than 30 cents a barrel—only slightly more than bare royalty and rental charges.

Government Takes Steps to Remedy the Situation

Believing that sales by independents tended to undermine the whole international price structure, deprived Libya of substantial revenues, and made difficult the application of the royalty-expensing agreement approved by OPEC and already made applicable to Saudi Arabia, Iran, and Qatar, the government took steps to remedy the situation.

In September 1965, the Libyan Council of Ministers established a special committee—the Petroleum Prices Negotiating Committee—to negotiate an agreement with the oil-exporting companies on prices and discounts. At its first meeting, on September 28, the committee, consisting of the undersecretary of the ministry of finance, the director of oil affairs, the secretary of the supreme council for petroleum affairs, the head of the oil ministry's company accounts department, a financial advisor to the Libyan Cabinet, and a consultant to the oil ministry, made clear to company representatives the government's intention to amend its petroleum law to apply OPEC's royalty-expensing formula with or without the companies' consent.

Following the committee's report on its negotiations to the Libyan Council of Ministers, the minister of oil notified all bidders for new concessions that the government was postponing a decision on bids for new acreage in order to provide sufficient time "for the preparation and issue of the required amendments to existing petroleum legislation with a view to safeguarding the country's interests."[26]

Thereafter matters moved rapidly toward a climax. While a parliamentary committee busied itself with drafting an amendment, the negotiating committee endeavored to persuade concession holders to agree to amend their concessions to permit the application of the royalty-expensing formula. When it became apparent that several concessionaires would not voluntarily comply with the government's request, the government promulgated by decree an amendment to its petroleum law (November 24, 1965). In doing so, it employed a London and a New York public relations firm to prepare and issue background memoranda explaining to all interested in international oil matters the reasons for its action. The amended law provided that concessionaires calculate taxable income on a basis of posted prices, subject only to the discounts provided in the OPEC approved

26. *Ibid.*, October 8, 1965.

formula (7.5 percent of 1965, 6.5 percent in 1966, and thereafter to such discounts as might be negotiated) and that royalties be treated as a cost rather than as a credit against income tax. The government estimated that the new fiscal regulations, if fully complied with, would increase its oil revenue in 1965 alone by approximately $135 million.

While wielding the stick, however, the government did not abandon the carrot. The law contained two provisions designed to induce voluntary compliance. It promised that the government would forego and forget all outstanding claims against any complying company that had arisen out of disputes over prices, rebates, discounts, allowances, and royalties growing out of operations before the end of 1964, and it provided that any subsequent dispute over terms of the concessions or their application be settled by arbitration under international law.

Government claims against concessionaires, which were in contention, involved large amounts. Against one group of independent companies they totaled approximately $60 million. A government spokesman characterized this provision of the law as a "generous gesture" designed to wipe the slate clean and invite voluntary acceptance of the "norm of taxation applied throughout the Middle East."[27]

Government Eventually Forces Compliance

It remained to be determined whether the oil companies which had thus far refused to negotiate such a settlement would comply with a decree, sweetened though it was by what Libyan officials regarded as generous concessions, that conflicted with their contractual obligations and rights as guaranteed under the 1955 law as amended in 1961. It soon became evident that the independent companies were unlikely to comply. Before the government promulgated its decree, they had presented a strong plea against the proposed amendment. In a conference with Libyan Prime Minister Husain Maziq, representatives of seven companies emphasized the role that independents had played in Libya's oil development. They pointed out a conflict of interests between the independents and the majors and between the major companies and Libya itself. They observed that the majors' interest in producing relatively high-cost Libyan crude was slight

27. See memorandum written by a special oil advisor to the Libyan government, reproduced in London and New York newspapers and summarized by the *MEES*, December 31, 1968, Supplement.

compared with their interest in low-cost Arabian crude. The increase in their Libyan costs, which the new royalty and tax provisions would exert on their total average costs, would be slight, but it would have a disastrous effect on the independents. They suggested that this would not be displeasing to the majors, who would welcome an opportunity to buy them out at bargain prices. And they argued that once the independents disappeared from the Libyan scene, the majors would have a cost incentive to increase Persian Gulf output at Libyan expense. Libya's gain in income from expensing royalties and increasing its tax revenues would be a "short-run gain but a long-run loss." And they hinted that they would not comply with a decree that emasculated acquired rights. Since under existing law no company would be compelled to comply unless all did so, the new legislation would prove futile.

These arguments did not convince the government, but they put to test a principle that Frank Hendryx, formerly legal council to the Standard Oil interests and later advisor on oil matters to Saudi Arabia, had eloquently expounded before the First Arab Petroleum Congress at Cairo in 1959 and that subsequently became the subject of hot debate: a government has not only the right but the obligation to modify unilaterally the terms of an oil concession when it believes it is in the public interest to do so.

While the Libyan case does not establish that a government has such a right, it does establish that it has the power. It soon became evident that Libya proposed to use it. In an interview with a correspondent of the Beirut daily *al-Hayah,* Libyan Prime Minister Maziq, commenting on his government's partially successful effort to win acquiescence by persuasion, issued a warning to companies that remained recalcitrant. He said, "If they continue this attitude, we shall be compelled to issue new legislation which will oblige them to abide by the new system."[28] And he indicated the character of the contemplated law—a prohibition of exports by recalcitrant concessionaires. Meanwhile OPEC at its Vienna conference in December passed a resolution recommending that all member countries refuse any new oil rights of any nature whatsoever "to any company or the subsidiary of any company refusing to comply with Libya's new oil policies."

Confronted by threat of compulsory legislation and the possibility of punitive discrimination of OPEC members, five of the seven recalcitrant companies signified their intention to comply (Marathon, Continental, Pan American, Grace, and Bunker Hunt) leaving only two holdouts, Phillips and Gulf. To bring them into line and to put the power of the state behind

28. *MEES,* December 31, 1965.

the policy announced by its spokesmen, on December 26, 1965, the Libyan Parliament unanimously passed a decree empowering the council of ministers to take such measures as might be necessary to insure compliance with the November law. These measures specifically included "the stopping of production or exportation" by companies not complying (Article 4) and, if necessary, "the expropriation of some or all" of their "physical assets," with the council determining how much and how they would be paid (Article 5). Confronted with such stern penalties, the holdouts surrendered.

Because the major oil companies in their negotiations with OPEC had already agreed to expense royalties as provided in the November decree, they, of course, were not reluctant to see the independent companies forced to comply. But the majors could scarcely have wished to see the principle of expropriation incorporated into a permanent statute. Whether they exerted pressure against the government to prevent this is not known. But its threat having brought compliance, parliament never published its decree, dubbed the "Law of Compulsion," in the Official Gazette and never promulgated it.

The Royalty Settlement

The royalty settlement permitted the companies to discount the posted prices by 8.5 percent in 1964, 7.5 percent in 1965, and 6.5 percent in 1966 in calculating their income tax obligations. This meant that for 1966 the government would receive about five cents per barrel more than they would have received under the pre-1964 arrangements. This compares with an 11-cent increase if OPEC demands had been met completely. The settlement provided, however, for consultation in 1966 between the governments and the companies on possible reductions in the future in the rate of discount. Any agreement reached is to be governed by the "competitive, economic and market situation . . . expected at the time of such consultation" to prevail during future years.

After it heard the report of its board of governors and examined the economic situation in the international oil industry at its eleventh conference, held in Vienna in April 1966, OPEC recommended that "each Member Country concerned take steps toward the complete elimination of the discount allowance granted to the oil companies."[29] In accordance with

29. *OPEC,* XI.71.

this recommendation the Iranian minister of finance, the oil ministers of Libya and Saudi Arabia, Qatar's director general, and OPEC's secretary general accompanied by his economic advisor met in Beirut on August 5, 1966, and decided to initiate negotiations with Aramco in Saudi Arabia. On September 28 they began negotiations. Negotiations began in a leisurely manner and adjourned rather promptly. After the conclusion of the Israeli-Arab war they increased in tempo. At OPEC's extraordinary conference in Rome during September 1967, the conference decided that the five member countries concerned with expensing royalties should meet for consultation on the issue on October 5 and 6. In December representatives of the governments and of the companies resumed negotiations and reported some progress. Negotiations were concerned not only with the elimination of the 6.5 percent tax allowance but with the closely related problem of the gravity differential allowance built into the original royalty-expensing settlement. At a two-day conference in Beirut in January the member countries concerned decided to accept a last minute offer the companies had made only two days before the conference convened. A decision by OPEC members unilaterally to inaugurate a discount phasing out perhaps led the companies to raise their previous offer. Under the agreement reached, the percentage discounts are to be phased out in four years, declining to 5.5 percent in 1968, 4.5 percent in 1969, 3.5 percent in 1970, 2 percent in 1971, and cease entirely in 1972. The gravity allowance is to be eliminated in seven years. Following the closing of the Suez Canal the companies temporarily eliminated all discounts on shipment from Mediterranean ports. Because of this Libya was not a party to the Beirut agreement but planned to enter separate negotiations regarding the issue.

18

OPEC and Oil Proration

THE price cuts of 1959 and 1960, which precipitated the founding of OPEC, had borne more heavily on Venezuela than on the Middle East countries. Although Venezuela was the world's largest exporter of crude oil, its growth in production had suffered to the benefit of the Middle East. While Venezuela's 1960 output increased by a meager 3 percent over 1959, total Middle East production increased by 14 percent. In 1958 and 1959 Venezuela's production was actually less than in 1957. Confronted by the declining importance of his country as an oil-producer, concerned with its loss of national revenue through oil price cuts, and familiar with the American experiment in price stabilization through oil proration, Venezuela's minister of mines and hydrocarbons Dr. Juan Perez Alfonzo was among the first to agitate for an oil proration program that would include all the major exporting countries and particularly the Middle East.

OPEC Sets Proration As a Goal

Venezuelan members have never allowed OPEC to forget its commitment to an output control program. Alfonzo and Tariki publicly advocated such a program at the first Arab petroleum congress in 1959. At its organizational meeting in Baghdad in 1960 the OPEC Conference recognized in its first resolution the need for its members to formulate a program for the regulation of production. At its second conference, held in Caracas in January 1961, OPEC acknowledged that a "just-pricing formula" supported by international proration would require detailed study[1] and, pursuing this idea, it employed an American firm, Arthur D. Little, Inc., to make a comprehensive analysis of profits, prices, and output in the international oil industry and to formulate recommendations for an output control program. The study, although completed, was never made public, but it is widely known that the Arthur D. Little analysis recognized serious obstacles to international proration as a means of increasing the revenue of the oil-exporting countries.

OPEC, not Venezuela, Loses Enthusiasm for Proration

The influence that the Arthur D. Little study exerted on OPEC's policy is a matter of conjecture, but OPEC's inaction is a matter of fact. Apparently the Little study dampened OPEC's enthusiasm for tackling this problem. At any rate, during its next four conferences OPEC remained silent on proration. But Venezuela did not. Alirio A. Parra, chief of the division of petroleum economics of Venezuela's Ministry of Minerals and Hydrocarbons, in a scholarly paper on "Oil and Stability" presented at the Third Arab Petroleum Congress, held in Alexandria in 1961, outlined the aims of an international proration program and analyzed the factors on which its success would depend. Parra's analysis assumed a constantly expanding demand for crude oil. The purpose of proration, as he conceived it, was not to limit current output in order to raise prices but to gear increases in output to increasing demand in order to insure stability of prices. As Parra put it, proration's aim should be each year to distribute equitably among OPEC members the estimated annual increase in production necessary to meet market demand. Under his plan an international co-ordinating committee would make periodic forecasts of market demand for oil in the principal markets similar to those made by the United States Bureau of Mines in the domestic market. After these were made it would be OPEC's task to devise a formula for distributing equitably among member countries the estimated increase in production. The Venezuela Ministry of Mines had already analyzed the criteria, direct and indirect, on

---

1. *Resolutions Adopted at the Conferences of the Organization of Petroleum Exporting Countries,* Caracas, January 1961, Resolution II.13. (Roman numerals indicate the number of the conference and arabic numerals indicate the number of the resolution.)

which equity might depend. OPEC should take account of the previous average growth rate of its members, their production potential, and the magnitude of capital invested in oil within their borders. It should gear allowables positively to these variables. But OPEC should also take account of certain variables having an indirect effect on the propriety of allowables. These should include the relative importance of oil in the national income of member countries, oil's contribution to the countries' governmental revenues, its role in their exports, their per capita income, and the number of employees in their oil industry. Here again OPEC should gear allowables to these variables in a positive manner.

Parra's analysis was intended to be suggestive, not definitive, but since he spoke as a representative of the Venezuelan government, it was presumably official. And it indicated the concern of his government with the problem of output control.

OPEC Moves under Venezuela's Urging

As OPEC's enthusiasm for devising a proration program waned, Venezuela's waxed. At one time the Venezuelan members reputedly threatened to resign if OPEC did not tackle the problem of output control. Tardily it did so. At its seventh conference held in Djakarta in November 1964, at which it referred the agreement reached on expensing royalties to its members for their action, OPEC took its first positive step toward the solution of the problem of overproduction. It stepped cautiously. Noting the continued deterioration in the relation between the price of crude oil, on which the revenues of its members depended, and the price of manufactured goods, on the importation of which the members must rely, it authorized the establishment of the "OPEC Economic Commission" and directed its secretary general to request the commission to formulate recommendations "to counteract continuing erosion of crude and product prices" including, if necessary, the formulation of a production control program as contemplated in its initial Baghdad resolutions.[2]

Not until its eighth conference, held in Geneva in April 1965, did OPEC formally establish the economic commission as a "permanent" and specialized unit of the Organization of Petroleum Exporting Countries and indicate the details of its composition, its functions, and its financing.[3]

2. *Ibid.*, VII.50.
3. *Ibid.*, VIII.55.

Meanwhile, however, the commission had worked out a proposed schedule of allowable output increases for the OPEC region as a whole and had suggested how the increases be divided among OPEC members. When these were presented for discussion at the Geneva conference, they apparently precipitated dissension and bitterness so sharp that OPEC took no formal action on the commission's recommendation.

At its ninth conference, held in Tripoli in July 1965, apparently in a calmer atmosphere, OPEC adopted as a "transitory measure" a production plan calling for a "rational" increase in production in the OPEC area, geared to an estimated increase in world demand, and voted to submit a production schedule for the approval of member-country governments.[4]

Not until August 23 did OPEC release for publication the official action of its April and July conferences, and it has never officially released the detailed production schedules which it submitted to its member countries.

Arab Press Criticises OPEC

While facts about the April meeting were scarce, rumors were plentiful, and they fed the appetites of the critics and propagandists. None was more voracious than the Iraqi *Review of Arab Petroleum and Economics*. In its May issue it editorialized,

The colossus in Geneva is now sick and possibly on its deathbed. Differences among its members are sapping its strength and reducing it to a skeleton. Such differences are no longer papered over but freely admitted by the Organization's spokesmen.

And it propagandized in a manner typical of those who customarily explain Middle East troubles in terms of Western imperialism:

A day will come when we shall hear no more of technical reasons preventing timely publication of [OPEC] resolutions. Instead we may be told in plain English that the United States Government or the oil companies or both have intervened in time to change the direction of OPEC's ship and prevent its reaching its destination.

It concluded,

4. *Ibid.,* IX.61.

Planning of supply is the only effective answer to the companies' cartelized lifting arrangements. If we fail to take away from the companies, through planned production, the dangerous offtake weapon by which they can play one OPEC member country against another, we can never succeed in negotiating a rational price structure or in solving any basic oil problem.

The recent OPEC fiasco is explicable only in terms of the majors' interference through whispering in some Middle East or North African ears that backtracking would be rewarded by increased offtake.

Future generations will curse our politicians for failure to clip the power of companies to decide our fate.[5]

OPEC Explains Its Action

In releasing its eighth and ninth conference resolutions OPEC also released a note on the deterioration of prices, giving its reasons for recommending limitations on production increases and the distribution of allowable increases among its members. It explained,

Resolution IX.61 calls for measures designed to eliminate one of the principal causes for the deterioration of prices: without reducing total supply, they are intended to apply a brake to cut-throat competition among the suppliers of crude who operate in the Member Countries. Such competition has only been possible because of the surplus producing capacity in existence. In effect, what the Resolution proposes is to define the proportions of the total offtake to be taken from a given area.

To allay uneasiness among consumer countries it emphasized that

the production programme envisaged in the Resolution will not in any way curtail supplies of oil to the consumer. On the contrary, as producers, the Member Countries of the Organization have an obvious interest in encouraging the expansion of demand for oil provided that such demand is met on an equitable basis that will benefit both producers and consumers. The Member Countries have ample reserves with which to do so.[6]

But in recommending quotas for output increase to its member countries, OPEC was concerned not only with stabilizing prices through a controlled expansion program but in preventing the international companies from using their control over output as a strategic weapon in dealing individually with its member countries. OPEC called attention to the

5. *Review of Arab Petroleum and Economics,* Baghdad, May 1965.
6. OPEC, Note on Resolution IX.61.

interlocking structure of the major companies which accounted for most of the production in member countries:

> The board of directors of the operating company in one country are nominees of the companies that have nominated the board in another country. It is therefore common practice for them to decide regularly or as the need arises what proportion of their requirements is to be taken from any given country. The existence of surplus producing capacity in the area allows these companies a wide measure of flexibility in allocating their total requirements as between countries. The offtake patterns—and hence the annual revenues—of the producing countries are therefore subject to the decisions of the oil companies.[7]

After presenting annual production figures for the several Middle East countries for 1959, 1960, and 1961 with yearly percentage changes, the OPEC memorandum stated,

> It is clear that although the average percentage increase for the Middle East as a whole did not vary from year to year, the individual offtake figures have fluctuated widely, year in year out. There are various explanations for this, which go beyond purely commercial criteria; but this is obviously too great a concentration of power in the hands of a small group of large international companies.[8]

OPEC proposed that its members limit this power by making joint decisions on the rate of production increase and on its distribution among themselves.

OPEC's Proposed Production Schedules and Their Meaning to OPEC Members

While OPEC has never officially released its proposed production schedule, others have revealed its details. The proposed plan provided for a 10 percent increase in members' total daily output for the twelve months ending June 30, 1966, to be distributed as indicated in the accompanying table.

Apparently these figures represented merely what OPEC members thought their increase in production would be, based on what they had been and what their concessionaires' plans, as they understood them, called for. They were more prognosis than proration. As such they apparently

7. *Idem.*
8. *Idem.*

Country	Allowable Increase	Percent over Previous Year
Iran	304,000 b/d	17.5
Iraq	125,000	10.0
Kuwait	157,000	6.0
Saudi Arabia	254,000	12.0
Qatar	67,000	32.0
Libya	210,000	20.0
Indonesia	48,000	10.0
Venezuela	115,000	3.3

represented a consensus. But different members attached different weights to them and to their significance. Members assessed them in the light of their country's interests. Venezuelan Oil Minister Perez Guerrero, OPEC's staunchest proration advocate, at a press conference hailed the plan as a "victory for Venezuela," which had long exercised "a self-imposed restriction program." And he counted on the "co-operation from the oil companies who will be able to see that this measure will also benefit them in halting the continued world oil price erosion."[9]

Libya, whose production was young, growth rapid, and hopes high, viewed the OPEC program differently. Oil Minister Kabasi characterized it as a transitory production plan based only on "actual real production expected." He declared,

So far as Libya's concerned, there's no production limit, none imposed, and we never accepted one. . . . Libya plans to go ahead to develop production until it reaches maturity. This was specifically accepted by OPEC which agreed that Libya is to have its normal development.[10]

But Libya was not the only OPEC dissenter. Saudi Arabia, whose reserves were rapidly mounting and whose development plans envisaged constantly increasing oil revenues, was loath to accept any arbitrarily imposed limitations on its output. In an exclusive interview with news editor Ian Seymour of the *Middle East Economic Survey* in February 1966 Saudi Arabia's Oil Minister Yamani pointed out that Saudi Arabia had ratified the proposed output schedule on two conditions: that Saudi Ara-

9. *Petroleum Intelligence Weekly,* August 30, 1965, p. 6.
10. *Idem.*

bia's commitment be for only the first six months of the proration period, and that approval for the second six months would depend on Saudi Arabia's receiving an increase in allowable production.[11]

Yamani expressed skepticism on the workability of any proration program that was limited to OPEC areas. Oil companies with alternate sources of supply could readily shift production from OPEC countries to other areas not encompassed by the OPEC program, and they would not hesitate to do so if they found their operations restricted by OPEC. Thus the OPEC plan might do more harm than good to member states.

Anxious to increase its oil revenues and to recapture its previous position as Middle East exporter, Iran was equally critical of the OPEC proration program. In an interview with Wanda Jablonski in the spring of 1965, Dr. Reza Fallah, an NIOC director, expressed the view that creating an artificial shortage of crude would merely speed expansion of other sources of supply, encourage the use of alternative fuels, and in all probability fail to stabilize prices. Oil proration, he said, is "definitely out."[12]

OPEC's Program Fails

Proration was a falcon of hope released by Tariki and Dr. Alfonzo in the spring of 1959. By January 1966 it was a dead bird. At the end of the plan's first year actual production figures for the OPEC countries revealed that Saudi Arabia and Libya had exceeded their allowable increases by substantial amounts. All other countries had fallen short of their allowables.

Since OPEC had sought to curb output increases as a means of stabilizing prices, it might have been regarded as a source of OPEC satisfaction that actual increases for the area as a whole had fallen short of OPEC's approved allowables by more than 2 percent. It was not. In countries where actual output increases fell far short of OPEC approved allowables, notably Kuwait and Iraq, both of which were engaged in controversies with their concessionaires, the suspicion grew that concessionaires were using their control over output as a punitive device. Other countries were troubled over the adverse effects that curtailed production increase would exert on their developmental programs. In particular, Iran,

11. *MEES,* February 11, 1966, Supplement.
12. *PIW,* May 17, 1966, pp. 6–7.

which had striven unsuccessfully since the disastrous days of Mossadegh and oil nationalization to regain her position as the Middle East's largest oil exporter, was disturbed over the consortium's failure to meet Iran's allowable increase, even though only by a slight margin.

As it became evident that 1965–66 production was falling below its approved output, OPEC at its eleventh conference, held in April in Vienna, took official notice of the failure of some concessionaires to meet their "allowables." In Resolution XI.73 it averred that the "unsatisfactory rate of increase of production" in member countries "cannot be ascribed to the lack of outlets for their crude in the international markets." It charged the oil companies with manipulating production "contrary to the national interest" of member countries, and it resolved that OPEC would give "full support" to the efforts of the "countries concerned to safeguard their legitimate national interests."

What began primarily as a program to curb the rate of increase in oil output became one to insure that each country's output would at least reach the level stipulated in OPEC schedules.

The American Experience: An Inappropriate Guide

Conceived as a program to establish a rational control of output by its members, OPEC's "proration" program failed. That it did so is not surprizing. It drew its inspiration initially from the proration experience in the United States, where conditions are quite different. The American program, though its unavowed object is the stabilization of prices, has been successfully defended as a conservation measure before the courts. Customarily in the United States, where the minerals belong to the owners of the land overlying them, property tracts have been the unit for oil exploitation. These have rarely been compatible with an efficient and economical exploitation of the underground oil deposits. Frequently the number of property tracts overlying a single oil reservoir, which is basically a geologic unit, has been large. Inevitably exploitation with the property tract as the developmental unit has meant waste, frequently on a gigantic scale.[13] It has engendered a race among the several owners of drilling rights to get as

---

13. See George W. Stocking, *The Oil Industry and the Competitive System: A Study in Waste* (Boston: Houghton Mifflin, 1925), particularly Chapters VII–X. See also Stocking, "Stabilization in the Oil Industry; Its Economic and Legal Aspects," *American Economic Review, Supplement,* March 1933.

much oil as quickly as possible before a rival producer gets it because once exploitation begins, the oil, actuated by the water or gas pressure, moves readily from place to place. This has not only made for periodic and serious overproduction but it has meant wasteful utilization of the underground forces which propel oil to the surface. The effort of each to get as much oil as possible in the short run has meant that all get less than would be available in the long run. Recurrent and wasteful overproduction has periodically depressed prices—at times almost to the vanishing point. Prices in East Texas fell from over a dollar a barrel when the field was first discovered to less than a dime, and oil went begging at that price. Despite the depressed condition in which the oil industry found itself, voluntary proration, inaugurated in the middle 1920s, failed. Enforced by state and federal statutes, upheld by federal courts, supplemented by the Bureau of Mines authoritative periodic estimates of demand, complemented by interstate oil compacts, and bulwarked by federal restrictions on imports, the proration program ultimately proved an effective device for stabilizing prices. Consumers tolerated it and oil companies supported it. But competition among rival states, each anxious to increase its own revenues and wealth, would probably have brought it down had not the State of Texas, which accounts for about 40 percent of both the nation's reserves and its annual output, played the role of a dominant producer recognizing that its interests are served by output restrictions even though other states pursue a less rigorous production control.

Middle East Conditions Not Conducive to Proration

In the Middle East no such institutional basis exists for an effective proration program. There subsurface minerals belong to the state. Oil concessions are large, embracing thousands of square miles and containing many separate oil reservoirs. To insure a program of development sound by engineering standards does not require the co-operation of the several oil-producing countries. The interests of the state as owner of the underground deposits and the interests of the concessionaires in exploiting them scientifically generally coincide. Where they do not, as sometimes in the conservation of gas, a state's ability and authority to remedy the matter do not depend on the action of other states. In short, oil proration in the Middle East lacks the justification of conservation. It is primarily an output and price stabilization scheme the successful operation of which depends on the effective co-operation of rival states and, more important, on the support of the concessionaires. The Middle East has neither. The conflict of interest among OPEC member states has already been noted. Only Venezuela, a mature producing country whose output when OPEC was born exceeded by far that of any Middle East country and equalled more than 50 percent of all, gave unrestrained support to an output control program. Libya and Iran, as already observed, wanted no restrictions on their output. Nor did Saudi Arabia.

Both Parra in his Arab Congress paper and Yamani in various public statements have recognized the difficulty of formulating a proration program that will divide a restricted output equitably among rival political units. The rate of population increase, living standards, social needs, the significance of oil revenue to gross national product and governmental revenues, the size of reserves, their potential production, historical growth rate—all affect a country's attitude toward output limitation. The problem of measuring and weighing these and other relevant factors makes it *impossible* to devise a formula that will *equitably* distribute allowable production and *difficult* to devise one that will *acceptably* do so. By contrast the American problem is simple. Texas allowables are geared to a maximum efficient rate of production in each field and to oil companies' nominations as to the amount of oil they are willing to buy. The state's regulatory commission holds hearings and determines allowable production at frequent intervals. But assuming that, despite the conflict of interests among them, the oil exporting countries of the free world could devise a formula acceptable to each of them, proration would still lack an essential ingredient for its success—the co-operation of the producing companies. And this OPEC's "transitory" measure has lacked.

Corporate executives zealously guard their prerogative of making independent decisions in matters affecting their corporate welfare. Oil executives were not consulted by OPEC in devising a "transitory" production control program and they at no time gave their approval to it. Some, apparently convinced of the plan's probable failure, thought the less said about it the better. Some made it clear to their host governments that under the concession contracts they alone have discretion on output decisions. And at least one, Aramco, apparently confident of its legal right under its concession agreement to produce and export as much oil as in its business judgment it thinks wise free from government intervention, and anxious to establish that right beyond all dispute, proposed that the issue be submitted to arbitration in accordance with procedures provided in the concession contract.

The Saudi government, noting that it had approved the OPEC quota only for a trial period of six months, demurred, suggesting that if at some future date the issue should become vital, it would explore the question of arbitration.[14]

OPEC Puts the Proration Program to Sleep

Without abandoning formally its proration program, OPEC quietly laid it to rest. Proceedings of the eleventh conference, held in Vienna in April 1966 make no mention of proration, and the only resolution bearing indirectly on the subject is that previously referred to in which OPEC chides concessionaires for having failed by producing too little oil to protect the national interests of some member countries. Without acknowledging in its proceedings that it had done so, it apparently approved production schedules for a second year virtually unchanged from those of its first. It seemed equally apparent that neither the companies nor the countries expected to abide by them.

14. *MEES,* Saudi Oil Minister Yamani's interview with Ian Seymour. Yamani said that oil companies operating in Iran, Libya, and Qatar had made similar requests that the issue be arbitrated. What the host governments' responses were, he did not know.

PART IV

THE ECONOMICS OF
MIDDLE EAST OIL PRICING

19

The Problem of Oil Pricing

As previously indicated, it was the August 1960 reduction in Middle East crude oil prices that occasioned the birth of the Organization of Petroleum Exporting Countries. Since its organization the restoration of prices has been continuously one of OPEC's major aims. OPEC has based its quest for higher prices on the assumption, sometimes implied, sometimes expressed, by OPEC officials or self-appointed spokesmen for the Middle East oil industry that Middle East oil companies acting as oligopolists or in collusion set prices at monopoly levels and that they reduced prices with no loss to their over-all integrated operations but with significant loss of revenues to the Middle East countries.

To evaluate OPEC's price-raising goal and the economic logic on which it is based, it is necessary to examine briefly the prewar corporate quest for stability in the oil industry and the pricing formula by which the international oil companies priced crude and its products.

The Quest for Corporate Stability

The international oil companies regarded the stabilization of international markets as an essential auxiliary to the domestic stabilization program they engineered with the help of both state and federal governments during the late 1920s and in the 1930s. By the close of the 1930s limitation of output and proration of allowable production had become essential features of the conservation programs of the leading oil-producing states. Spokesmen for the industry had come to regard oil proration as a statesmanlike program protecting the interests of all concerned.

In 1928 oil men took steps to translate their concern about price instability in international oil markets into a program of action. It has already been related how Sir John Cadman of Anglo-Persian, Sir Henri Deterding of Royal Dutch-Shell, and Walter C. Teagle of New Jersey Standard Oil, while grouse shooting on Sir John's estate at Achnacary, talked about the problem of overproduction and what to do about it and how their conversation eventuated in the so-called "As-Is Agreement." One of the agreement's major objectives was to maintain existing relationships among international oil marketers and at the same time insure a more efficient utilization of existing facilities. To achieve the latter, participants were to supply petroleum products from the most economical source. Marketers with inadequate supplies in any area were to obtain their needed supplies from companies having surplus productive facilities in that area instead of shipping them from an area where the marketers might have surplus facilities of their own. In this way producers would retain the advantage of geographic locations, and needless transportation costs would be eliminated. To insure identical pricing among rival sellers, one of the basic elements in the agreement provided that

each group shall be paid f.o.b. port of shipment for each product on the basis of Gulf (Texas) price; or if the goods are supplied c.i.f. port of import, the marketing organization shall pay to each group this price plus the freight rates scheduled for the port of import.[1]

Because the company might use its own tankers, ship by tankers on long-term lease, or by tankers hired in a spot market for particular cargoes, its costs varying with the method used, and because freight costs involved in the several methods of shipment would fluctuate over time, the question of what freight rate to be used in calculating c.i.f. price posed a problem. To solve it the agreement provided for the preparation of a schedule of "relative freight rates from each port of shipment to each port of import."[2]

1. *The International Petroleum Cartel,* staff report of the Federal Trade Commission, submitted to the Subcommittee on Monopoly of the Select Committee on Small Business, U.S. Senate, Government Printing Office, Washington, D.C., 1952, p. 205.
  2. *Idem.*

Although the "As-Is Agreement" never became fully operative, it set a pattern of pricing to which the international oil companies conformed in a generally effective way for more than two decades. To say this is not to contend as does the Federal Trade Commission that the "As-Is Agreement" represented an illegal cartel operating more or less effectively until the Department of Justice brought suit against the participants in the agreement. Nor does it conflict with the historical fact that delivered prices for crude oil were based on Gulf Coast prices for many years before 1928 nor with the fact that this method of pricing found its logic in the early structure of the oil industry. As long as Texas was the principal oil-producing state and virtually the only place where foreign importers could obtain adequate supplies, prices tended to equalize at the port of export and delivered prices represented the Gulf Coast price plus ocean transportation costs. But in the As-Is Agreement we find the first evidence of a conspiratorial arrangement to perpetuate a pricing system that was breaking down under the impact of surplus world production and increasing competition.

A Free Market for Crude Oil Lacking

Until after World War II a market for Middle East crude in the economist's conception of the term had not developed. Fully integrated companies owned the concessions and produced mainly for the benefit of their subsidiary or affiliated refinery and marketing units. Buyer-seller bargaining relationships of a free market were lacking. Subsidiary refining and marketing units were not free to choose alternative sources of supply or exercise their bargaining power to secure the most favorable terms possible. The producing units made few if any normal commercial sales. As we have noted, several of the companies had negotiated long-term contracts with each other which more closely identified their common interests. As the oil moved from the production units to its ultimate market, the transfers that took place were not for the most part in the arms-length transaction of a free market, but were frequently identified as intracompany billings and were essentially bookkeeping transactions. The company operations were worldwide but the management was highly concentrated with executives in New York or London making decisions on price and production, sources of shipment and the like. Whether to supply a particular market from Venezuela, the Middle East, or the United States was frequently a matter of intracompany or even intercompany convenience depending on local as well as worldwide factors. Petroleum markets were in some cases more centralized than the control of production; ordinarily not more than four and sometimes only two of the major companies marketing in any one country. For example, at the close of the war only three companies—Royal Dutch-Shell, Jersey Standard, and Socony—marketed in the whole of the Eastern Hemisphere east of the Suez. Until after the close of the war two companies accounted for almost the whole of Middle East oil production. Anglo-Iranian, Iran's sole producer, refined its oil at the Abadan refinery or transported it to refineries located at various points in the British Empire where it conducted marketing operations. Anglo-Iranian was the principal supplier of bunker fuel oil to the British Navy and negotiated prices under a long-term contract. When exploitation of Iraq oil began in 1934 with the completion of a pipeline from Kirkuk to the Mediterranean, the owners of IPC—Anglo-Iranian, Jersey Standard, Socony, and CFP—used the Iraqi output primarily to supply their subsidiary refining and marketing facilities.

Basing-point pricing on the pattern suggested in the As-Is Agreement offered a convenient technique for insuring that the integrated companies would have similar bookkeeping costs and a common delivered price in the transport of crude oil or refined products to their subsidiaries. It facilitated a common pricing policy in the sale of products and of crude as and when it was sold outright. As one noted authority described it, "There we have all the paraphernalia of a full-fledged system of regional pricing based on the somewhat theoretical Gulf quotations."[3] Its successful operation depended on identical freight-rate schedules available to all shippers. The International Association of Tanker Owners provided these before the war; the United States Shipping Administration and a committee of London brokers after the war.[4]

During World War II the British Navy became an important buyer of bunker oil, much of which was delivered at various points in the Eastern

3. P. H. Frankel, *Essentials of Petroleum* (London: Chapman & Hall, 1944), p. 116.
4. *Ibid.*, p. 162. For a detailed treatment of the logic of basing-point pricing in general, consult Fritz Machlup, *The Basing-Point System* (Philadelphia and Toronto: The Blakiston Press, 1949). For the logic of the system and its significance to pricing for iron and steel, consult George W. Stocking, *Basing Point Pricing and Regional Development* (Chapel Hill: The University of North Carolina Press, 1954). For a critical analysis of basing-point policy in international oil markets and of price movements, 1947–57, consult M. A. Adelman, *The World Petroleum Market*, Chapter V. (Manuscript scheduled for publication in 1970 or 1971.)

Hemisphere. Under the system of pricing the British Government paid the Gulf Coast price plus transportation cost to the Persian Gulf even though the delivery was made at Abadan on the Persian Gulf. The director general of the British Ministry of Transport, disturbed by this situation, began an investigation of oil prices. The oil companies failed to supply him with cost data, choosing apparently to reduce the price of bunker oil at Abadan to that prevailing at the Gulf Coast.[5] In effect, they adopted the Persian Gulf area as a second basing point with prices identical with prices at the Gulf Coast.

Middle East Surplus Threatens Domestic Stabilization Program

By the close of the war the international pricing pattern for Middle East petroleum products was well established. The pattern of crude oil pricing was less well defined. During the war it became increasingly apparent that in the Middle East the international oil companies would soon have an unlimited source of low-cost crude. Within a year after the war's close the Middle East was producing more than three times its 1940 production and its reserves had barely been tapped. Controlling oil reserves destined to become primary sources of supply for Western Europe and much of the rest of the world, the oil companies recognized the significance to the world oil industry of the production and pricing policy that they would use in pricing Middle East oil. To have permitted oil to flow into world markets without restriction, with price governed only by production costs, would have been disastrous to the stabilization program that the American regulatory agencies had so meticulously constructed during and before the war and to the Middle East companies themselves whose operations were worldwide and whose capital investments elsewhere were predicated on relatively high and stable prices. Having accepted during the war the principle of a dual basing-point system with the Persian price equal to Texas Gulf Coast prices, it is not surprizing that in their postwar price they accepted Texas Gulf prices as a guide.

European Co-operation Administration's Concern with Oil Prices

The year in which the United States launched its European recovery program witnessed a significant shift in the flow of international petroleum

5. *Petroleum Times,* May 13, 1944, p. 298.

into world trade. It was not only the first year since 1921–22 that the United States was a net importer of crude oil but it marked the emergence of the Middle East as a significant supplier of crude oil to the United States and signaled the end of its historical reliance on Latin America as the only source of imported crude.

It also focused the attention of the European Recovery Administration on oil prices. The most important single commodity to be financed under the recovery program was petroleum and its products. Total purchases by ECA eventually amounted to 11.4 billion dollars, more than 10 percent of which was petroleum.

As an economic consultant to ECA and a member of its panel set up to advise on petroleum prices, I had occasion to familiarize myself at first hand with ECA regulations, with the various petroleum price moves, and with the arguments with which the oil industry justified them. In the interval between the close of the war and the inauguration of the European recovery program prices advanced at the Gulf Coast more rapidly than on the Persian Gulf where the pressure of surplus capacity tended to restrain price advances. A differential had developed between the two sets of prices. When the recovery program became operative in April 1948, an f.o.b. price of $2.22 a barrel for 36° gravity oil at the Persian Gulf prevailed on virtually all shipments. The Economic Co-operation Act of 1948, as amended and the regulations of ECA established standards by which to determine the reasonableness of the prices charged on sales to ECA. They were to be no higher than the market price prevailing in the United States at the time of purchase adjusted for differences in the cost of transportation, quality, and the terms of payment; they were to be no higher than the price charged by the supplier on comparable sales to any other purchaser; and they were to approximate as nearly as practicable the lowest competitive price at the source of supply.

International Oil Companies Lower Prices and Standard of New Jersey Explains Pricing Formula

After consultation with the international oil companies ECA explained the price of $2.22 as equaling the crude oil price f.o.b. United States Gulf plus freight at the United States Maritime Commission rates to the United Kingdom less freight rates from the Persian Gulf to the United Kingdom. Since the United States was a net importer of oil and Venezuela was the source of large shipments to Europe, this afforded only an artificial explanation of the price. Under pressure from ECA the companies lowered the price at Ras Tanura in the Persian Gulf to $2.02. In response to a congressional investigation of oil prices, Jersey Standard explained and justified this price as follows:

> To our knowledge there is no uniform "world pricing system" for petroleum. The company's prices are established independently and reflect competitive market conditions world wide. . . . For your information I am outlining herewith the way in which our prices to Europe and ECA countries are established.
>
> Our prices for supplies from the various sources are arrived at as follows: crude oil from the Caribbean areas for all destinations sold f.o.b. Caribbean supply points at the competitive Caribbean price . . . are announced f.o.b. prices for crude oil supplies at the eastern Mediterranean or Persian Gulf and equated to the Caribbean price for crude plus freight at established U.S.M.C. rates from the Caribbean to Western Europe less freight on the same basis from Western Europe to the Persian Gulf.[6]

Other sellers of crude financed by ECA priced oil by the same formula.

Shipments from Middle East to United States Increase

Meanwhile the major oil companies had begun shipping oil in increasing quantities to the United States. Because of the longer freight haul the net back or realized f.o.b. Persian Gulf price on such shipments was considerably below that of ECA financed shipments to Western Europe. This discrepancy became a matter of discussion and controversy between the companies and the ECA administrators and ECA sought the advice of its panel of experts.

Meanwhile ECA had adopted a general policy of not financing purchases at prices higher than those regularly charged by a supplier on comparable transactions to other customers. Because of the urgency of the supply situation ECA in September 1948 at the request of the oil companies had waived this requirement on oil purchases. It now decided to revoke its waiver. Thereupon, effective July 15, Gulf Oil reduced its Kuwait price for 36° gravity crude to $1.75. In September Jersey, Caltex, and Socony made similar reductions.

6. Mutual Security Agency, *Petroleum Price History*, p. 4.

Standard Oil of New Jersey justified to ECA and its consultants the $1.75 price for its Persian Gulf crude as follows:

Middle East [crude oil] production is [now] sufficient to cover Eastern Hemisphere crude requirements. Therefore, the price of that crude might be expected normally to fluctuate between a high (which would just permit Western Hemisphere crude to be shipped to Western Europe) and a low (which would just permit Middle East crude to be shipped to the Western Hemisphere). The competition between companies selling Middle East crude has brought the present price near the low of this range. Western Hemisphere crude oil—except for specialty crude—is not now a substantial competitor for European business.

Although there was disagreement within the panel, it gave its tentative sanction to the $1.75 price.

ECA Discontinues Financing Oil for Europe

When in the latter half of 1950 freight rates substantially increased and prices realized on deliveries to the United States (which had not only continued but increased) declined, ECA became uneasy over the $1.75 price at which it had continued to finance sales to Europe. It accordingly requested that the companies supply it with documents covering sales to the United States. Relying on this documentary evidence, ECA alleged that the companies all along had been billing shipments to the Western Hemisphere at a uniform price of $1.45 while charging $1.75 on its ECA-financed shipments to Western Europe. Thereupon ECA began negotiations with the supplying companies for reimbursement of the alleged overcharge, totaling in excess of $66,000,000. On June 21, 1952, the Mutual Security Administration (ECA's successor) suspended new procurement authorizations and announced that after August 31, 1952, it would discontinue all financing of nonspecialty crude oil for European recovery. Thereafter ECA policy presumably exerted no influence on the price of Middle East oil.

During the next two years the price of 36° gravity Middle East oil remained unchanged. In the latter part of 1952 it moved upward, as did oil of similar grade on the Gulf Coast. Thereafter for the next several years Persian crude prices moved in unison with United States Gulf prices. They remained so geared until after the Suez Canal crisis.

20

The Breakdown of Stable Prices

GEARING world oil prices to United States Gulf prices depended on two factors: a concentration of ownership in the principal areas of production outside the United States adequate to insure common pricing policies and a method of pricing that would insure common pricing practices. The structure of the international oil industry and the contractual relationships existing among the major companies supplied the one; basing-point pricing supplied the other. With only a few firms controlling production both in Venezuela and the Middle East, they apparently realized that the best interests of each could be served by promoting the welfare of all. In basing-point pricing they found a convenient device for translating a policy of mutual betterment into uniform market practice.

Prices at the Persian Gulf have only an arbitrary relationship to production costs. They contained a substantial element of what may be called monopoly profits. Equally important, they afforded a price structure convenient for the "profit-sharing" program which the oil companies had worked out with the Middle East governments. The structure began to collapse in 1959, and with the price reduction of August 1960, Middle East prices were freed from dependence on Caribbean prices. It was this breakdown that the Middle East governments protested and that gave birth to OPEC.

American Independents Gain Foothold in Middle East

By the end of the 1950s the pressure of prolific Middle East reserves was mounting and the conflict of interest of the companies holding reserves was becoming more apparent. Moreover, significant changes were taking place in the structure of the international oil industry itself. At the beginning of the 1950s seven integrated companies, operating on a worldwide basis, together with Compagnie Française des Pétroles (which sold its oil in a protected market), produced all the Middle East's output. Selling their low-cost oil at prices geared to the Gulf of Mexico or Caribbean ports gave them profits that were the envy of oil producers less favorably situated. With United States output rigorously curtailed under proration programs and with a political environment at home encouraging to foreign exploration, American oil men, attracted by the profit potentialities of low-cost oil, went abroad in a search for it that was unparalleled in scope and intensity. When World War II ended, twenty-eight American companies in seventy-eight foreign countries engaged in oil exploration or production abroad.[1] By 1958, 190 American companies were exploring for or producing oil in 91 foreign countries. Perhaps the most significant of these ventures for Middle East oil were those of the Getty Oil Company (formerly the Pacific Western Oil Corporation) and the American Independent Oil Company. The Getty Oil Company, controlled through an intricate corporate structure by J. Paul Getty, one of America's thriftiest, most successful, and truly independent oil men, obtained in February 1949 a concession covering King ibn Saud's interest in the Kuwait-Saudi Arabian Neutral Zone. King ibn Saud had previously granted drilling rights to Aramco, but Aramco, apparently attaching more importance to Saudi Arabian offshore areas, had relinquished its concession in the Neutral Zone in 1948 in return for an extension of its concession to the offshore areas. With Aramco out, Getty promptly moved in.

Meanwhile Ralph K. Davies, vice-president of the Standard Oil Company of California and distinguished deputy administrator of the Petroleum Administration for War, organized the American Independent Oil Company (Aminoil), a Delaware corporation capitalized at $100 million, owned by himself and J. S. Abercrombie and eight American oil companies

1. Leonard M. Fanning, *The Shift of World Petroleum Power away from the United States* (Pittsburgh: Gulf Oil Company, 1958), p. 7.

which for the most part had hitherto confined their activities to the American continents and which, with the exception of Phillips Petroleum Company, were relatively small companies. In June 1948, Aminoil obtained a concession covering Kuwait's half interest in the Kuwait–Saudi Arabian Neutral Zone and, under an arrangement later worked out with the Getty Oil Company, began exploring for oil. After drilling five dry holes and conducting extensive exploratory operations, Aminoil brought in the discovery well on April 13, 1953. Early in 1954 Aminoil shipped its first 100,000 barrels of crude to Japan—an event that Davies with prophetic foresight hailed as a "milestone in the history of the independent oil enterprise abroad."[2] By the close of 1959 the concessionaires were producing 116,000 barrels of oil daily from 122 wells and had opened an estimated reserve of 6.5 billion barrels.

The agreement of August 1954 by which the controversy between the Anglo-Iranian Oil Company and the Iranian government was settled contributed similarly to a change in the structure of the international oil industry. The Anglo-Iranian Company, sole owner of the concession before the controversy, received only a 40 percent interest in the consortium. As previously noted, the remainder went to the Royal Dutch-Shell group, Compagnie Française des Pétroles, and five American companies already operating in the Middle East—Jersey Standard, Socony-Vacuum (later Socony Mobil Oil), Standard of California, Gulf Oil, and Texaco. Later each of the American companies surrendered a 1 percent participation which was allocated to Iricon Agency, Ltd., a corporation wholly owned by Aminoil, certain owners of Aminoil, and five relatively small American companies hitherto confining their operations to the Western Hemisphere. These two developments brought thirteen new companies into the production of Middle East oil. More important than their number was their position in the oil industry. They were all relatively small, and all had hitherto confined their marketing operations to the United States.

Arabian Oil Company Obtains Concession

Not only were independent American companies entering the Middle East fields in increasing numbers but the nationals of other countries were striving to obtain a foothold there. Notably successful was the Arabian Oil

2. *Oil and Gas Journal*, December 21, 1953, p. 102.

Company, backed by Japanese capitalists. As previously indicated, the Arabian Oil Company in December 1957 signed a concession agreement with Saudi Arabia granting it exploration and production rights covering Saudi Arabia's interest in the offshore area of the Neutral Zone in the Persian Gulf. In July 1958 it obtained a similar concession covering Kuwait's interest. Although both concessions were more liberal in their terms than the traditional concession of the international oil companies, whose spokesmen regarded them as unworkably onerous, the Arabian Oil Company in July 1959 spudded in its well No. 1; by January 1960 at a depth of 5,000 feet brought in a 6,000-barrel well; by the close of 1962 had completed 34 producing wells with an average of more than 6,000 barrels daily; and by the spring of 1964 had a daily productive capacity of 240,000 barrels. With his company's reserves estimated at between 5 and 10 billion barrels and with an assured place in the rapidly growing Japanese market, AOC's president could report to his shareholders in the spring of 1964 that their company "has a high future full of big dreams and hopes."[3]

By these several developments Middle East oil producers were not only becoming more numerous but more diversified in their interests and outlook. Unlike the established international companies, whose long-run concern is with the availability of underground reserves in relation to probable future demand and which manifest a farsighted concern over stability in international markets, the newcomers were more interested in what oil could do for them today than what it might do for them tomorrow.

Developments in Venezuela

While these developments in the Middle East were changing the structure and character of the international oil industry, developments in Venezuela were working to a like end. Shortly after President Medina assumed office in 1941 he announced his intention of revising Venezuela's petroleum legislation in an effort to increase the government's share in the profits of the industry.[4] Thereafter the government, by a persuasive use of the carrot and the stick, induced concessionaires to convert two-thirds of their

---

3. For a detailed discussion of the Arabian Oil Company consult George W. Stocking, "Arabian Oil Company: Progress and Prospects," *Middle East Economic Survey, Supplement,* September 10, 1964.

4. Edwin Lieuwen, *Petroleum in Venezuela* (Berkeley: University of California Press, 1954), pp. 91–93.

old concessions (about 6 million hectares) to the new law and to surrender the remainder. By this procedure concessionaires who had obtained their original grants under a variety of terms conforming to the requirements of ten different laws enacted between 1910 and 1938—laws expressing varying degrees of concern or indifference to the country's welfare—began to operate under uniform standards. In 1944 the Venezuelan government invited bids on new concessions under terms better calculated to protect Venezuela's interest in its most important natural resource.

More rigorous terms did not dampen the enthusiasm of bidders. In ten months they had contracted for exploitation rights to 6,500,000 hectares and some had agreed to royalties as high as 33⅓ percent and substantial exploration taxes as well. Although in granting concessions the government showed a preference for newcomers, established companies were among the successful bidders.

With the overthrow of the Medina administration in 1945 the government reversed its policy on concessions. But the basis had already been laid for the greatest oil boom that Venezuela had yet experienced. Because the 1945 law required concessionaires to release one half of their acreage within three years after the granting of the concession, concessionaires began exploration on a scale and a scope hitherto unknown in Venezuela. In 1945 oil prospectors had twenty-seven seismograph and nine gravity-meter parties at work.[5] As a result of these intensified activities Venezuela's production increased by approximately 100 percent in the seven years from 1945 to 1950.

Although Venezuela's oil output more than doubled in the ten years following the exploration boom under Medina's administration, reserves did not keep pace with mounting production. Moreover, Venezuela's production was concentrated largely in Venezuela in the Lake Maracaibo area of the State of Zulia, where virtually all of the oil was of low specific gravity, and in eastern Venezuela in the States of Anzoateguri and Monagas, where lighter oils were found but production was less prolific.

Venezuela Again Invites Bids for Concessions, and Companies Eagerly Respond

To improve its reserve position, insure a better geographic distribution of production, a greater diversification of oil grades, and an expansion of domestic refining, and to obtain higher exploration taxes and royalties with

5. *Oil and Gas Journal,* December 29, 1945, p. 265.

which to finance an expansive development program, the government invited bids in early 1956 on eleven designated blocks in the Lake Maracaibo area. Despite the more onerous terms exacted of them, prospectors responded with avidity. In its invitation for bids the government set a minimum price of 4,000 bolivars per hectare (about $484 per acre). The successful bidders on the first eight concessions paid approximately twice that amount. In the language of the *Oil and Gas Journal,*

The established companies were as eager to expand their operations as newcomers were to get a toehold in Venezuela, and their desire was whetted by a decade of waiting.

    The situation shot bids sky high, well beyond the reach of many companies which are now watching for the government's next move.[6]

But not beyond the reach of all. Some newcomers combining their efforts obtained choice tracts and paid well for them. The Signal Oil and Gas Company, representing a four-company combine consisting of itself, the Hancock Oil Company (merged with Signal in 1958), Sohio Petroleum Company (Standard Oil Company of Ohio), and Pure Oil Company, paid more than $33 million for a Lake Maracaibo concession covering 27,614 acres, the highest sum paid by newcomers for any tract. The Sun Oil Company, the Atlantic Refining Company, and Texaco Seaboard, Inc., together paid more than $18 million for a 24,710-acre tract and almost $2 million for a 2,079-acre tract in Lake Maracaibo. As a bonus or "special advantage to the nation" they made a gift of $500,000 to Venezuela's program for the development of a petrochemical industry. The San Jacinto Oil Company, joining forces with the Lion Oil Division of the Monsanto Chemical Company, the Murphy Corporation, the Tennessee Gas Transmission Company, the Union Oil and Gas Corporation of Louisiana, the Sharples Oil Corporation, and American Petrofina, Inc., paid $3,873,787 for 1,969 acres in Lake Maracaibo. To the Superior Oil Company went the distinction of paying the highest per acre price. For 7,776 acres it paid $22,019,834 or more than $2,846 per acre.[7]

    Gratified by the industry's response to its 1956 invitations and in keeping with the policy inaugurated by Jimenez, the government invited bids in the spring of 1957 to 89 separate tracts totaling 2,500,000 acres. While the 1957 concession program was not so successful as its predeces-

    6. *Ibid.,* July 2, 1956, p. 77.

    7. *Summary of Concessions,* Venezuelan Oil Scouting Agency, supplied by Venezuelan Embassy, Washington, D.C.

sor, it brought additional newcomers to Venezuela. The Continental Oil Company, joining with Texaco, the Ohio Oil Company, Cities Service, and Richfield, through Paria Operations, Inc., paid $101,637,699 for the privilege of exploiting 152,004 acres in the Paria Gulf area of Venezuela. This was the largest single bid in the second offering. The Phillips Petroleum Company, itself a newcomer to Venezuela, joined forces with Sunray Mid-Continent Oil Company, Ashland Oil & Refining Company, Kerr-McGee Oil Industries, Inc., El Paso Natural Gas Products Company, Western Natural Gas Company, Pacific Petroleums, Ltd., and Canadian Atlantic Oil Company, Ltd., newcomers all, to pay $44,695,793 for three tracts—one in Lake Maracaibo, one in Monagas, and one along the Colombian border—totaling 98,840 acres. Venezuelan Sun Oil Company was again a successful bidder, this time in partnership with Pan American Petroleum Corporation (a Standard Oil Company of Indiana subsidiary), Atlantic Refining Company, and Seaboard Oil Company. The combine paid more than $32 million for a Lake Maracaibo tract and $485,437 for a border tract.

Neither were the established major companies, which had learned over the years just how profitable Venezuelan oil could be, indifferent to the new opportunities that the 1956–57 invitations presented. Mene Grande Oil Company (Gulf Oil) paid $136,413,365 for 225,685 acres; Creole Petroleum Company (Jersey Standard) paid $76,913,318 for 471,746 acres; and Shell paid $80,661,822 for 183,157 acres. As "special advantages" Mene Grande agreed to furnish certain technical services to the government, and Shell agreed to supply 2,110,000,000 cubic feet of gas annually for general distribution and for the country's budding petrochemical industry.[8] Together the newcomers and old-timers paid $685,221,900 for the privilege of exploring and exploiting the oil rights on more than 2 million acres of land.

Significance of the 1956–57 Concessions

The invitation bidding wrought substantial changes in the structure of the Venezuelan and eventually the international oil industry. It brought some thirty new operators into the world market at a time when the United States through its import policy was rapidly curtailing that market.

8. *Oil and Gas Journal,* March 25, 1957, p. 118.

The holders of new concessions lost little time in ascertaining through the drill just how profitable their investments would prove. The results were gratifying. By June 1958, Superior Oil Company and Venezuelan Sun Oil Company had each brought in a well with a daily flow of more than 10,000 barrels. Before the close of 1958 Superior had produced as much as 67,000 barrels daily from its Lake Maracaibo tract and Venezuelan Sun had completed fifteen wells, only three of which had an initial production of less than 3,000 barrels a day. The San Jacinto group was producing as much as 22,000 barrels a day from five wells. Altogether the newcomers were producing more than 100,000 barrels a day in the Lake Maracaibo area alone.

The significance of the newcomers is indicated by the shift in distribution of Venezuela's annual production over the years. In 1952 Creole, Shell, and Mene Grande produced 90 percent of Venezuelan oil. In 1958 they produced only 82 percent. Oil output in Venezuela was rapidly passing from the hands of the international companies with established markets, which had long concerned themselves with the protection of the price structure, into the hands of newcomers concerned with gaining a place in world markets. As their production mounted, the newcomers found themselves hard pressed to find buyers for their crude at existing posted prices. They did not hesitate to sell for less.

Russia Increases Oil Exports

While Venezuela producers were clamoring for customers, Russia was conducting what oil men referred to as "the Soviet oil offensive." Before World War II Russia's crude production ranked second only to that of the United States. In 1932 with a crude production of 154,775,000 barrels Russia exported more than 25 percent of her total supply of crude oil and refined products. By 1940 its crude production had increased to 218,600,000 barrels, virtually all of which it consumed within its own boundaries. During the war production decreased substantially and exports dried up. In the postwar decade Russia increased her crude output in every year and when the Suez crisis occurred was producing at an annual rate of more than 600 million barrels, four times its 1944 output. As its output increased, Russia began an intensive campaign to increase its share of world markets. This campaign had a twofold aspect—one political, the other commercial. On the one hand it represented an effort to expand

Russia's political influence in neutral countries or to undermine United States' influence among those countries committed to the West. On the other hand it represented an effort to improve Russia's trading position in world markets. Whether the one or the other objective, the techniques were pretty much the same. Russia offered oil at lower prices than the international oil companies customarily charged, arranged bilateral barter agreements of a character not ordinarily resorted to by the international oil companies, and offered credit terms that the oil companies were unwilling or unable to match.

In its courtship of Egypt following the Suez crisis, in the fisheries dispute between Iceland and the United Kingdom, in its effort to build up good will in India, in its 1961 trade agreements with Ceylon and Pakistan, Russia either supplied oil on favorable barter terms or offered to sell at lower prices than customarily charged by the international oil companies. Nor were its efforts confined to the underdeveloped countries. The USSR penetrated the markets of Western Europe and of Japan by barter agreements, by direct government to government deals, by sales to national oil companies and independent oil brokers. At the Second Arab Petroleum Congress, held in Beirut in 1960, the president of the Soviet oil trust frankly acknowledged that Russia's goal was the recapture of its previous market share.

North Africa Becomes an Important Producer

Meanwhile producers of the free world had begun an assiduous search for oil in North Africa. Algeria, which in 1950 reported a total production of only 24,000 barrels, had increased its output to 3,420,000 barrels in 1958. Developments in Libya were more significant. Under Libya's mineral law of 1953 nine international oil companies had begun geological reconnaisance in Libya. By 1958 Libya had granted 77 concessions to 17 oil companies or groups of companies, some of which were taking their first significant step in the international oil arena. By 1959 concessionaires had completed six discovery wells; by mid-1961 they had completed 109 wells with a tested daily production of 185,000 barrels. Libya's oil boom, which by 1966 was to account for 1,500,000 barrels daily, was in full swing.[9]

9. Petroleum Commission, *Petroleum Development in Libya,* 1959 through mid-1961, pp. 17–22.

Argentina Seeks Self-Sufficiency

While the number of oil producers and their annual output were increasing in the Middle East and North Africa, similar developments on a smaller scale were taking place in South American countries. Argentina, for example, had begun a program designed to eliminate reliance on foreign oil. By the end of 1961 this program had paid off. Argentina's oil production was triple that of 1958 and the cost of oil imports had been reduced from $280 million in 1958 to $97 million in 1961.[10] Argentina had even begun to export a little fuel oil.

United States Restricts Imports

As Argentina was striving to cut down on its crude imports, the United States was conducting a more significant program of import curtailment. Mounting oil imports during the 1950s disturbed greatly independent oil producers, who alleged that oil imports threatened to impair the national security. In February 1955 the President's Committee on Energy Supplies and Resources reflected this concern in reporting that if imports should exceed significantly their 1954 ratio to domestic production, the national security would be jeopardized. Responding to mounting pressure, the president, under authority granted him by Congress in the Trade Agreements Extension Act of 1955, instituted a voluntary import program following the Suez crisis in 1957. On the failure of the voluntary program to reduce total imports of crude and refined products, the president on March 10, 1959, issued Proclamation 3279. With subsequent amendments this proclamation established a mandatory restrictive import program, tying imports to domestic production and dividing allowable imports among domestic refineries under a transferable quota system which permitted them to share in the profits to be derived from the use of low-cost Middle East oil.[11]

The effect of these several developments—an increase in foreign oil

10. *Petroleum Press Service,* April 1962, p. 135, Table II.
11. *An Appraisal of the Petroleum Industry of the United States,* U.S. Department of the Interior, 1965, pp. 7–11.

reserves, particularly in the Middle East, an increase in the number of producers of foreign oil with more of the production falling into relatively weak hands, Russia's effort to recapture her foreign markets, and a limitation of the market for foreign oil through import restrictions—was an intensification of rivalry in international oil markets.

Conflict of Interests Among Major Internationals

As the solidarity of the industry was weakened by the process of accretion, the community of interest among the firms was also greatly weakened. Mounting oil production in more numerous hands eventually engendered a more vigorous rivalry among them.

Although the corporations that originally controlled Middle East production were all integrated companies producing crude oil, transporting most of it to their own refineries and marketing its products throughout the world, their integrated operations were not equally well balanced. Some refined more oil than they produced, buying to meet their refining and marketing needs from others or from independent producers, largely in the United States. Companies that were net buyers of crude oil had conflicting interests within themselves: as buyers they wanted cheap oil; as sellers of petroleum products they wanted stable prices. Throughout most of the 1950s the integrated companies customarily billed the crude with which they supplied their refineries at posted prices, even though they produced it at much lower cost or bought it on long-term contracts at less than posted prices. All the major companies followed common costing practices, and all were interested in stable prices for their refined products at levels adequate to yield a reasonable profit on their refining and marketing operations. As reserves mounted, those companies with refining and marketing facilities inadequate to dispose of the crude oil they produced, the short-run marginal cost of which was near zero and the average cost of which was far below posted prices, eagerly sought buyers even though to find them necessitated their making price concessions.

Of the major Middle East producers only Jersey Standard had in both its regional and worldwide operations an approximate balance between production of crude oil and natural gas liquids on the one hand and its refinery runs on the other. During the 1950s it refined a little more crude than it produced. In the Eastern Hemisphere it produced a little more crude than it refined. In its total world operations it was a net buyer of crude, refining about 5 percent more crude than it produced.[12]

The Royal Dutch-Shell group, on the other hand, was a deficit producer in both hemispheres, its gross production in the Eastern Hemisphere being only about 60 percent of its refinery runs and in the Western Hemisphere about 94 percent. It made up the deficiency in crude largely by long-term purchase contracts from its Middle East associates.

Although Socony Mobil produced more oil than it refined in the Eastern Hemisphere, it was a heavy net buyer in the Western Hemisphere and its total worldwide refinery runs exceeded its worldwide production by a substantial margin.

The total worldwide production of Standard of California and Texaco was only slightly larger than the total amount of oil they refined, but their Eastern Hemisphere production exceeded their Eastern Hemisphere refinery runs by more than 50 percent. Their Western Hemisphere production was only about 82 percent of their refining, the balance being supplied by imports or purchases from domestic producers. With restrictions imposed on United States' imports they had to depend more heavily on foreign markets to dispose of their surplus Middle East production.

In Gulf Oil and British Petroleum, co-owners of the fabulous Kuwait reserves, the imbalance was most marked. The gap between their gross oil production and the crude processed at their refineries steadily widened during the 1950s. For the entire decade Gulf Oil produced 65 percent more oil than it refined, and BP's excess was only slightly smaller. In the first half of the decade demand for petroleum and its products increased so rapidly that neither Gulf Oil's nor BP's productive capacity constituted a problem. In truth, when the negotiations between BP (then Anglo-Iranian Oil Company) and the Iranian government regarding modification of the concession agreement collapsed and Iran repudiated the concession and nationalized the industry, BP's great productive resources stood it in good stead. When production and refining in Iran came to a standstill in 1951 and BP found itself in short supply, it made good its deficiency by increasing its Kuwait, Iraq, and Qatar production, increasing its output from refineries outside Iran and making temporary processing agreements with other companies. In 1951, when the Iranian dispute culminated, BP's crude production in Kuwait, Iraq, and Qatar totaled 16 million tons. By 1952 it had increased to 25,600,000 tons. The Kuwait production had increased by

12. These percentages and those that follow were calculated from the various company annual reports covering the years 1950 through 1959.

almost 50 percent. In the following year production from these sources increased to 32,200,000 tons. When the consortium agreement was concluded in 1954 and BP's old Iranian properties went back into production, BP produced some 36 million tons from its Kuwait, Iraq and Qatar concessions. Its Kuwait production had just about doubled.

With the resumption of Iranian operations in October 1954 under consortium management the industry's problem was to fit Iranian production into markets that had learned to live without it. By 1955 BP's total production, despite the consortium agreement under which it surrendered 60 percent of its Iranian production to rival companies, was considerably in excess of its 1950 total, which included all of Iran's output. In truth, with the exception of 1957 when the Suez crisis curtailed shipments from the Persian Gulf, BP's production with that of the other Middle East companies surged upward. Toward the end of the decade it became increasingly difficult for Middle East producers to dispose of all their oil, particularly for those companies whose integrated operations were not well balanced. By August 1960, Jersey Standard in announcing price reductions on its Middle East crude stated that such reductions were necessary "because of competitive offers and sales made at sizable discounts below posted prices by major companies."[13]

Thus had the stability of Middle East prices weakened, and OPEC had come into existence.

13. Platt's Oilgram *Price Service,* Vol. XXXVIII, No. 155, p. 1-A.

21

Costs, Prices, Profits, and Revenues

A MAJOR concern of the Middle East oil countries has been with the amount of revenue they derive from their oil resources. With the adoption of the 50-50 profit-sharing agreement these revenues more than doubled, and with minor interruptions they have since steadily increased. Despite the fact that the president of the Standard Oil Company of New Jersey characterized the 50-50 agreement as a "tested principle for maintaining an equality of interest through all aspects of an inevitably complicated relationship intended to endure for many years," Middle East governments have come to regard it as unfair and inadequate. As their oil revenues have grown, their resentment has increased. Outsiders familiar with the tales of the profligate and selfish use that the royal family of Saudi Arabia have made of their oil revenues or the affluence that they have brought to the little Sheikdom of Kuwait and the other small sheikdoms of the Persian Gulf area are not apt to appreciate this resentment. They may likewise find ground for criticising the inefficiency and corruption that have sometimes characterized the more politically mature countries in the application of oil revenues to their economic development.

But this is beside the point. Both those who seek to hold or capture political power in the Middle East and a growing group of intelligent, educated, public-spirited, articulate younger citizens who help to mould public opinion have come to appreciate the significant role that oil plays in their country's economy and the extent to which their country's prosperity and growth depend on it. They resent the control that the oil companies have exercised over their economic welfare. They believe that the oil companies' primary interest is in maximizing their profits without due regard to the country's welfare.

Profits and Prices

Their concern about profit has largely focused on the price of oil. The price cuts of August 1960 precipitated OPEC's organization. One of OPEC's chief goals has been the restoration of precut prices. As previously related, OPEC's early ambition was the inauguration of an output-control program that would subject price-making to its discipline. As its members came to appreciate the obstacles to this goal, they directed their efforts toward increasing their revenues by persuading the companies to treat royalties as a cost rather than a contribution toward their 50 percent share of profits. This they have largely achieved. But they have never abandoned hope of transferring decision-making on prices from the foreign companies to the domestic government or at least insuring that the government would play a greater role in the decision-making process.

Prices in a Free Market

OPEC's ambition warrants a more detailed consideration of the factors determining prices, profits, and the division of profits between the governments and the oil companies. In a free market with no single buyer or seller able to influence the price of oil, production in any area or by any supplier will be carried to the point where the cost of producing the most expensive unit is just equal to the price it can command on the market. Such a price may yield a very considerable profit or economic rent to areas and producers with low average costs. The division of this surplus between the resource owner and the producer in the absence of free competition may properly be subject to negotiation. The most that any owner can hope to get and, assuming competition, the least that he need take is that share that will leave for the producer a competitive return on the capital invested in the enterprise. That is the least that will justify any producer's continuing or undertaking to produce oil over the long run. Depending on the price of oil and the cost of producing it, that share may be considerable.

Characteristics of Middle East Oil Production that Bear on Costs

What are the costs of producing oil in the Middle East oil fields? More specifically, how does the average per barrel cost of Middle East oil compare with costs in other regions and with the price of oil? Before coming to grips with these specific questions it may prove helpful to examine some general characteristics of oil production in the Middle East that have a direct bearing on production costs. As has been indicated in an earlier point in this study, nowhere else in the world have men found such large and rich deposits over so great an area. As the chief geologist for the Anglo-Iranian Oil Company put it, "The occurrence of so many giant oil fields in one area is unique in oil-field experience throughout the world."[1]

But not only have the major oil companies had the opportunity to develop the world's largest and richest oil fields, they have been able to do so under conditions ideally suited for maximizing production and minimizing costs. The oil fields have been large, but the concessions have been larger. In hunting for oil, in finding it, in producing it, oil technicians have been able to utilize an accumulated body of scientific knowledge and technical procedures that have done much to reduce the element of chance in an otherwise unusually speculative enterprise. Once they have located an underground reservoir, in determining its geographic extent, the depth and thickness of the producing horizon, the permeability and saturation of the underground formation, the nature of the expulsive forces, they can proceed with a deliberation and prudence that contrast sharply to the haphazard, reckless fashion which the law of capture made unavoidable in much of American development. In developing a field operators can locate and space wells in a pattern calculated to utilize with efficiency and economy the natural expulsive forces on the effective use of which the ultimate recovery of the oil so largely depends.

Aramco's Saudi Arabian Experience

Aramco's experience in Saudi Arabia illustrates the efficiency with which an oil concession may be developed under a unified management.

1. Wallace Pratt and Dorothy Good, eds., *World Geography of Petroleum,* American Geographic Society, Special Publication No. 31, (Princeton: Princeton University Press, 1950), "The Middle East," by G. M. Lees, pp. 159–206.

Aramco drilled its first six wells, all of which proved unproductive, on what appeared to be an anticlinal formation disclosed by surface outcropping on the hills surrounding Dhahran. Since the completion of its seventh well with a large initial daily production, Aramco has systematically explored much of its concession in a search for structural formations favorable to the accumulation of oil in commercial quantities. This involved plane table mapping of the geological outcrops in the mountainous area of western Arabia and the occasional outcrops east of the mountains extending to the Persian Gulf. Subsequently, Aramco conducted a systematic and carefully planned seismographic and core-drilling program designed to reveal the stratigraphy of underground formations. With these studies completed, Aramco's task was to determine by exploratory wells whether a particular underground structure contained oil in commercial quantities and, if so, to develop a well location and spacing program designed to insure an efficient exploitation of the underground oil reservoir.

By such methods by 1960, when I first visited the company's operations, Aramco had discovered and was exploiting some nine or ten oil fields, most of them lying along the Persian Gulf. Among them was the Ghawar field, one of the world's greatest. It extends in a north-south direction for more than 140 miles some 50 to 100 miles inland from the Persian Gulf. The Ghawar field illustrates both the magnitude of Saudi Arabia's oil reservoirs and the economy with which Aramco is exploiting them. In mapping surface structures in 1935, Steinke and Koch, Aramco geologists, detected the first clue to the existence of the En Nala anticlinal axis. Before World War II, Aramco followed this discovery with gravity mapping, and in 1941 by drilling widely spaced structural holes Aramco confirmed the existence of a major anticline. Immediately after the war with further structural drilling and by gravity mapping, company geologists defined the location for two promising wildcats in the Ain Dar and Haradh areas. By drilling to a depth of approximately 7,000 feet, Aramco completed a productive well at Ain Dar in June 1948 and a second well at Haradh in February 1949. These two wells, approximately 140 miles apart, determined roughly the limits of the underground structure. By the end of 1957 Aramco had drilled 129 wells extending over the length of the field in an area roughly 15 miles wide. As of that date Aramco was producing oil from 88 wells, had 27 wells shut in temporarily, was using 8 wells for observation purposes, had suspended drilling in 2 wells and had abandoned only 4. The first well completed had an estimated daily capacity of 19,000 barrels. By 1957 with the southern half of the field only roughly defined by widely spaced wells and shallow-structure drilling and the northern limits of the Ain Dar field not yet determined, the Ghawar field averaged nearly 600,000 barrels a day with each producing well yielding 7,000 barrels.[2]

The East Texas Experience

A comparison of the East Texas field and its development with Aramco's exploitation of the Ghawar field indicates the economic significance of the exploitation of an underground oil reservoir by a single producer. A lonely wildcatter, C. M. (Dad) Joiner, completed the drilling of the discovery well in the East Texas field on October 3, 1930. Joiner, who had more faith in God than in geology, had drilled two dry holes adjacent to the discovery well on lands which geologists had characterized as unlikely to produce oil. When Joiner's well came in with an initial production of only 300 barrels a day, skeptics "wrote it off" as an insignificant pocket discovery not likely to produce much oil and not indicative of a large oil reservoir. Two months later when another wildcatter completed the No. 1 Crim well about ten miles north of the Joiner well with an initial daily production of 20,000 barrels, skepticism gave way to frenzied excitement. With every farm a potential oil bonanza, a hectic leasing and drilling program exploded overnight. Within thirty days eager oil men had drilled three more wells. Within a year they had drilled 3,600 and produced more than a million barrels of oil. A quarter of a century later 559 operators owning 2,567 leases were producing more than 220,000 barrels of oil a day from 20,200 wells. During the interim they had drilled approximately 29,000 wells over an area approximately forty miles long and about a half dozen miles wide. They had produced approximately 3,200,000,000 barrels of oil on leases estimated still to contain about the same amount of recoverable oil.[3]

2. For a more detailed account of the development of the Ghawar field consult *Bulletin of the Association of Petroleum Geologists,* February 1959, pp. 434–454, by Aramco's staff.

3. East Texas data largely from the *Oil and Gas Journal,* October 3, 1955, pp. 78–81. A. M. Haikail, of Aramco, draws a parallel between the development of East Texas and Abqaiq in which he explains the more economical development of Abqaiq in terms of the advances in science and technology. (Paper, the Sixth Oil Petroleum Congress, Baghdad, March 6–13, 1967, "Modern Concepts of Oil Field Development.") In the development of Abqaiq Aramco utilizes a technology which did not exist when East Texas was developed, but the wastes of East Texas are to be explained primarily by the system of ownership under which it was developed.

By any criteria, East Texas has proven a great oil field. With a combination of favorable factors—size of the underground reservoir, sand thickness and saturation, bottom hole pressure, the depth of the productive horizons—it is rare if not unique in the industry's history. It has been one of the domestic industry's lowest cost fields and under a unified control which would have permitted the efficient utilization of geological knowledge and engineering techniques might have been one of the world's lowest cost producers.

But the scramble for oil under the law of capture, with surface property lines determining the number and location of wells, inevitably wasted the reservoir pressures, led to rapid water encroachment, and threatened the early ruin of the field. Martial law, together with legislative and court action at both the state and federal level supporting a proration program, eventually brought the East Texas field under control. But not until incalculable wastage had occurred and East Texas oil was selling at 10 cents a barrel. Moreover, the proration program which eventually stabilized production encouraged excessive drilling with wasteful capital expenditures. East Texas, with all of its waste and neglected opportunities, affords a dramatic illustration of how cheaply oil might have been recovered from a rich reserve.

But the United States has had only one East Texas, while the Middle East has dozens of Ghawar's. A rough notion of the economic superiority of Middle East production is reflected in the daily average production of wells in the two areas. In 1966 in the United States 583,000 wells produced an average of 14.2 barrels of oil a day. In the four major Middle East oil-producing countries (Iran, Iraq, Saudi Arabia, and Kuwait) 1,146 wells produced an average of more than 7,500 barrels a day. Iran's 184 wells averaged 11,586 barrels a day; Kuwait's 503 wells averaged 4,938 barrels.

These data suggest what oil men acknowledge—that the Middle East is the world's lowest cost oil producer. They do not prove it. They fail to take account of the greater cost of drilling—an Aramco vice-president testified before a subcommittee of the Senate Public Lands Committee in 1947 that to drill wells in Saudi Arabia cost from $250,000 to $300,000 each, more than twice as much as wells of comparable depth in the United States—or of the millions that Middle East concessionaires have spent on road construction, schools, hospitals, medical services, training programs, and numerous other facilities provided by the state in western countries.

Adelman's Cost Study

Fortunately several studies show the approximate per-barrel cost of producing oil in the Middle East including such expenditures. The most thorough of these is M. A. Adelman's "Oil Production Costs in Four Areas."[4] Adelman's study compares the average and the marginal costs of producing a barrel of oil in the free world's most important producing areas. It is a bold and meticulous study—bold in its assumptions and meticulous in its analysis.

Adelman recognizes that a precise answer to such a complex problem is fraught with inadequacy of data, numerous pitfalls of accounting, challenging assumptions, and daring estimates. But he brings to his task rare analytical and critical ability, unusual professional competence, and painstaking thoroughness.

What Adelman is trying to get at is the necessary price of oil if it is to continue to be produced in the areas under consideration. He defines cost as "identically equal to break-even price" or the economist's "word of art 'supply price,' . . . if the market price just barely repays cost, a barrel is just barely worth producing."[5] He breaks his analysis down according to the components of cost: operating costs, development costs, and finding costs. He first calculates operating costs in the United States by dividing total operating expenditures by total operating capacity. He then derives estimates for the other world areas by assuming that operating costs per well increase as the square root of capacity—an assumption that one of his critics characterizes as "a leap into the unknown . . . [that] . . . reminds . . . [one] of the Daedalus and Icarus legend."[6]

Adelman finds that in the early 1960s the United States average operating costs for nonstripper wells operating at capacity were 17 cents a barrel. Per barrel operating costs for Venezuela were 6.5 cents. In Africa the Libyan operating costs were 2.2 cents; in Algeria, 3.9 cents; in Nigeria, 2.7 cents. In Iran operating costs were only 1 cent a barrel; in Iraq, 1.2 cents; in Kuwait, 1.8 cents; in Saudi Arabia, 1.5 cents. He calculates the

4. M. A. Adelman, "Oil Production Costs in Four Areas," reprinted, *Proceedings of the Council of Economics,* Annual Meeting of American Institute of Mining, Metallurgical, and Petroleum Engineers, February 28–March 2, 1966.

5. *Ibid.,* p. 4.

6. *Ibid.,* p. 9.

United States developmental costs per barrel at $1.34 and in Venezuela at 55 cents. In Africa the comparable costs in Libya were 13 cents; in Algeria, 42 cents; in Nigeria, 28 cents. Comparable Middle East costs were 6 cents in Iran, 3 cents in Iraq, 8 cents in Kuwait, and 8 cents in Saudi Arabia. He calculates a total per barrel cost, exclusive of funding costs, in the United States of $1.51; in Venezuela, 62 cents; in Libya, 15 cents; in Algeria, 46 cents; in Nigeria, 31 cents. In the Middle East his total costs are 7 cents for Iran, 4 cents for Iraq, 10 cents for Kuwait, and 10 cents for Saudi Arabia. Although he recognizes that predicting finding costs is impossible, he concludes that "Persian Gulf development operating cost today fixes the supply price of oil, *including necessary finding cost,* for the whole world,"[7] and he calculates this cost as between 15 and 20 cents a barrel for each of the four Persian Gulf countries. For the United States and Venezuela comparable figures are 60 to 65 cents; for Libya, 45 to 50 cents; for Algeria, 49 to 54 cents; for Nigeria, 36 to 40 cents.[8]

To thus state Adelman's conclusions bare of his intricate and scholarly analysis may do an injustice to them. But to summarize his analysis, extending through 62 closely-spaced pages accompanied by 112 references and 31 tables, would lose more in accuracy than it would gain in brevity. His cost figures, although varying in detail, are consistent with others available. They justify the inference to be drawn from my own qualitative analysis that nowhere else in the world can so much oil be produced at so little cost as in the Persian Gulf.

The Chase Manhattan Bank Study

Other quantitative approaches to the problem of comparative costs lend support to this conclusion. The Chase Manhattan National Bank of New York compiles annual expenditures made by oil companies in developing crude oil reserves in the United States and foreign producing areas. When related to the total annual production they show the per barrel cost of maintaining and expanding oil production in these areas. These data show that in the period 1951–60 to expand production in the United

7. *Ibid.*, p. 60.
8. *Ibid.*, Table XXXI, pp. 114–115. Adelman has recently revised his calculations, with the result that Iran shows a total per barrel cost of 12 cents, Iraq 7 cents, and Kuwait 9 cents. Saudi Arabia's total cost remains unchanged at 10 cents. His revisions do not change his conclusions.

States oil companies spent from $1.32 to $1.94 for each barrel of oil produced or an average of $1.73. To maintain and expand Venezuelan production during the same period they spent from 24 to 95 cents or an average of 51 cents for every barrel. To maintain and expand production in the Middle East they spent from 11 cents to 21 cents or an average of 16 cents for every barrel produced. Extending the calculations through 1965 by using data presented by the bank's annual reports on "Capital Investments by the World Petroleum Industry" gives an average per barrel expenditure over the 15-year period of $1.67 for the United States; 38 cents for Venezuela, and 13 cents for the Middle East.[9]

About its calculations the bank states that they are significant and valuable "because they—(1) can be compiled from available information, (2) are not dependent on estimates of new reserves added each year, and (3) illustrate with some degree of accuracy the trend in the costs of expanding production."[10]

Profits in Middle East Oil

With its production costs the lowest in the world and its posted prices some eight or ten times its production costs, we would expect profits and the rate of return on investment in Middle East oil to be large. Spokesmen for the industry reject this expectation. B. A. C. Sweet-Escott of the British Petroleum Company in a paper presented to the Fourth Arab Petroleum Congress in Beirut on November 6, 1963, warned that the industry might face serious difficulty in finding funds to finance the expansion necessary to meet demand if profit ratios continued at the "low rates" of the 1950s. He alleged that the net income of the seven international oil companies expressed as a percentage of average capital employed in their worldwide integrated operations had fallen from 13 percent in 1957 to 9.1 percent in 1959 and that it had recovered to only 9.6 percent in 1962. He estimated that if current forecasts were correct, the gap between funds generated by the major companies (retained profits and depreciation) and the funds

9. Chase Manhattan Bank, *Capital Investments by the World Petroleum Industry*, November 1961; November 1962; *Capital Investments of the World Petroleum Industry*, 1962, 1963, 1964, 1965, 1966.

10. *Ibid.*, *Capital Investments by the World Petroleum Industry*, November 1961, p. 8.

needed for fixed assets alone would rise from an average of £150 million a year over the last five years to £225 million a year in the period 1961–75 and he questioned the ability of the capital market to supply such amounts.

A Royal Dutch-Shell Managing Director Analyses Profits

L. E. J. Brouwer, a managing director of the Royal Dutch-Shell group, sounded a similar note in an address before the Iraqi Engineers Association in Baghdad on May 26, 1964. He called attention to increasing competition in international oil markets arising primarily from newcomers in the industry. Although the major companies have handled a far larger volume of oil and oil products in recent years than previously, their relative importance has declined. Whereas in 1954 they accounted for 93 percent of the free world production, 79 percent of the refining capacity, and 76 percent of the market trade outside the United States, by 1962 competition had reduced their share to 83 percent, 67 percent, and 66 percent respectively. In order to put these figures in "proper perspective" he insisted one must take account of the "over-all rate of return on investment." This showed a disturbing decline. Whereas in 1956 the seven major international companies averaged 15.9 percent on their total net assets, in 1963 they averaged only 11.6 percent. And he echoed Sweet-Escott's concern in pointing out that the bulk of the industry's capital expenditures must be financed out of companies' earnings. He implied that a further decline in profits would jeopardize the companies' ability to provide the capital essential to their functioning as efficient and effective instruments in safeguarding the sales and revenues of the producing countries.

The First National City Bank Study

The First National City Bank of New York has struck an even more doleful note on oil company earnings. It has published annually since 1960 data bearing on the earnings on so-called "downstream" operations of seven international oil companies. It takes as its point of departure the total annual net earnings of these companies from their integrated operations and from them deducts the annual payments to the Middle East governments. Assuming that the earnings of the international oil companies'

Eastern Hemisphere production operations equals the payments to the governments, by subtracting these from total net earnings it arrives at earnings from downstream operations. These show a loss in every year beginning with 1957. The loss increases annually until by 1967 it has reached $1,693 million. That is to say the bank's study indicates that in every year the companies operated their transportation, refining, and marketing business at a loss.

Leaving aside the question of the accuracy of the bank's calculations, the loss can be explained largely by their cost-accounting procedures. Customarily until quite recently the companies have charged their European refining and marketing subsidiaries the posted prices for Middle East oil which they have used in calculating payments to the governments. During this period the supply of oil moving into world markets was so great that it could not be sold at posted prices. To meet the competition of independent producers the majors were forced to sell at discounts from posted prices on sales to unaffiliated concerns. The competition of independent refiners who bought their oil at discounts was reflected in weak refinery and marketing prices throughout Western Europe. It is not surprising, therefore, that the subsidiaries of the major companies, charged by their parents the full posted prices, operated in some cases and at certain times at an accounting loss.

The bank calls attention to other accounting problems that might invalidate its findings: failure to prorate head office expenditures among the various geographical regions which the head office serves, failure to prorate tanker earnings between the areas in which crude or products are loaded and the areas to which they are delivered, etc. But while the bank finds that these reservations obscure precise measurements of downstream losses, it concludes that they do not appear to invalidate its conclusion that downstream losses in the Eastern Hemisphere yield a negative result.

The willingness of the companies to expand greatly their investments in downstream operations raises grave doubts about its conclusion. Nevertheless it seems certain that the payments to the governments have increased far more rapidly than have the net earnings of companies on the whole of their integrated operations. The bank reports earnings as having remained relatively constant for the last five-year period—$1,428 million in 1963 compared with $1,472 million in 1967—while payments to the governments showed an increase of 60 percent during the same time, from $1,908 million to $3,165 million. The bank's findings are set forth succinctly in Table I.

TABLE I

Eastern Hemisphere Earnings and Payments to Governments by Seven
Major International Oil Companies 1957–1967
(Millions of Dollars)

	1957	1958	1959	1960	1961	1962	1963	1964	1965	1966	1967
Earnings	1,056	953	987	1,089	1,120	1,230	1,428	1,256	1,371	1,492	1,472
Payments to govern- ments	1,070	1,210	1,309	1,381	1,454	1,637	1,908	2,167	2,471	2,841	3,165
Indicated downstream profit (+) loss (−)	−14	−257	−322	−292	−334	−407	−480	−911	−1,100	−1,349	−1,693

Source: First National City Bank, *Profit Problems in the Eastern Hemisphere*, October 1968

The Significance of Rates of Return on Integrated Operations

The findings of Sweet-Escott, Brouwer, and the First National City
Bank are alike in that they assume that the significant profit figure for the
international oil companies is the rate of return on their total investment
throughout the Eastern Hemisphere in oil production, transportation, refining, and marketing.

Although with far different implications, Dr. Edith Penrose, a competent economist who has distinguished herself by her scholarly studies of the
international oil industry, takes substantially the same position.[11] Dr. Penrose's immediate concern with the rate of return which the international oil
companies make on their integrated operations grew out of the demand of
representatives of the Middle East governments that their governments
share in the profits of the major oil companies derived from the refining
and distribution of Middle East oil in world markets.

Dr. Penrose observes that the total cost of any given amount of the
finished product is the sum of the costs incurred at every stage of production, and profits for the firm as a whole arise when the sales receipts exceed
costs. This is certainly true for every individual corporate enterprise engaged in the several separate operations by which a raw material is converted into and sold as a finished product. There are good accounting and
commercial reasons, however, why such a firm should calculate, as far as
feasible, costs at each stage of operations, even though its chief concern as

11. Edith Penrose, "Middle East Oil: The International Distribution of Profits and
Income Taxes," *Economica*, August 1960, pp. 203–213.

a business enterprise is its rate of return on the whole of its investment. Where the several stages of production are conducted by different corporate entities, all owned or controlled by a parent company, sound accounting procedures would seem to require that costs and rates of return on capital invested in the separate business enterprises be calculated for each. Only by such accounting can a business firm, that is, the parent company, make sound decisions as to where and when investments should be expanded or contracted, whether products at one stage of the operation should be sold wholly or in part to an affiliated company at another stage or to independent firms or bought wholly or in part by affiliated companies from independents. Such independent accounting may be even more essential where integrated firms compete in their over-all operations but jointly own an important productive unit at some stage. This is the situation of Aramco, jointly but not equally owned by Standard of New Jersey, Mobil Oil, Standard of California, and Texaco.

There are particular reasons why in the oil industry calculation of costs and rates of return at each stage of the process are imperative. Pipeline transportation in the United States, for example, is subject to a regulatory agency that limits the rate of return. In the Middle East both Aramco and IPC transport oil to the Mediterranean by pipelines crossing Syria and Lebanon, and each of these countries shares in the profits of such operations. It is essential, therefore, to calculate costs and revenues separately for their pipeline operations.

The Low Rate of Return in Refining and Marketing

The low or negative rate of return which the international oil companies have shown in their downstream operations is owing in part, as has been indicated, to their having charged the affiliated refining companies the posted price for Middle East oil when competitive refiners were buying oil at discounts. But in general the lower rate of return in refining and marketing operations is occasioned by the greater severity of competition in these stages. Because they require relatively little capital, independent refiners and marketers have entered the European markets in increasing numbers in recent years. Their decisions to do so have been influenced by the relative advantages of alternative opportunities open to them. Rates of return in oil refining and marketing generally have been far less than in Middle East oil production but sufficiently high to attract capital in increasing amounts. What the international oil companies can earn in these stages of production is set by what others can earn. The rate has been relatively low.[12]

Dr. Penrose's analysis, supplemented by the First National City Bank's and Escott-Sweet's and Brouwer's studies, may warrant the conclusion that the Middle East governments have benefited more by a 50-50 sharing of production profits calculated on posted prices than they would have by sharing profits on the whole of the international companies' operations.

Middle East Production Highly Profitable

However, neither Sweet-Escott's nor Brouwer's fear that declining profits in the industry would deprive the industry of the funds with which to finance an oil output essential to meet demand has much merit. The worst that could happen would be a readjustment through the market mechanism whereby world markets get along with less oil at higher prices. Such a contingency is not likely within the calculable future.

Capital has been flowing into Middle East oil production in increasing amounts both by the majors and independent companies despite low rates of earnings in marketing and refining. It has done so because investments in oil production in the Middle East have been highly profitable.

Competent studies confirm this conclusion. Shortly after its organization OPEC commissioned Arthur D. Little, Inc., industrial consultants, to conduct a study of the structure of crude prices and the rates of return enjoyed by the major oil companies on their investments in OPEC countries. Experts familiar with the pitfalls of corporate accounting and with the nature of oil production conducted the study. They had at their command not only such published corporate reports as were available but company cost and earning figures which constituted a basis for calculating the annual taxes due the several governments.

12. According to Solomon and Laya, the relatively low rates of return in marketing and refining may result in part from the practice of these divisions of capitalizing high fractions of their outlay rather than expensing them. They state that it is not "Surprising that the production division of the petroleum industry, which expenses a high fraction of investment outlays, tends to have higher yields than refining or marketing divisions." "Measurement of Company Profitability: Some Systematic Errors in the Accounting Rate of Return," Ezra Solomon and Jaime C. Laya. Presented at the Conference on Financial Research and Its Implications for Management, Graduate School of Business, Stanford University, June 9, 10, 11, 1966.

Although neither Arthur D. Little nor OPEC has released the study to the public, Sheik Abdullah Tariki, Saudi Arabian minister of petroleum when the study was under way, now an oil consultant with offices in Beirut, in a lecture delivered before the Iraqi Engineer's Association in Baghdad on June 1, 1963, on "Arab Oil" disclosed in detail the net profits, the capital investments and the rates of return on investment for the five years 1956 to 1960 inclusive. These figures are rather generally known in Middle East oil circles.

With complete annual data on net assets, gross income, and total costs for Aramco, the experts calculated rates of return varying from a low of 57 percent in 1958 to a high of 71 percent in 1960 for a five-year average of 61 percent. With complete data on Iraq on net assets and total costs, the experts estimated gross income on a basis of posted prices and going discounts and arrived at rates of return varying from 36 percent in 1957 to 75 percent in 1960 with an average return on annual net assets of 62 percent.

With complete data on net assets for the Iranian consortium group and estimates of gross income from the sale of oil and oil products, but with incomplete costs of operation, the experts estimated rates of return varying between a low of 63 percent in 1956 to a high of 78 percent in 1957 for an average return of 71 percent for the five-year period.

For the Qatar Petroleum Company owned by IPC from complete data on assets and total costs and gross income estimated from posted prices and going discounts, they calculated annual rates of return ranging from 83 percent in 1960 to 150 percent in 1958 for an average of 114 percent.[13]

Not having sufficient data for the Kuwait Oil Company, the experts made no estimates of rate of return.

Discounted Cash Flow Returns

Annual rates of return on investments in petroleum made over a long period do not reflect the profitless investments that the companies may have experienced during the early years of exploration and development. Accountants have recently developed concepts and procedures that permit calculating more realistic rates on long-term investments. Commonly known as the discounted cash flow method, it makes allowance for varia-

13. Data from Tariki's Baghdad lecture, mimeographed, Table VIII, p. 29.

tions in the time pattern of cash outlays and cash earnings. In the words of Joel Dean, one of its originators,

it computes the rate of return as the maximum interest rates which could be paid on capital tied up over the life of the investment without dipping into earnings produced elsewhere in the company. The mechanism of the cash flow method consists essentially of finding the interest rate that discounts future earnings of a project down to the present value equal to the project cost. This interest rate is the return on that investment.[14]

As Solomon and Laya have pointed out, the discounted cash flow concept corresponds to the financial analyst's concept, "effective yield to maturity," used in measuring the income from bonds and to the economist's concept, "marginal efficiency of capital."[15] Although used customarily to guide executives in making decisions on the comparative advantage of various investment alternatives, it may be useful in determining a more realistic rate of return on investments already made than does the more conventional accounting method. In nontechnical terms, it makes allowance for two facts: that the early life of an investment may have been barren of earnings and that when decisions were made the present value of future earnings should be subject to time discounts.

Mikdashi's Study

With the co-operation of IPC, Zuhayr Mikdashi of the faculty of Business Administration, American University of Beirut, compiled annual disbursements and receipts for the IPC group from 1925 through 1963.[16] These data indicate that ten years elapsed before the group's net income after taxes and depreciation equaled annual disbursements—in short, before the investment yielded any net profits. From the data he assembled Mikdashi calculated discounted cash flow rates of return. About this he says, "They came out to about 13–15 percent on a basis of certain

---

14. Joel Dean, "Measuring the Productivity of Capital," *Harvard Business Review*, January–February 1954. Reproduced in Ezra Solomon, ed. *The Management of Corporate Capital*, pp. 21–34.

15. Solomon and Laya.

16. See Mikdashi, Zuhayr, *A Financial Analysis of Middle East Oil Concessions: 1901–1965* (New York: Frederick A. Praeger), Appendix II, pp. 273–292. Mikdashi's *Analysis* is a painstaking and comprehensive study based on several years of research and consultation with company executives throughout the Western world.

figures and assumptions. The accuracy of the DCF figures may very well be disputed." But his "computations show strikingly that historical profitability is much lower than current accounting rates of return suggest."[17]

On data from a variety of sources and on a basis of various assumptions, Mikdashi's calculations indicate that Aramco's owners got "DCF rates of roughly 20 to 28 percent, . . . significantly below the OPEC rate from 1956 to 1962."[18]

Similarly, Mikdashi calculates a DCF rate received by the Iranian Consortium from the date it took over operations following the nationalization of the industry through 1964. About it he says, "The 1954–64 discounted cash flow rate of return is deemed to be close to 70 percent, using one appraised value of the Consortium in the terminal year 1964, and adjusting roughly for changes in the purchasing power of money (e.g., sterling)."[19] This approximates the OPEC average rate for 1956–60 of 71 percent. The significantly higher DCF rate for the consortium than for Aramco and IPC is to be explained in Mikdashi's words by the fact that "(a) there has been practically no lag between disbursements and receipts, and (b) the DCF rate takes into consideration expected future returns whereas the accounting rate does not."[20]

Judged in either historical or contemporary perspective IPC, Aramco, and the consortium's investments have been highly profitable to their owners. There is no reason to believe that the Kuwait Oil Company's investment has proven otherwise.

Middle East Governments Want Larger Shares

The Middle East governments are fully aware of this and equally determined to enlarge their share of the profits. Neither the current rates of return on the companies' investments nor the gap between the per barrel cost, including the contributions to the governments, and the current price of oil afford an index to a "fair" distribution of profits. It does indicate a range within which collective bargaining may shift the governments' share without disaster to either party. It is extremely unlikely that the government will be content with less. It is more likely that as the government

17. *Ibid.*, p. 194.
18. *Ibid.*, p. 183.
19. *Ibid.*, pp. 221–222.
20. *Ibid.*, p. 222.

brings greater pressure on the companies that they will have to accede in one way or another to government demands for more.

Meanwhile it seems not unlikely that declining prices may narrow rather than broaden the area within which collective bargaining may be carried on. Should such prove true, the governments' clamor for an increase in their shares will likely meet increased resistance by the companies.

Any prognostication about the course of prices is, of course, hazardous. But with the major companies continuously expanding their reserves, with new concessions being granted on lands previously monopolized by the original concessionaires, with national companies determined to participate directly in the development of relinquished areas, with Middle East costs still below the price necessary to bring its oil on to world markets, it is unlikely that, despite OPEC's effort, except for short periods the downward trend of prices will soon be reversed.

PART V

THE PROSPECTS FOR MIDDLE EAST OIL

OPEC Broadens Its Program

THE Organization of Petroleum Exporting Countries was created to gain through collective action control over the concessionaires of its members which they had alienated by the broad terms of their concessions. Collective bargaining was a temporary substitute for national sovereignty. Through prolonged negotiations over the royalty issue OPEC in effect established its right to speak for its members. More recently it has assumed a more pretentious role. Paradoxically, by asserting the sovereign right of each of its members to speak and act for itself, OPEC may weaken the power it has gained to speak for them through collective action.

The Right of Permanent Sovereignty

In 1952 the United Nations first invoked the right of permanent sovereignty by a country in the development of its national resources. It subjected the doctrine to refinement from time to time and expressed it in its most complete form in a resolution adopted by the General Assembly on November 25, 1966.

At its sixteenth conference held in Vienna in July 1968, OPEC adopted with only minor changes in language the General Assembly's preamble to its 1966 resolution as a "Declaratory Statement of Policy in Member Countries." OPEC's former secretary general characterized OPEC's declaratory statement as representing "a powerful surge forward in the process of unification of the Organization's policies." It was not adopted lightly by member countries, but was "the result of many months of effort and long discussions." It represents the "crystallization of current and recent trends of thought as well as a chart for the future."[1]

The Preamble to the Declaratory Statement and Its Four Major Goals

The preamble to the "Declaratory Statement of Petroleum Policy" reads in part as follows:

The Conference, recognizing that hydrocarbon resources in Member Countries are one of the principal sources of their revenues and foreign exchange earnings and therefore constitute the main basis for their economic development; bearing in mind that hydrocarbon resources are limited and exhaustible, and that their proper exploitation determines the conditions of the economic development of Member Countries, both at present and in the future; bearing in mind also that the inalienable right of all countries to exercise permanent sovereignty over their natural resources in the interest of their national development is a universally recognized principle of public law and has been repeatedly reaffirmed by the General Assembly of the United Nations, most notably in its Resolution 2158 of November 25, 1966; considering also that in order to ensure the exercise of permanent sovereignty over hydrocarbon resources, it is essential that their exploitation should be aimed at securing the greatest possible benefit for Member Countries; considering further that this aim can better be achieved if Member Countries are in a position to undertake themselves directly the exploitation of their hydrocarbon resources, so that they may exercise their freedom of choice in the utilization of hydrocarbon resources under the most favorable conditions; taking into account the fact that foreign capital, whether public or private, forthcoming at the request of the Member Countries, can play an important role, inasmuch as it supplements the efforts undertaken by them in the exploitation of their hydrocarbon resources, provided that there is government supervision of the activity of foreign capital to ensure that it is used in the interest of national development and that returns earned by it do not exceed reasonable levels; recommends that the following principles shall serve as basis for petroleum policy in Member Countries.

Within this framework the member countries declare their intent as far as feasible to explore and develop their hydrocarbon resources directly. When this is not feasible, they may enter into contracts of various types for

1. Francisco Parra, in a press statement, July 24, 1968, quoted in *Middle East Economic Survey,* July 26, 1968.

a reasonable remuneration. Such contracts are open to revision when changing circumstances justify it. Changing circumstances may also justify the revision of existing contracts.

The Declaratory Statement of Policy outlines four major goals for OPEC members in the exercise of their sovereignty over existing contracts: (1) an accelerated relinquishment program with the government participating in the choice of acreage to be relinquished; (2) the requirement that all contracts provide for the payment of taxes based on posted or reference prices with the government determining the prices; (3) the guarantee of fiscal stability to operators for only a reasonable length of time, with the government exercising the right to recapture excessively high earnings through taxation; (4) the acquisition by the government of a reasonable participation in the ownership of existing concessions.

Such is the program contemplated through the exercise of their sovereignty by OPEC's members.

The National Oil Companies

Regarding the economic consequences of this program, one can only speculate. Accelerating the relinquishment program with government participation in the choice of acreage to be relinquished will place a vast acreage under the supervision of the state oil companies. All OPEC major countries now have their national oil companies. In Iran it is NIOC; in Indonesia it is Permina; in Venezuela it is CVP; in Kuwait it is KNPC; in Iraq it is INOC; in Saudi Arabia it is PETROMIN; in Libya it is the Libyan General Petroleum Corporation. It is the avowed purpose of each of these companies to so expand their activities that they will become significant factors in international markets. Some are in a hurry to accomplish this.

Iran's NIOC has gone furthest and fastest. It will be recalled that Iran's Petroleum Act of July 1957, under whose general provisions NIOC operates, expresses a sense of urgency on the part of its framers. It was enacted "with a view to extending as rapidly as possible . . . research, exploitation, and extraction of petroleum throughout the country" and bringing all "phases of the industry within the nation's control—refining, transportation, and marketing," at home "as well as abroad."

NIOC's joint ventures reflect the same sense of urgency. As previously indicated, the NIOC–Pan Am agreement obliges NIOC and Pan Am to exercise their "utmost" effort to sell the "maximum quantity of petroleum economically justifiable" and each party to pay taxes on a basis of realized prices. Each of Iran's more recent joint ventures places similar obligations on the parties to the contract. ERAP's contract with NIOC provides that ERAP will market for NIOC three million tons of crude each year for five years and four million tons yearly for a second five-year period. For this oil it will pay taxes on realized prices less a 2 percent brokerage fee. NIOC still has large uncommitted areas for the development of which it will create other joint ventures. The recent consortium agreement will increase this area by 25,000 square miles. Apparently NIOC intends to develop them. M. Farmanfarmaian in an address before the Second Seminar on the Economics of the International Oil Industry at the American University of Beirut in May, 1968, estimated that by 1972 NIOC's joint ventures in the Persian Gulf will be producing 700,000 barrels daily, about one-fourth of Iran's 1968 production.

As pointed out, the General Petroleum and Mineral Organization (PETROMIN) is Saudi Arabia's national instrument for getting into world markets. Although PETROMIN's board chairman, Oil Minister Yamani, is aware of the threat that state oil companies constitute to the stability of international markets, he is determined that Saudi Arabia, whose oil reserves are now estimated to be the world's largest, will not lag behind other countries in its development and production programs. PETROMIN's joint-venture contract with AUXIRAP and its more recent agreement with AGIP are more than straws in the wind.

Kuwait is a rather different story. Its instrument for getting into world markets is the Kuwait National Petroleum Company (KNPC). Owned 60 percent by the government and 40 percent by private Kuwaiti investors, it is authorized by its charter to "engage in all phases of the petroleum industry, natural gas, and other hydrocarbons, refining, manufacturing, transporting . . . and distributing, selling, and exporting such substances."

To get into world markets promptly it has chosen to engage in oil refining on a large scale. As planned, its Shuaiba refinery was to have been one of the most modern in the world and the world's largest. Its unique function was to have been its ability to utilize low-cost hydrogen derived from the Kuwait Oil Company's nearby Burgan oil field and to convert its oil throughput largely into middle distillates then in short supply. It was to have a flexibility permitting it to vary product output to meet changing market conditions. In the winter of 1967 I visited the plant then being constructed under the supervision of an American technician. When questioned about the plant's future, he expressed fear that the Kuwait government might find it difficult to man the rather complicated refinery with a competent staff.

Whether because of manpower problems, "mechanical failure" in some units of the refinery, or because the process is in an "experimental stage," as alleged by Tariki, who served as an adviser to Kuwait's ministry of finance and oil, the refinery operated at a loss during 1968. With refinery costs of 70 cents a barrel, compared with 30 cents at KOC's similar plant at nearby Ahmadi, KNPC's refinery losses totaled ID 4,813,285 for the year.[2] With adequate finances, however, which KNPC can command, its manpower problems and mechanical difficulties should prove temporary.[3]

In the immediate future KNPC has contracted to obtain crude supplies for its refinery from KOC's hitherto shut-in Umm Gudair field. In the long run it hopes to get its crude requirements from the development of the areas that KOC has relinquished. Though it does not expect a bonanza oil discovery in these areas, it does have a reasonable expectation of finding oil in commercial quantities.

But whether it relies on its own sources of crude or buys it from existing producers, KNPC basically represents one more seller in an over-crowded oil market. Its influence on the structure of the international oil industry is of the same sort as if it were an oil producer. The oil it buys will occasion an increase in output by its suppliers. One cannot expect them to deny oil to their present customers nor to slow their efforts to find new customers because they agree to supply KNPC with its requirements.

In the long run KNPC seems destined to become a crude producer by way of the recently concluded joint-venture deal for the relinquished acreage with Cia. Hispania de Petroles, S. A. (Hispanoil). Hispanoil represents a combination of four Spanish refining companies, three of which are state owned, the fourth being owned by Spanish banking interests. A significant aspect of the proposed joint venture is that the Spanish government will guarantee it 25 percent of Spain's crude oil imports for 15 years at prevailing world prices.

Since Qasim's expropriation of 99.5 percent of the IPC concession, political instability has interfered with the development of a consistent oil

---

2. *MEES*, October 24, 1969.
3. *MEES*, November 21, 1969, reports that all units of the Shuaiba refinery are "now fully operational" and that the difficulties of start-up troubles have now been overcome. *MEES*, April 6, 1970, reports, however, that the Shuaiba refinery operated at a substantial loss during 1969.

policy. The basic charter of such a program has been set, however. INOC is to become the agency for developing the north Rumaila field and for making contracts for exploring and developing the vast acreage to which it has access. The French ERAP agreement and the contract with Russia will facilitate the process.

Although the production of any one of the national oil companies is likely to be relatively small for some time, in the long run the output promises to be substantial. And they will be selling oil for what it will bring. If the production hopes of the national oil companies are realized, it will probably bring less in the future than it now brings.

Taxes Based on Reference Prices or Posted Prices

OPEC has already achieved recognition of the principle of basing tax payments on posted or reference prices. Until the royalty issue was settled, the Middle East governments had calculated tax payments on the prices as established in the price cuts of August 1960. Because the oil companies have sold increasing amounts of oil at discounts, they have paid to the governments in taxes including royalties more than 50 percent of their net revenue. The royalty settlement authorized a discount from posted prices in the calculation of taxes. But the agreement provides for the fading out of the discounts and thereafter the calculation of taxes on a basis of posted prices.

Oil concession contracts of Venezuela and Indonesia, and Libya until its 1965 legislation, specifically authorized their concessionaires to use actual realized prices as a basis for calculating their taxable income. A concessionaire's willingness and ability to grant discounts from posted prices are obviously enhanced as long as doing so reduces his tax obligations. When a concessionaire is required to base his tax calculations on posted prices even though he sells his oil for less, his costs in effect are raised, as is the level of prices at which he can profitably sell his product. Under these circumstances he may be no less eager to find buyers but will be more cautious in pricing his product. A stable reference price not only tends to lessen the frequency and amount of price-cutting but to increase the revenues of the state. And this is its ultimate object.

At its eleventh conference, in Vienna in April 1966, OPEC in Resolution XI.72 designated the use of posted or reference prices for tax calculations as a specific objective. Venezuela had established a Co-ordination

Committee for the Conservation and Commerce of Hydrocarbons as early as April 1959. Its objective was to prevent sales of petroleum at abnormal discounts. It did not work well. After OPEC's 1966 resolution, Venezuela began a more determined drive to improve its tax realization. After extensive negotiations with the major oil companies, accompanied by threats of legislative action, it succeeded in establishing as of January 1, 1967, a schedule of reference prices midway between realized prices and posted prices as a basis for calculating oil taxes.

OPEC's "Declaratory Statement of Public Policy" adds something new to this principle. It declares not only for the payment of taxes on a basis of posted or reference prices but contemplates the state's determining the prices. In short, its objective is to obtain for OPEC members the authority to increase tax revenues by designating prices on which taxes are to be calculated regardless of prevailing market prices. The posted prices under prevailing market conditions have no real meaning except for tax purposes. For the governments to obtain the authority to adjust this price is to recognize their authority to raise taxes at their discretion, despite the provisions of their concessions to the contrary.

The Recapture of Excessive Earnings and Participation in Ownership

The requirement that all contracts provide for the payment of taxes on the basis of reference prices fixed by the government will increase the cost of producing oil. The recapture through taxation of excessively high earnings, with the government deciding what constitutes excessive earnings, places the risks on the concessionaire, with the government's taking the profits in good years with no guarantee to the companies of minimum earnings in bad.

The economic consequences of government participation in ownership of existing concessions are less clear. What is clear is that OPEC members have dedicated themselves to the achievement of this objective at the earliest possible date. Even before OPEC had spoken officially, Saudi Arabian Oil Minister Yamani, at the 1968 American University seminar on the economics of the international oil industry, indicated that the major companies must revise their whole system of operation by enlisting the partnership of their host governments. Calling attention to the recognition of this principle in recent agreements, he insisted that psychological repercussions of the Israel-Arab war had made it "absolutely essential" that the majors "and not the least Aramco" follow suit if they wish to operate peacefully in the area. "Partnership with the host government is a must; any delay will be paid for by the oil companies concerned," he said.[4] Yamani's ultimate objective is a 50–50 partnership, but he emphasizes that it is the principle of partnership that counts.

On October 7, 1968, at a meeting of the U.S.–Arab Chamber of Commerce at New York's St. Regis Hotel, the Kuwait minister of finance and oil expressed a similar sentiment in these words:

> One desirable means of establishing long-lasting and appreciable collaboration between a host government and a concession-holding company is the participation of the governments in the ownership and operation of such a company. This would not only give the government a more active role in the management of the most important source of its national income, but it would place the foreign company on a firmer ground. In our country, as indeed anywhere else, a partner is more welcome than an exploiter.[5]

At OPEC's seventeenth conference, held in Baghdad, November 9–10, 1968, apparently ownership participation was the chief topic of a closed session. The Iraqi oil minister is said to have placed it formally on the agenda, seconded by Kuwait.

Participation in Ownership a Substitute for Nationalization

Participation in ownership is a substitute for nationalization. It would raise many complex problems—first and foremost of which would be the terms on which the government would acquire participation. Middle East oil is currently selling in excess of its cost. The concessionaires are making noncompetitive profits on their investment. If the government should pay a price commensurate with the earning power of the investment, in theory it would be a matter of indifference to the companies whether or not they sold. But it would be an extremely unprofitable investment for the governments to make. On the capital invested they could expect only a normal rate of return, whereas with no investment they now share in the monopoly profits of their concessionaires.

But the assumption that they pay a price commensurate with the

4. *MEES,* June 7, 1968, *Supplement.*
5. *Ibid.,* October 18, 1968, *Supplement.*

earning power of the investment is wholly impractical. If the governments participate in ownership, the most they will expect to pay for their share will be a price based on the depreciated value of the original investment, or on the cost of reproduction, or some criterion to be worked out by negotiation.

Assuming that the governments acquire participation on what they regard as reasonable terms, what effect would this have on the governments' income and the companies' costs? What would be the effect on the volume of oil output and the stability of prices? These questions may be considered on the basis of two different assumptions: that the governments behave as silent partners, leaving decision-making entirely to the present owners; that the government actively participate in making decisions.

Governments Behave As Silent Partners or Active Managers

The margin between the cost of producing Middle East oil including the government tax and the price is still large. The concessionaires continue to make noncompetitive profits. With the government a part owner their monopoly profit would be lessened. The larger the government's share, the less would be their profit. Presumably, the companies have adjusted their output among the several countries in relation to aggregate demand and their own several supply requirements in such manner as to maximize their profits. Government participation increases the government's share in the profits and increases the companies' per barrel cost. The upper limit to the government's take will be set by the relationship between costs and prices. When they are equalized, there will be nothing more for the government to take. Participation in ownership has the same effect as would an increase in the rate of sharing profits, now nominally 50–50.

The governments have made clear their intent to participate actively in managerial decisions if and when they participate in ownership. As part owner, each country is likely to exert pressure on the concessionaire to increase output. Iran, which before its nationalization venture was the Middle East's most important producer, is now launched on a program of recapturing its former premier position. Saudi Arabia became first in output in 1966 and is unlikely to surrender its premier position without a struggle. Iraq has resented its slow expansion and is determined that its production be speeded up. Kuwait, although blessed with government revenues beyond its fiscal needs and with the world's largest per capita income, has manifested no disposition to surrender output for the benefit of its neighbors.

The instrument through which government ownership will be exercised is the national oil company. It may elect to take its share of output and to sell it. It will pay it to do so as long as it receives a price which yields a profit in excess of the rate of tax it would have gotten from the concessionaires with appropriate allowance for its capital investment. If the concessionaires choose to buy the government's share, their costs will be increased unless they act merely as selling agents valuing the oil for tax purposes at the realized price.

But whether the national companies sell on the market or to their partners, the result will be an increased flow of oil onto world markets. The rate of increase will depend on how aggressive the national oil companies are. They are not unaware of the problem which their uncoordinated production will create, but they have evinced no capacity for solving it.

Companies Will Resist Participation

Regardless of the governments' motives in demanding participation in ownership, the companies will resist it. While OPEC representatives have made clear that they are prepared to support vigorously the demand to share in ownership, Aramco has made clear its opposition. In the *Middle East Economic Survey* for December 20, 1968, Aramco's president, R. I. Brougham, released the legal opinion of Dr. Saba Habachy on the issue of "changing circumstances." Dr. Habachy, a distinguished scholar and practitioner of international law, at present an adjunct professor of law and lecturer in legal institutions of the Middle East at Columbia University, develops at length the significance of changing circumstances under Arab legal principles. In recognizing the binding character of the law of contracts, he concludes that the two fundamental sources of Islamic law, the Koran and the Sunna, do not support the doctrine of changing circumstances.

Dr. Habachy's opinion brought a quick response from an official of the Kuwait minister of finance, who challenged the applicability of Dr. Habachy's ideas, and on January 13, Saudi Arabian Oil Minister Yamani, lamenting the legalistic wrangling in which the issue might become involved, stated,

We did not expect Aramco and the other oil companies to accept the principle of participation the moment it was announced and its application demanded. The road is long and difficult, but the day will come when the companies will share our view that participation between them and ourselves is the only way to establish a stable oil industry and to protect both parties against the danger of new trends that have manifested themselves in consuming countries.[6]

OPEC's Program Will Squeeze Out the Residue of Oligopoly Profits

The program outlined in the OPEC's declaratory statement in its entirety is calculated to drive up costs, increase the flow of oil on world markets, and lower the price of oil. It will tend to squeeze out the residue of monopoly profits, or economic rent, which the oil companies now make. Dr. Penrose suggests that it may result in the oil companies' surrendering their concessions before they expire and becoming contractors for the purchase of crude oil in a buyers' market.[7] To such a contingency Yamani is bitterly opposed. "In the long run it could result in a drastic reduction in oil prices, and the only beneficiaries would be the consuming countries."[8]

If the program engenders lower prices, OPEC is apparently prepared to fall back on a proration program. Yamani, who is now opposed to such practice, would "become a prime supporter of such a measure."[9] For reasons already discussed, such a program is unlikely to work. In addition to the fundamental obstacles which confront it, it is likely to meet the opposition of consuming countries, some of which are already resisting the high prices of imported crude.

The goals outlined in OPEC's declaratory statement of policies are long-run goals, the road to which is likely to be fraught with bitter controversy between individual governments and their concessionaires. Their ultimate objective is the capture by the Middle East governments of the whole of the economic rent which their low-cost oil has engendered. There is danger that they may wipe it out entirely.

6. Yamani's statement was made in an interview published in the Kuwait daily al-Siyasah, MEES, January 17, 1969.

7. Edith Penrose, "Government Partnership in the Major Concessions of the Middle East: The Nature of the Problem," MEES, August 30, 1968, Supplement.

8. Ibid., June 7, 1968, Supplement.

9. Idem.

23

Political Instability and the
Security of the Concessions

THE security of the oil concessions may depend ultimately on the political stability of the host countries. Political stability of the Middle East countries depends in part on the forces operating within a given country, in part on external forces which affect the whole region. Stability as it derives from internal forces is a reflection of economic and political factors. Political maturity makes for political stability. Iraq, Kuwait, and Saudi Arabia as independent nations are creations of the twentieth century. Iran in contrast has been an independent nation since ancient times. Much of its history has been characterized by political rivalry among those who have sought the power to govern. Since the failure of the Mossadegh revolution and the return of Shah Muhammad Reza, however, the government has been characterized by increasing stability and by remarkable economic development. The oil industry has supplied the fuel for Iran's economic progress.

Iran Gears Five-Year Plan to Oil Revenues

In negotiations with the consortium in the spring of 1968 Iran made clear its intention of gearing its oil output to its needs as envisioned in its development program. On March 1, 1968, Prime Minister Hoveyda in an address before the Iranian Parliament stated,

448

The oil income figures contained in the 1347 [Iranian calendar] budget reflect neither Iran's true needs nor her demands; they merely represent the bare minimum required by Iran during the five-year period of the Fourth Development Plan. We cannot stand idly by while our own oil resources are kept unexploited underground and not utilized for the country's development.[1]

In an address at the inauguration of the Soviet-aided steel project at Isfahan on March 14 the shah put it even more bluntly:

What we must emphasize is that . . . decisions cannot be taken unilaterally. We as the owner of this oil and the master of this land, must have a say in the production of this wealth because the needs of the country are clear. It is clear to what purpose the country's revenues are being spent. No firm, no company, no organization can tell us, merely because it has an agreement with us, "We will produce and export so much of your national wealth, but you cannot touch the rest as we do not wish to exploit it." What this means is that they want to deprive Iran of this wealth which rightfully belongs to us.[2]

The owners of the consortium, as the leading producers in other Middle East countries, view the matter differently. They have the delicate task of maintaining an equilibrium of satisfaction among the several host governments. They feel compelled to dampen excessive demands of any one country. They take the position that the increase in output of any one country is limited by the general level of output for the Middle East as a whole. As judged by world demand, they estimate the annual rate of growth for the Middle East at six or seven percent. Applying these figures to Iranian output for the revenue year 1969–70 left a gap of $100 million between the Iranian contemplated expenditures and the realized income. Expenditures under the plan (63.1 percent of which are scheduled to come from oil revenues) are slated to increase more rapidly than the consortium's projected increase in output. The gap will grow.

The Shah Carries His Propaganda Abroad

In an endeavor to mobilize Western sentiment in support of Iran's demand for an increase in oil output sufficient to close the gap, the shah sponsored a brochure written by David Missen and published in Britain in the spring of 1969 under the title *Iran: Oil in the Service of a Nation*.[3] The

1. *Middle East Economic Survey*, March 8, 1968, published the full text of the prime minister's remarks.
2. *Ibid.*, March 22, 1968.
3. (London: Transorient Books, Ltd., 1969).

brochure constitutes a detailed account of Iran's recent progress under the shah's leadership: progress in industry, in agriculture, in the redistribution of feudal estates, in social and economic reform, in the status of women, in the introduction of the co-operative movement, and in improvement in living standards. Alleging that Iran has entered the critical stage of "economic take-off on the transition path from a less developed to a developed and diversified economy"[4] and that it has attained a "political stability [that] is a byword in Asia and indeed throughout the less developed countries,"[5] the author mobilized arguments in support of the shah's demands. He calls attention to Iran's population of approximately 28 million, far larger than the combined populations of Iraq, Saudi Arabia, Kuwait, and the oil-producing sheikdoms of Qatar and Abu Dhabi, and its comparatively meager oil revenue per capita: £9.5 for Iran in 1967, £15 in Iraq, £50 in Saudi Arabia, £243 in Qatar, £519 in Kuwait, and £2,187 in Abu Dhabi.

The author draws a contrast between the uses to which Iran has devoted her oil revenues and those of other Middle East countries. "Unlike her smaller neighbors Iran does not accumulate vast sums of unspent foreign exchange in the hands of a ruling elite"[6]—unspent sums so large that they constitute a threat to international monetary stability. "At no time in the past four years have the reserves of this one statelet [Kuwait] with hardly half a million people amounted to less than 50 percent of the available foreign exchange of the Bank of England."[7]

In Iran "none of the oil earnings have gone into private fortunes, and none has been wasted in either doctrinaire projects, military attacks on Iran's neighbors, or on forms of conspicuous personal consumption too common in some neighboring states."[8] No other "country in the Middle East can claim the potential for balanced growth or anything approaching the success and efficiency with which oil revenues have been used in the past."[9] The author concludes with the observation that the shah has made it ominously clear that "he regards the situation where another country or, still more serious, commercial companies from other countries, can dictate to a sovereign state the rate at which its resources shall be developed and used as intolerable and unjustifiable."[10]

4. *Ibid.*, p. 4.
5. *Ibid.*, p. 13.
6. *Ibid.*, p. 3.
7. *Ibid.*, p. 23.
8. *Ibid.*, pp. 19–20.
9. *Ibid.*, p. 25.
10. *Ibid.*, p. 29.

The Consortium's Compromise Settlement Brings Spirited Arab Responses

With this exercise in international propaganda designed to soften consortium resistance, the shah's representatives began negotiations on May 10, 1969, with consortium representatives. On May 14 the negotiations reached a compromise settlement. Covering only a single year, it can be only a prelude to further negotiations. The agreement as reported in the *Middle East Economic Survey* for May 16 provided that the consortium make available during 1969–70 the full amount of revenue that Iran demanded. This will be provided in part by an increase in production of lighter, higher-priced crudes, in part by short term advances against future revenues. It does not meet Iran's demand that the consortium gear oil revenues to expenditures under the five-year plan. This is left for future consideration.

Meanwhile the neighboring Middle East countries have made it clear that they will not tolerate an increase in Iran's oil revenues at their expense. Kuwait's crown prince and prime minister, Sheik Jabir al-Ahmad al-Sabah, warned the oil companies that "an increase in their production in other areas at the expense of . . . oil production" in his country "will make it necessary to reconsider their concession agreements."[11] And Iraq's minister of oil and minerals, Dr. Rashid al-Rifa'i, in a statement to the Iraqi News Agency on May 17, 1969, declared, "We shall not tolerate an increase in production in any country at the expense of Arab oil output in the region as a whole or at the expense of Iraqi oil in particular.[12] Oil Minister Yamani a year earlier had made clear Saudi Arabia's intention of not sitting quietly by while the governments brought pressure on the oil companies to increase their output. Alleging that Saudi Arabia had never interfered with Aramco's production plans, he stated that "If . . . we see that the oil companies have allowed other governments to interfere with the free play [of economic forces] vis-à-vis Saudi offtake levels, we will at once move to safeguard our interests."[13]

As long as Iran remains prosperous, it is quite possible that the consortium, by giving a little and Iran, by bending a little, can maintain the status quo. But a severe economic recession would intensify the govern-

11. *MEES,* May 23, 1969.
12. *Idem.*
13. *Ibid.,* April 5, 1968.

ment's demands on the consortium and harden the consortium's response. That Iran, whose democratic roots are so shallow, could survive the stress of an oil crisis is doubtful. It could not in 1952. Policy under the condition of crises may well be governed by emotion rather than reason. The life of the concession may be at stake.

Iraq and the Security of IPC

This study of the oil controversy in Iraq (Chapters 9–14) indicates the turmoil that has characterized continuously that country's entire political life. The political winds have changed direction almost with the frequency of the seasons if not with their regularity. And they have blown with increasing intensity. With a single brief exception, since the Revolution of 1958 military men have headed the government. Although all have aspired to the creation of a united Arab nation, each has jealously guarded the power he has won by coup and countercoup. Each has made his contribution toward the insecurity of the IPC concession. Revolution that brought the confiscation of 99.5 percent of the concession ultimately led to Iraq's turning to France and Russia for aid in exploiting the confiscated lands. The Arab-Israeli six-day war and the postwar activity of the fedayeen have confounded the chronic confusion which has engulfed Iraq. They have created an insecure environment for the operation of an oil industry based on governmental sanction.

Kuwait and the Security of KOC

With its governmental departments and its elected national assembly Kuwait has the paraphernalia of a democratic state but neither the experience nor the disposition to operate it as such. The sheik and his counselors are well satisfied with the concession arrangement, which until recently had given this small tribal state the world's highest per capita income and has made it the banker of the Middle East. Dissident elements in the national assembly, primarily Arab Nationalists, have not hesitated to criticize government oil policies and from time to time have challenged the validity of its concession agreement. Their influence has led to Kuwait's assuming a positive and aggressive policy in protecting the country's oil interests. Her poverty-stricken neighbors look with jealousy on the riches of Kuwait. It seems doubtful that they will tolerate indefinitely so much going to so few. The political viability of a nation, so small and so rich, but basically undemocratic, in the midst of so much misery and poverty is problematical.

Stability in Saudi Arabia

Saudi Arabia ostensibly has the most stable government of the Middle East. Its oil reserves are the largest. Its oil revenues afford funds adequate to its investment opportunities. Its oil minister is a common-sense individual who realizes that Saudi Arabia has a good thing going. While he wants to get out of the oil industry as much as he legitimately can, he would be reluctant to support any measure which would threaten Aramco's concession. King Faisal is respected as a competent ruler, dedicated to a far more intelligent use of oil revenues than was his predecessor.

But the recent history of Saudi Arabia brings in question its viability. Saudi Arabia is one of the political anomalies of the twentieth century. Created as a dynastic state in the first quarter of this century, it has remained free of the modern democratic paraphernalia which limit the power of a sovereign and give expression to the will of the people. This has not guaranteed its political stability. The years since the death of the old King Saud in 1953 have been characterized by a continuing feud between Prince Faisal and King ibn Saud. This resulted in the surrender of power by the king to Prince Faisal in 1958, its resumption by the king in 1960, its surrender to Faisal again in 1962. It culminated in 1964 in Faisal's being acclaimed king and ibn Saud being banished from the country. The feud ended with the death of ibn Saud in Greece in February 1969.

The Problem Confronting King ibn Saud

Each shift in power not only represented a cleavage in the royal family, but was accompanied by internal and external problems that threatened the life of the kingdom. The problems confronting King ibn Saud when he ascended the throne on the death of his father were new; the institutional equipment with which to cope with them was outworn. Suited to a relatively primitive society guided by religious tradition and tribal custom in administering justice, dispensing charity, and handling social relationships, it was ill-suited to channel large and rapidly growing governmental revenues into a social welfare program characteristic of modern states or into an economic development program to which industrially backward countries have come to aspire.

Traditionally the whole income of the state was the personal property of the king. But he was expected to hold it in trust for his people and administer it for their welfare. It was inevitable as the king and his counselors created government departments to assist in the process that they should be manned by members of the royal family. They were presumed to have a greater awareness and sense of responsibility in regard to state affairs than others, and the supply of alternates was strictly limited. But the royal family, educated in accordance with the Koran and presumably guided by its principles, had had no experience in modern governmental administration. They found themselves the willing occupants of posts for which they had neither aptitude nor training and they staffed departments with friends and relatives similarly inept in modern matters of state. They were ill equipped to fulfill their social responsibilities. To grasp their personal opportunities required a simpler discipline and it proved adequate to the occasion. Philby stated the matter with restraint and tolerance in pointing out that the official hierarchy was soon shot through with a "low standard of responsibility and altruism."[14]

The Government Faces Bankruptcy and Faisal Becomes Prime Minister

Despite an increase in annual oil revenues from $169,800,000 in 1953 when King Saud ascended the throne to $296,300,000 in 1958 when the king temporarily surrendered the powers of government to Crown Prince Faisal, the government was on the verge of bankruptcy. It owed approximately 1,250,000,000 Saudi riyals abroad and more than 600,000,000 at home. The gold, silver, and other foreign currency holdings of the Saudi Arabian Monetary Agency had fallen to a bare 14 percent of the note issue. Meanwhile the riyal had declined on the free market from an official rate of 3.75 to the dollar to an all-time low of 6.25 to the dollar. Not only was the government virtually bankrupt, but the king, with a personal

14. H. St. J. B. Philby, "The New Reign in Saudi Arabia," *Foreign Affairs,* April 1, 1954, pp. 446–450 at p. 449.

income equaled by few if any other individuals, was himself in debt. Such irresponsible conduct of governmental affairs had been accompanied by an elaborate building program both by the government and by a limited number of Saudi citizens whose affluence could be traced directly or indirectly to the expanding oil revenues and the prosperity that their expenditures had generated. During this boom the king attained worldwide renown for his conspicuous consumption and the extravagance of his living.

Not only was the government virtually bankrupt, but Saudi Arabia's position in the Arab world had seriously worsened as a result of an alleged plot by King Saud to prevent the union of Egypt and Syria and to bring about the assassination of President Nasser.[15] Revelations with regard to the alleged plot created a wave of resentment against the king in Saudi Arabia. To rescue the government from bankruptcy and to ride out the political storm, the king called on Crown Prince Faisal to head the government as prime minister and turned over to him control of state affairs.

Shortly after assuming office, Faisal inaugurated a stabilization program designed to put the government on a sound fiscal basis. His task was to stabilize the currency, to balance the budget and retire the outstanding debts. To achieve these ends the government instituted a rigorous program of import, exchange and budgetary controls.

When Faisal inaugurated his austerity program unaccompanied by any significant political reforms, the royal family split into three feuding groups. Faisal and the members of the royal family whose support he had enlisted in putting through his fiscal reforms; King ibn Saud, his sons and those brothers jealous of power and smarting under Faisal's economy program; and a small group of so-called "free" or "liberal" princes who under Prince Talal's leadership agitated for far-reaching political reforms.

Faisal's austerity program was accompanied by falling prices, a slump in business activity, and dissatisfaction within the business community. The

15. The plot was revealed by President Nasser in Cairo on March 5, 1958, and the plot's details were publicized by 'Abd al-Hamid Sarraj, Syrian chief of military intelligence and later vice-president of the U.A.R., at a press conference in Damascus on the same date. At this conference Sarraj not only revealed the details of the conspiracy but released documentary evidence to support them. On the day following Sarraj's press conference King Saud officially denied playing any part or having any knowledge of such a conspiracy and announced that he had ordered a special committee to inquire into the whole affair. Despite the king's denial his father-in-law, As'ad Ibrahim, alleged to be the go-between in the conspiracy, was later tried in absentia and convicted by a Syrian court.

king, who had grown restless with the loss of his governmental power, enlisted the support of the young liberals and in December 1960 forced Faisal into temporary retirement.

King ibn Saud Resumes Control

Before lending their support to the king and accepting cabinet positions, the young liberals had obtained from the king assurance that he would institute a reform program consistent with their aspirations and the country's welfare. Not only had the king elevated to cabinet position the leaders in agitating for reform but he had surrounded himself with like-minded advisers, chief of whom were 'Abd al-'Aziz ibn Mu'ammar, and 'Abd Allah ibn Hujailan. Mu'ammar, well educated and widely traveled since his student days at the American University of Beirut, had been deeply concerned about the social and economic problems of his country. Early in his reign King Saud had appointed him director general for labor in the Eastern Province. Later, accused of activity on behalf of the "National Front for Liberation," one of the two or three reform groups operating sub rosa in Saudi Arabia, and conducting a more open conspiracy outside its borders, Mu'ammar was imprisoned. When released, he spent some time abroad but had returned to his native land before the king regained control of governmental affairs. In the 1960 Faisal-Saud struggle, lacking confidence in Faisal, he cast his lot with the king.

The new cabinet had scarcely been seated and the king's liberal advisers designated than it became evident that the king was more interested in enlisting the support of the "liberals" than in carrying out their program. It also became evident that the so-called "bloodless coup" was merely a critical episode in a continuing struggle for power within the royal family. Within less than two months the princes supporting Faisal were conducting a campaign for his reinstatement.

Meanwhile the king's illness, which led to his widely publicized trip to Peter Brent Brigham Hospital in Boston and a visit with President Eisenhower, brought about some sort of a reconciliation with his feuding brother. Before leaving Saudi Arabia, the king issued a royal decree designating Prince Faisal as regent and acting prime minister during his absence.

However, not all the brothers were reconciled. Prince Talal, who had been serving as minister of finance, resigned and left the country. From

Beirut he tried to mobilize liberal sentiment against the king and Faisal by conducting a campaign of vigorous and bitter publicity. The conservative brothers, feeling their strength, in March 1962 forced the resignation of the Commoners in the cabinet, including the minister of petroleum, Sheik Abdullah Tariki.

Faisal Made King

In the fall of 1962 following the Yemen crisis the king dismissed his cabinet and charged Prince Faisal with the task of forming a new government. As the Yemen crisis deepened, the ineptness of the king to cope with domestic and external problems became apparent to all. On November 2, 1964, in response to a unanimous decision of the royal family (Prince Talal after a public apology had returned penitent) and the Ulama, the council of ministers, deposed ibn Saud and acclaimed Faisal as king.

King Faisal affirmed his determination to institute reforms in the political, administrative, and economic fields. He has utilized the Ford Foundation to advise in administrative organization and the training of personnel and has curbed payment to the royal family for the benefit of other departments. An economic development program, educational reform, and improvement in communications are central to his administration. He has won the support, at least temporarily, of a new generation of political dissidents, educated in the West, with conflicting political ideals, who before Faisal's being made king accentuated dissension among the several factions within the royal family. Apparently they have become convinced that a stable political regime under Faisal's leadership offers more for the country's welfare than would the confusion of revolution.

This brief discussion of the political history of Saudi Arabia should indicate that no enduring foundation has been laid for a program characterized by stability and progress. Power rests in the hands of the king and his council. The king has come to recognize that the survival of the royal family as rulers of Saudi Arabia depends primarily on a wiser use of oil revenues. The regime is essentially nonlibertarian, supported by no enlightened free press, intolerant of dissent.[16]

---

16. Faisal hushed forever the dissent of Mu'ammar by causing him, after he had served as ambassador to Switzerland, to be put in prison, an imprisonment the rigors of which his friends say he has not survived.

Israel and the Oil Concessions

Meanwhile Israel and the cold war have made insecure the king's tenure and that of all the governments of Middle East Arab countries. The most disturbing factor in the Middle East is the existence of Israel and the refusal of the Arabs to accept it. The war of 1948, the refugee problem, the war of 1956, and the war of 1967, as all the world knows, have left the Arabs embittered. The six-day war left Israel determined to relinquish none of the conquered territory until a permanent peace settlement is achieved. The war exacerbated Arab hostility toward the West and temporarily stopped the flow of oil to it. The Arabs in their anger threatened to seize all oil properties.

On June 4 and 5, 1967, the ministers of the several Arab producing states met in Baghdad to consider the crisis and to develop a common policy on the uses of oil as a weapon against the "aggression" of Israel. The conference, attended by representatives of the U.A.R., Iraq, Algeria, Libya, Kuwait, Saudi Arabia, Syria, Lebanon, Qatar, Bahrain, and Abu Dhabi, unanimously approved a resolution to stop the flow of oil from the Arab producing countries with a view to preventing its sale and delivery to any state that commits or takes part in an aggression directly or indirectly against any Arab state. Immediately thereafter Saudi Arabia, Kuwait, Iraq, and Libya stopped all oil exports. Saudi Arabia resumed exports at midnight on June 13, on written guarantee that Aramco would ship no oil to "those states named by his Majesty's Government as having taken part in the aggression on Arab states." A spokesman for Aramco identified the embargoed countries as the United States and Great Britain. By June 19 all Arab states except Iraq had resumed oil exports to countries other than the United States and Great Britain. During the last week in July Iraq authorized IPC to ship Iraqi crude from its Tripoli terminal to all destinations except the United States, West Germany, Rhodesia, and South Africa.

As the limited embargo continued, it became evident that the Arab states were the chief sufferers. According to the Kuwait daily, *al-Ray al-'Amm,* Kuwait stood to lose $145,000,000 annually in oil revenue.[17] Oil Minister Yamani estimated Saudi Arabia's loss for the twenty-four days of June at $30,264,900 and the annual loss at $122,600,000.[18] Meanwhile Iran and Venezuela had stepped up their oil shipments.

17. *MEES*, August 4, 1967.
18. *Ibid.*, July 14, 1967.

When the Arab summit conference convened in Khartoum on August 29, the futility of the embargo had become evident. The conference rejected Iraq's proposal for a total ban on oil shipments and voted that oil be used as a positive weapon to strengthen the economies of the Arab states most directly affected by the "aggression." The result was eventual lifting of the embargo and agreement by Saudi Arabia, Kuwait, and Libya to contribute $378,000,000 annually to Jordan and the U.A.R., which had borne the brunt of the Israel-Arab War.

Oil Minister Yamani had questioned the wisdom of the embargo as early as mid-July when in an interview he said, "injudiciously used, the oil weapon loses much if not all of its importance and effectiveness. If we do not use it properly, we are behaving like someone who fires a bullet into the air, missing the enemy and allowing it to rebound on himself."[19]

Whether or not occasioned by a recognition of the failure of the embargo and the need of insuring that future policy be kept in the hands of the leading oil exporting states, Saudi Arabia, Kuwait and Libya, which were the contributors to the fund for rehabilitating the chief sufferers of the June war, in January 1968 formed the Organization of Arab Petroleum Exporting Countries (OAPEC). Oil Minister Yamani in a statement to the press emphasized the essentially economic and commercial nature of the new enterprise and stated that its objective is "to maximize the benefits derived by its members from their natural resources."[20] Its restricted membership, however, may be significant, as may be the fact that Iraq, consistently following a more belligerent policy toward the major concessionaires, refrained from membership. Significant also may be Yamani's statement of the reasons for OAPEC's organization. In an interview with the Kuwait newspaper *al-Siyasah* on September 12, 1968, he stated that one of the three purposes that gave rise to OAPEC's organization was "To keep oil activity within the organization with a view to protecting the member states from precipitous decisions and making oil a genuine weapon to serve the interests of the producing countries and the Arab countries in general."[21]

While policy was thus placed in the hands of the more conservative Arab states and the ones that have the most to lose if oil is to be used as a weapon to "liquidate the consequences of Israeli aggression," it is unlikely that these states could guarantee the preservation of the international oil companies' investments if all-out war breaks out again.

19. *Ibid.,* July 21, 1967.
20. *Ibid.,* January 12, 1968.
21. *Ibid.,* September 13, 1968.

The Cold War and Middle East Oil

American policy in the Middle East has been designed not only to prevent the encroachment of Russia on her neighboring states but to minimize Russian influence throughout the area. It has failed. By subordinating ideology to pragmatism the Soviet Union has effectively cultivated the friendship of most Middle East countries. Russia has supplied them with military equipment and technical and financial assistance in their economic development while blinding itself to their avowed hostility toward communism and their treatment of Communists. It has echoed their plaint against the "imperialist" ambitions of the oil companies and the West. It has encouraged have-not nations to covet the wealth of nations that have. The United Arab Republic, committed to national socialism, with Soviet approval has conducted an ideological war against Saudi Arabia's royal family and the wealth that they control. Cairo's Yemen adventure was in part a face-saving response to Syria's withdrawal from the United Arab Republic and in part an effort to subvert Saudi Arabia and gain access to the Persian Gulf oil region. Faisal, influenced by institutional affinity with the Yemen monarch and aware of the threat that Nasser's success would constitute to his rule, was drawn into the struggle. The early Arab summit conferences were unsuccessful in obtaining the withdrawal of Egyptian troops from Yemen. Not until the Khartoum conference was Nasser persuaded to abandon his Yemen adventure and then only in response to Saudi Arabia's agreement to contribute financial aid with Libya and Kuwait to Egypt as long as the Suez Canal remains closed.

Meanwhile Saudi Arabia has become an important source of financial aid to al-Fatah, whose sole aim is the "liberation" of Palestine, and Faisal has given verbal dedication to a suicidal war against Israel. This serves to divert Nasser's attention from the longrun task of subverting Saudi Arabia.[22] Although it contributes to the danger of precipitating anew an all-out war between the Arab states and Israel, it dampens internal opposition to the rule of the royal family. It dampens but does not extinguish it.

September 1969 dispatches from Beirut, Lebanon, indicate that twice within two months Faisal with a tipoff from the American Central Intelligence Agency has aborted a military coup and executed thirty conspirators.

22. See J. Gaspard, "Faisal's Arabian Alternative," *The New Middle East,* March 1969, pp. 15–19.

This indicates how tenuous is the royal family's hold on the throne. Faisal is recognized as the ablest member of the royal family for whom no successor is in sight. His death might herald the death of the dynasty.

The overthrow of the monarchy would not necessarily mean any abrupt change in the status of Aramco's concession. After their swift and decisive coup in Libya, the military announced their intention to honor all current oil agreements. But Libya's transition to the radical Arab camp with its slogan of "unity, freedom and socialism" seems certain to change the environment within which the oil industry operates.

Neither Arab socialism nor Russian influence, however, is apt to prevent the flow of Middle East oil to the West. Middle East oil continues to grow in abundance and at the world's lowest cost. Recent estimates indicate that Saudi Arabia's reserves have reached the 100 billion-barrel mark. John K. Jamieson, Standard of New Jersey's new chief executive officer, placed the significance of the giant new Alaskan Prudhoe Bay field with a potential of 5 to 10 billion barrels into proper perspective by stating that last year Aramco routinely added 8.5 billion barrels to its Saudi Arabian reserves.[23] "Tension in the Middle East," Jamieson observed, "seems to be getting worse and worse, and we view this as a most critical, dangerous area." This study furnishes a solid basis for Jamieson's apprehension but, barring war, neither Russian encroachment on the Middle East nor the growth of Arab nationalism is apt to impede the flow of Middle East oil to the West more than temporarily.

The apparatus for its distribution in the Western world is complex and expensive and firmly under the control of the international oil companies. They have their worries. The terms under which they take the oil may grow progressively more burdensome. But neither the Arabs nor Russia can long deny a market to Middle East low-cost oil.[24]

23. *Wall Street Journal*, August 28, 1969.
24. In early 1970 the likelihood that the Israeli-Arab conflict may develop into a full-scale war grows greater, as does the possibility that Russia and the United States may be involved in it. Under these circumstances any prognostication regarding the future of Middle East oil is hazardous.

Appendix

TABLE II

Annual Production of Crude Oil for Indicated Countries 1946–1968
(Thousands of Barrels)

Year	Iran	Iraq	Saudi Arabia	Kuwait
1946	146,819	37,168	59,944	5,928
1947	154,998	37,595	89,852	16,228
1948	190,384	28,025	142,853	46,547
1949	204,712	33,080	174,009	89,790
1950	242,475	50,972	199,547	125,722
1951	127,600	65,981	277,962	204,910
1952	10,100	142,378	301,861	273,433
1953	9,800	212,215	308,294	314,592
1954	22,400	232,198	351,044	350,294
1955	120,035	254,394	356,449	402,828
1956	198,289	234,610	367,037	405,696
1957	262,742	164,086	376,254	427,627
1958	301,526	269,939	386,343	524,070
1959	338,810	312,772	420,733	525,995
1960	390,766	355,829	480,734	619,193
1961	438,804	367,851	540,237	633,280
1962	487,084	368,358	599,666	714,598
1963	544,325	424,090	651,890	765,150
1964	626,107	459,403	694,302	842,160
1965	696,520	479,099	905,190	861,527
1966	778,109	508,141	850,059	906,702
1967	950,180	448,239	1,024,263	912,427
1968	1,039,367	553,131	1,114,177	956,549

Source: Annual Statistical Bulletin, Organization of the Petroleum Exporting Countries, Dr. Karl Lueger-Ring 10, 1010 Vienna, Austria

TABLE III

Annual Revenue from Oil for Indicated Countries 1946–1968

Year	Iran Million £	Iraq Million £	Saudi Arabia Million $	Kuwait Million £
1946	7.1	2.7	10.4	———
1947	7.1	2.7	18.0	———
1948	9.2	2.1	52.5	———
1949	13.5	3.1	39.1	———
1950	16.0	6.7	56.7	4.9
1951	7.0	15.1	110.0	6.7
1952	———	33.1	212.2	20.7
1953	0.1	51.3	169.8	60.2
1954	7.4	57.7	236.3	69.3
1955	32.2	73.7	340.8	100.5
1956	53.9	68.9	290.2	104.3
1957	76.0	48.9	296.3	110.2
1958	88.3	79.9	297.6	128.5
1959	93.7	86.6	313.1	159.8[g]
1960	101.8	95.1	333.7	158.6
1961	104.0	94.8	377.6	167.1[h]
1962	122.3	95.1	409.7	173.0
1963	135.7	110.0	607.6[b]	190.6
1964	172.2	126.1	523.2	206.2
1965	190.7	131.4	662.6[c]	216.1
1966	216.8	140.8	789.7[d]	231.7
1967	277.5	131.7	909.1[e]	242.7
1968	324.7	203.0[a]	932.4[f]	239.2

a. Including £17 million as special payments for oil shipped from Mediterranean ports since June 1967.

b. Including special payments by Aramco of $152.5 million.

c. Including $46 million additional tax.

d. Including $29.4 million additional tax.

e. Including $23.46 million additional tax.

f. September 1967–September 1968 (budget estimate).

g. Revenue from 1/1/1959 to 3/31/1960 (15 months).

h. Revenue for 1961–69 is for budget years.

Source: Annual Statistical Bulletin, Organization of the Petroleum Exporting Countries, Dr. Karl Lueger-Ring 10, 1010 Vienna, Austria

Selected Bibliography

Baldwin, George B. *Planning and Development in Iran.* Baltimore: Johns Hopkins Press, 1967.

Barrows, Gordon H. *The International Petroleum Industry.* New York: International Petroleum Institute, 1965.

Binder, Leonard. *Iran.* Berkeley: University of California Press, 1962.

Brown, Edward H. *The Saudi Arabia Kuwait Neutral Zone.* Beirut: Middle East Research and Publishing Center, 1963.

De Chazeau, Melvin G., and Alfred E. Kahn. *Integration and Competition in the Petroleum Industry.* New Haven: Yale University Press, 1959.

East, W. G. *The Geography of Energy.* London: Hutchinson University Library, 1964.

Engler, Robert. *The Politics of Oil.* New York: Macmillan, 1961.

Fanning, Leonard M. *Foreign Oil and the Free World.* New York: McGraw-Hill, 1954.

Finnie, David H. *Desert Enterprise: The Middle East Oil Industry in its Local Environment.* Cambridge, Mass.: Harvard University Press, 1958.

Ford, Alan W. *The Anglo-Iranian Oil Dispute of 1951–1952.* Berkeley: University of California Press, 1954.

Frank, Helmut J. *Crude Oil Prices in the Middle East.* New York: Praeger, 1966.

Frankel, P. H. *Essentials of Petroleum.* London: Chapman and Hall, 1940.
———. *Oil: The Facts of Life.* London: Weidenfeld & Nicolson, 1962.
———. *Mattei: Oil and Power Politics.* London: Faber & Faber, 1966.

Hamilton, Charles W. *Americans and Oil in the Middle East.* Houston: Gulf Publishing Company, 1962.

Hartshorn, J. E. *Oil Companies and Governments.* London: Faber & Faber, 1962.

Hirst, David. *Oil and Public Opinion in the Middle East.* London: Faber & Faber, 1965.

Issawi, Charles, and Mohammed Yeganeh. *The Economics of Middle Eastern Oil.* New York: Praeger, 1962.

Kubbah, Abdul Amir G. *Libya: Its Oil Industry and Economic System.* Beirut: Rihani Press, 1964.

Leeman, Wayne E. *The Price of Middle East Oil.* Ithaca, N.Y.: Cornell University Press, 1962.

Lenczowski, George. *Oil and State in the Middle East.* Ithaca, N.Y.: Cornell University Press, 1960.

464

Lieuwen, Edwin. *Petroleum in Venezuela*. Berkeley: University of California Press, 1954.

Longrigg, Stephen Hemsley. *Oil in the Middle East*. London: Oxford University Press, 1964.

Lubell, Harold. *Middle East Oil Crises and Western Europe's Energy Supplies*. Baltimore: Johns Hopkins Press, 1963.

Lutfi, Ashraf T. *Arab Oil: A Plan for the Future*. Beirut: Middle East Research and Publishing Center, 1960.

————. *OPEC Oil*. Beirut: Middle East Research and Publishing Center, 1968.

Mann, Clarence. *Abu Dhabi*. Beirut: Khayat's, 1964.

Martinez, Anibal R. *Our Gift, Our Oil*. Vienna, 1966.

Mikdashi, Zuhayr. *A Financial Analysis of Middle Eastern Oil Concessions: 1901–1965*. New York: Praeger, 1966.

Mikesell, Raymond F., and Hollis B. Chenery. *Arabian Oil: America's Stake in the Middle East*. Chapel Hill, N.C.: University of North Carolina Press, 1949.

Mughraby, Muhamad A. *Permanent Sovereignty over Oil Resources: A Study of Middle East Oil Concessions and Legal Change*. Beirut: Middle East Research and Publishing Center, 1966.

O'Connor, Harvey. *World Crisis in Oil*. New York: Monthly Review Press, 1962.

Odell, Peter R. *An Economic Geography of Oil*. London: G. Bell and Sons, 1963.

Penrose, Edith T. *The Large International Firm in Developing Countries*. London: Allen & Unwin, 1968.

Philby, H. St. J. B. *Arabian Oil Ventures*. Washington, D.C.: Middle East Institute, 1964.

Rowe, J. W. F. *Primary Commodities in International Trade*. Cambridge: Cambridge University Press, 1965.

Shamma, Samir. *The Oil of Kuwait: Present and Future*. Beirut: Middle East Research and Publishing Center, 1959.

Shwadran, Benjamin. *The Middle East, Oil and the Great Powers*. New York: Council for Middle Eastern Affairs Press, 1959.

Siksek, Simon G. *The Legal Framework of Oil Concessions in the Arab World*. Beirut: Middle East Research and Publishing Center, 1960.

Taher, Abdulhady Hassan. *Income Determination in the International Petroleum Industry*. London: Pergamon Press, 1966.

Votaw, Dow. *The Six-Legged Dog: Mattei and ENI—a Study in Power*. Berkeley: University of California Press, 1964.

Index

467

71
74
75
76
77
79
81
83
85
88